T0233952

Lecture Notes in Computer Science 11790

More information about this series at http://www.springer.com/series/8379

Maciej Koutny · Lucia Pomello ·
Lars Michael Kristensen (Eds.)

Transactions on
Petri Nets
and Other Models
of Concurrency XIV

Editor-in-Chief
Maciej Koutny
Newcastle University
Newcastle upon Tyne, UK

Guest Editors
Lucia Pomello
Università degli Studi di Milano
Milan, Italy

Lars Michael Kristensen
Western Norway University
of Applied Sciences
Bergen, Norway

ISSN 0302-9743 ISSN 1611-3349 (electronic)
Lecture Notes in Computer Science
ISSN 1867-7193 ISSN 1867-7746 (electronic)
Transactions on Petri Nets and Other Models of Concurrency
ISBN 978-3-662-60650-6 ISBN 978-3-662-60651-3 (eBook)
https://doi.org/10.1007/978-3-662-60651-3

Preface by Editor-in-Chief

The 14th issue of LNCS *Transactions on Petri Nets and Other Models of Concurrency* (ToPNoC) contains revised and extended versions of a selection of the best papers from the workshops held at the 39th International Conference on Application and Theory of Petri Nets and Concurrency (Petri Nets 2018, Bratislava, Slovakia, June 24–29, 2018), and the 18th International Conference on Application of Concurrency to System Design (ACSD 2018, Bratislava, Slovakia, June 24–29, 2018).

I would like to thank the two guest editors of this special issue: Lars Michael Kristensen and Lucia Pomello. Moreover, I would like to thank all authors, reviewers, and organizers of the Petri Nets 2018 and ACSD 2018 satellite workshops, without whom this issue of ToPNoC would not have been possible.

October 2019   Maciej Koutny

LNCS Transactions on Petri Nets and Other Models of Concurrency: Aims and Scope

ToPNoC aims to publish papers from all areas of Petri nets and other models of concurrency ranging from theoretical work to tool support and industrial applications. The foundations of Petri nets were laid by the pioneering work of Carl Adam Petri and his colleagues in the early 1960s. Since then, a huge volume of material has been developed and published in journals and books as well as presented at workshops and conferences.

The annual International Conference on Application and Theory of Petri Nets and Concurrency started in 1980. For more information on the international Petri net community, see: http://www.informatik.uni-hamburg.de/TGI/PetriNets/.

All issues of ToPNoC are LNCS volumes. Hence they appear in all main libraries and are also accessible on SpringerLink (electronically). It is possible to subscribe to ToPNoC without subscribing to the rest of LNCS.

ToPNoC contains:

- Revised versions of a selection of the best papers from workshops and tutorials concerned with Petri nets and concurrency
- Special issues related to particular subareas (similar to those published in the *Advances in Petri Nets* series)
- Other papers invited for publication in ToPNoC
- Papers submitted directly to ToPNoC by their authors

Like all other journals, ToPNoC has an Editorial Board, which is responsible for the quality of the journal. The members of the board assist in the reviewing of papers submitted or invited for publication in ToPNoC. Moreover, they may make recommendations concerning collections of papers for special issues. The Editorial Board consists of prominent researchers within the Petri net community and in related fields.

Topics

The topics covered include: system design and verification using nets; analysis and synthesis; structure and behavior of nets; relationships between net theory and other approaches; causality/partial order theory of concurrency; net-based semantical, logical and algebraic calculi; symbolic net representation (graphical or textual); computer tools for nets; experience with using nets, case studies; educational issues related to nets; higher level net models; timed and stochastic nets; and standardization of nets.

Also included are applications of nets to: biological systems; security systems; e-commerce and trading; embedded systems; environmental systems; flexible manufacturing systems; hardware structures; health and medical systems; office automation;

operations research; performance evaluation; programming languages; protocols and networks; railway networks; real-time systems; supervisory control; telecommunications; cyber physical systems; and workflow.

For more information about ToPNoC see: http://www.springer.com/gp/computer-science/lncs/lncs-transactions/petri-nets-and-other-models-of-concurrency-topnoc-/731240.

Submission of Manuscripts

Manuscripts should follow LNCS formatting guidelines, and should be submitted as PDF or zipped PostScript files to ToPNoC@ncl.ac.uk. All queries should be addressed to the same e-mail address.

Preface by Guest Editors

This volume of ToPNoC contains revised versions of a selection of the best workshop papers presented at satellite events of the 39th International Conference on Application and Theory of Petri Nets and Concurrency (Petri Nets 2018) and the 18th International Conference on Application of Concurrency to System Design (ACSD 2018). In addition, this volume contains three papers based on the advanced tutorials held in conjunction with Petri Nets 2017/2018 and ACSD 2017/2018 as well as one paper submitted directly to ToPNoC.

As guest editors, we are indebted to the Program Committees of the workshops and in particular to the chairs. Without their enthusiastic efforts, this volume would not have been possible. We are also indebted to the lecturers of the advanced tutorials who took the time and effort to transform the material presented at the tutorial into self-contained papers aimed at a broad audience.

The workshop papers considered for this special issue have been selected in a close cooperation with the workshop chairs. Members of the Program Committees have participated in reviewing the new versions of the papers eventually submitted. We have received suggestions for papers for this special issue from:

- ATAED 2018: Workshop on Algorithms and Theories for the Analysis of Event Data (chairs: Wil van der Aalst, Robin Bergenthum, and Josep Carmona)
- PNSE 2018: International Workshop on Petri Nets and Software Engineering (chairs: Ekkart Kindler, Daniel Moldt, and Heiko Rölke)

The authors of the suggested papers have been invited to improve and extend their results where possible, based on the comments received before and during the workshops. Each resulting revised submission was reviewed by at least two referees. We followed the principle of asking for fresh reviews of the revised papers, also from referees not involved initially in the reviewing of the original workshop contributions. All papers have gone through the standard two-stage journal reviewing process, and eventually eight have been accepted after rigorous reviewing and revising. In addition, we invited the organizers of advanced tutorials to coordinate a paper presenting the material from the tutorial. The papers based on the advanced tutorials have also undergone a review process prior to inclusion in this special volume.

The tutorial paper "A Tour in Process Mining: From Practice to Algorithmic Challenges" by Wil van der Aalst, Josep Carmona, Thomas Chatain, and Boudewijn van Dongen gives an introduction to the field of process mining which in recent years has shown to be a highly successful approach for bridging the gap between traditional model-based process analysis and data-centric analysis techniques. The paper describes a range of techniques central to the discovery of process models from event logs, and in addition covers conformance checking techniques that can be applied for investigation how process models deviates from reality.

The tutorial paper by Karsten Wolf entitled "How Petri Net Theory Serves Petri Net Model Checking: A Survey" shows how the performance of verification based on reachability graphs can be improved by exploiting the rich structure theory developed for Petri nets. The paper provides a highly interesting demonstration of how different branches of Petri net research can be integrated. Specifically, the paper discusses how place and transition invariants, siphons, and traps make it possible to eliminate or simplify atomic propositions in the properties to be verified; provide more compact representation of markings (states); and detect behavioral cycles and diamonds. It is shown how this can be exploited in CTL and LTL model checking and in the stubborn sets and sweep-line methods.

Many concurrent and distributed systems contain multiple copies of the same components. The tutorial paper "Parametric Verification: An Introduction" by Étienne André, Michał Knapik, Didier Lime, Wojciech Penczek, and Laure Petrucci gives an introduction to the field of research exploring how to efficiently verify systems with such charecteristics. The paper concentrates on approaches based on parametric timed automata, parametric interval markov chains, parametric Petri Nets, and action synthesis. The paper also surveys software tools supporting the practical application of techniques for parametric verification.

The development of domain specific models requires appropriate tool support for modeling and execution. The paper "Integrated Simulation of Domain-Specific Modeling Languages with Petri Net-based Transformational Semantics" by David Mosteller, Michael Haustermann, Daniel Moldt, and Dennis Schmitz presents an approach to visually execute a simulation on a DSML, which semantics have been defined using a transformation to Petri nets, using the Renew Meta-Modeling and Transformation framework. To illustrate the approach, the authors present the integrated simulation of a selected subset of Business Process Model and Notation (BPMN) and refer to a model transformation to Petri nets.

Machine to Machine (M2M) communication and Internet of Things (IoT) are becoming more pervasive with the increase of use of communicating devices. Publish-subscribe protocols such as the Message Queuing Telemetry Transport (MQTT) protocol are widely used in this context. The paper "Formal Modelling and Incremental Verification of the MQTT IoT Protocol" by Alejandro Rodríguez, Lars Michael Kristensen, and Adrian Rutle presents a Coloured Petri Net (CPN) model of the MQTT protocol logic using CPN Tools. The modeling approach is incremental: the functionality of the protocol is gradually introduced using a set of CPN modeling patterns and properties are verified in each incremental step. As a main advantage, the effect of the state explosion problem in model checking is reduced.

The paper "Kleene Theorems for Free Choice Automata over Distributed Alphabets" by Ramchandra Phawade deals with the characterization of languages generated by free choice nets. It considers safe, state machine decomposable labeled free choice nets, with acceptance conditions. The labels are taken over an alphabet which is distributed consistently with the sequential components of the net. Syntactic expressions are given for the corresponding accepted languages by using synchronous products and Zielonka automata.

Raymond Devillers, Evgeny Erofeev, and Thomas Hujsa in their paper "Synthesis of Weighted Marked Graphs from Constrained Labelled Transition Systems: A Geometric Approach" extended recent work on investigating the problems of analyzing Petri nets and synthesizing them from labeled transition systems (LTS). They provide new conditions for the synthesis of Weighted Marked Graphs (WMGs), a well-known and useful class of weighted Petri nets in which each place has at most one input and one output. They also tackle geometrically the WMG-solvability of finite, acyclic LTS with any number of labels.

The paper "Evaluating Conformance Measures in Process Mining using Conformance Propositions" by Anja F. Syring, Niek Tax, and Wil M. P. van der Aalst provides a rich collection of conformance propositions and uses these to evaluate current existing conformance checking measures that are available in the process mining area. This includes the recall (fitness), precision, and generalization. Many recall, precision, and generalization measures have been developed and the paper formulates the challenges and requirements related to these measures.

The paper "Relabelling LTS for Petri Net Synthesis via Solving Separation Problems" by Uli Schlachter and Harro Wimmel, submitted directly to ToPNoC, deals with the problem of finding an unlabeled Petri net with a reachability graph isomorphic to a given usually finite labeled transition system.

As guest editors, we would like to thank all authors and referees who have contributed to this issue. The quality of this volume is the result of the high scientific value of their work. Moreover, we would like to acknowledge the excellent cooperation throughout the whole process that has made our work a pleasant task. We are also grateful to the Springer/ToPNoC team for the final production of this issue.

September 2019

Lucia Pomello
Lars Michael Kristensen

Organization

Guest Editors

Lucia Pomello   University of Milano-Bicocca, Italy
Lars Michael Kristensen   Bergen University College, Norway

Workshop Co-chairs

Wil van der Aalst   RWTH Aachen University, Germany
Robin Bergenthum   FernUniversität in Hagen, Germany
Josep Carmona   Universitat Politecnica de Catalunya, Spain
Ekkart Kindler   Technical University of Denmark, Denmark
Daniel Moldt   University of Hamburg, Germany
Heiko Rölke   DIPF, Germany

Reviewers

Kamel Barkaoui   Yann Ben Maissa
Robin Bergenthum   Alfonso Pierantonio
Didier Buchs   Pascal Poizat
Josep Carmona   Adrián Puerto Aubel
Maciej Koutny

Contents

A Tour in Process Mining: From Practice to Algorithmic Challenges

Wil van der Aalst[1], Josep Carmona[2](✉), Thomas Chatain[3],
and Boudewijn van Dongen[4]

[1] Process and Data Science Group, RWTH Aachen University, Aachen, Germany
[2] Computer Science Department, Universitat Politècnica de Catalunya,
Barcelona, Spain
jcarmona@cs.upc.edu
[3] LSV, ENS Paris-Saclay, CNRS, Inria, Université Paris-Saclay, Cachan, France
chatain@lsv.fr
[4] Eindhoven University of Technology, Eindhoven, The Netherlands
b.f.v.dongen@tue.nl

Abstract. Process mining seeks the confrontation between modeled behavior and observed behavior. In recent years, process mining techniques managed to bridge the gap between traditional model-based process analysis (e.g., simulation and other business process management techniques) and data-centric analysis techniques such as machine learning and data mining. Process mining is used by many data-driven organizations as a means to improve performance or to ensure compliance. Traditionally, the focus was on the discovery of process models from event logs describing real process executions. However, process mining is not limited to process discovery and also includes conformance checking. Process models (discovered or hand-made) may deviate from reality. Therefore, we need powerful means to analyze discrepancies between models and logs. These are provided by conformance checking techniques that first align modeled and observed behavior, and then compare both. The resulting alignments are also used to enrich process models with performance related information extracted from the event log. This tutorial paper focuses on the control-flow perspective and describes a range of process discovery and conformance checking techniques. The goal of the paper is to show the algorithmic challenges in process mining. We will show that process mining provides a wealth of opportunities for people doing research on Petri nets and related models of concurrency.

1 Introduction to Process Mining

This tutorial paper is based on a tutorial given at Petri Nets 2017 in Zaragoza (Spain) on Tuesday, June 27th, 2017 and a tutorial given at Petri Nets 2018 in Bratislava (Slovakia) on Tuesday, June 26th, 2018. The goal of these two tutorials was to introduce the topic of process mining for people with a Petri nets background and to show the exciting algorithmic challenges provided by

© Springer-Verlag GmbH Germany, part of Springer Nature 2019
M. Koutny et al. (Eds.): ToPNoC XIV, LNCS 11790, pp. 1–35, 2019.
https://doi.org/10.1007/978-3-662-60651-3_1

the various process mining tasks. Although process mining is widely used in industry, there are still many open research questions that require knowledge of both process science (e.g., formal methods and concurrency theory) and data science (e.g., data mining, machine learning, and statistics). To limit the scope, we focus on control flow and Petri nets as a representation.

1.1 Opportunities Provided by Event Data

Modeling behavior is valuable, but is often based on (unrealistic) simplifying assumptions. Simulation models and specifications can be used to get process insights, but these insights depend on the assumptions and abstraction used while modeling. Fortunately, when it comes to existing processes and systems, we no longer need to rely on modeling only. There is an abundance of event data. In the book [1], the term *Internet of Events* (IoE) is used to refer to the different types of event data readily available. These include: the *Internet of Content* (traditional databases, web pages, e-books, newsfeeds, movies, music, etc.), the *Internet of People* (e-mail, Facebook, Twitter, forums, LinkedIn, etc.), the *Internet of Things* (smart homes, high-tech systems, Industry 4.0, etc.), and the *Internet of Locations* (smartphones, wearables, etc.). Events may take place inside machines, enterprise information systems, hospital information systems, social networks, and transportation systems. Events may be "life events", "machine events", or "organization events". Process mining aims to exploit event data in a meaningful way, for example, to provide insights, identify bottlenecks, anticipate problems, record policy violations, recommend countermeasures, and streamline processes. Event data are like the breadcrumbs in the fairy tale of Hansel and Gretel. When stored properly, we can use them to reconstruct the real behavior.

Table 1 shows a small fragment of a larger *event log* describing 16218 events related to 1266 cases. The events in this log correspond to the handling of orders. Each row corresponds to two *events*, i.e., the *start* of an activity instance and the *completion* of an activity instance. Each *activity instance* (i.e., row) describes the execution of an activity for a particular order. The first column refers to the order number. The second column shows the activity name. The next two columns show the start time and end time of an activity. There are also additional attributes such as the person executing the event.

Event data, as shown in Table 1, can be used to *discover process models* best describing the behavior observed. This is similar to discovering a decision tree based on labeled instances, i.e., instances having descriptive attributes and a class label. However, in process mining, the instances correspond to events referring to cases (i.e., process instances), activities, timestamps, etc. Process discovery starts from the situation without a model and just event data. Given a process model, it is also possible to *check conformance*. Conformance checking uses as input both modeled and observed behavior. Discrepancies can be detected by replaying the event data on the process model. The process model may be hand-made or discovered using some process discovery technique. Next to process discovery and conformance checking, there are many additional process mining

Table 1. Fragment of an event log with 16218 events related to 1266 cases.

Order number	Activity	Start time	End time	Resource	Product	Prod-price	Quantity	Address
...	...	...	...	...	...	...	...	...
989	Place order	10/07/2015 05:10	10/07/2015 05:25	Sophia	APPLE iPhone 6s 64 GB	858.0	2	NL-7742XG-17
984	Send invoice	10/07/2015 08:47	10/07/2015 08:57	Jack	SAMSUNG Core Prime G361	135.0	2	NL-9468HG-14
810	Prepare delivery	10/07/2015 08:50	10/07/2015 08:53	Emma	APPLE iPhone 6s Plus 64 GB	969.0	4	NL-7875EH-26
898	Confirm payment	10/07/2015 08:56	10/07/2015 09:02	Lily	SAMSUNG Galaxy S4	329.0	4	NL-7944BB-6
858	Prepare delivery	10/07/2015 09:01	10/07/2015 09:04	Emma	SAMSUNG Galaxy J5	219.99	1	NL-7823JJ-7
990	Place order	10/07/2015 09:41	10/07/2015 09:51	Sophia	MOTOROLA Moto G	199.0	4	NL-9514BV-16
838	Confirm payment	10/07/2015 09:56	10/07/2015 10:03	Jack	SAMSUNG Core Prime G361	135.0	3	NL-7826GD-9
869	Prepare delivery	10/07/2015 10:39	10/07/2015 10:42	Emma	SAMSUNG Galaxy S4	329.0	1	NL-7742XG-17
875	Prepare delivery	10/07/2015 10:42	10/07/2015 10:46	Aiden	SAMSUNG Core Prime G361	135.0	4	NL-9407EM-35
898	Make delivery	10/07/2015 10:44	10/07/2015 11:06	Aubrey	SAMSUNG Galaxy S4	329.0	4	NL-7944BB-6
914	Send reminder	10/07/2015 11:38	10/07/2015 11:58	Abigail	MOTOROLA Moto G	199.0	3	NL-7828AM-11a
991	Place order	10/07/2015 11:41	10/07/2015 15:32	Lucas	SAMSUNG Core Prime G361	135.0	4	NL-7905BC-40a
888	Pay	10/07/2015 11:49	10/07/2015 11:54	Lily	APPLE iPhone 6 16 GB	639.0	3	NL-7821AC-3
977	Send invoice	10/07/2015 11:56	10/07/2015 12:01	Madison	SAMSUNG Galaxy S4	329.0	3	NL-7908XB-46
889	Pay	10/07/2015 12:03	10/07/2015 12:09	Lily	SAMSUNG Galaxy S4	329.0	2	NL-7948BX-10
883	Make delivery	10/07/2015 12:09	10/07/2015 12:37	Aubrey	MOTOROLA Moto G	199.0	1	NL-7751AR-19
912	Send reminder	10/07/2015 12:34	10/07/2015 12:46	Abigail	APPLE iPhone 6 16 GB	639.0	1	NL-9402NV-25
823	Confirm payment	10/07/2015 12:56	10/07/2015 13:04	Lily	SAMSUNG Galaxy S4	329.0	7	NL-7943MC-4
863	Prepare delivery	10/07/2015 12:59	10/07/2015 15:21	Lucas	APPLE iPhone 6 16 GB	639.0	2	NL-7905AX-38
860	Make delivery	10/07/2015 13:04	10/07/2015 13:32	Aubrey	SAMSUNG Galaxy S6 32 GB	543.99	3	NL-7823JJ-7
992	Place order	10/07/2015 13:41	10/07/2015 13:51	Jacob	SAMSUNG Galaxy S4	329.0	3	NL-7948DN-12a
900	Send reminder	10/07/2015 13:45	10/07/2015 13:59	Abigail	APPLE iPhone 6 16 GB	639.0	2	NL-9521KJ-34
905	Send reminder	10/07/2015 13:52	10/07/2015 14:08	Luke	APPLE iPhone 5s 16 GB	449.0	4	NL-7942GT-2
867	Prepare delivery	10/07/2015 14:36	10/07/2015 14:43	Sophia	SAMSUNG Galaxy S6 32 GB	543.99	2	NL-7908XB-46
...	...	...	...	...	...	...	...	...

techniques using both event data and models. For example, events data can be used to *enrich* or *repair* process models. It is also possible to *predict* performance or compliance. See [1] for a more complete overview of process discovery. The book [2] focuses on conformance checking.

1.2 Control Flow as the Backbone of Any Process

In this paper, we focus on *control flow*. This means that we ignore most of the attributes in Table 1. Initially, we are only interested in the ordering of activities within a *case* (i.e., process instance). To illustrate this, consider a particular order, e.g., order 889. There are six activity instances in Table 1 related to this order. Each activity instance corresponds to two events: start and complete. Suppose we focus on the start events. This means that order 889 is described by six events (rather than twelve). Table 2 shows these six events. Note that we only show the first three columns. Table 2 describes a sequence of activities for one case, namely: ⟨*place order, send invoice, pay, prepare delivery, make delivery, confirm payment*⟩. Note that the timestamps are only used to order the events.

Table 2. The events of type "start" for case 889.

Order number	Activity	Timestamp
889	Place order	23/06/2015 19:45
889	Send invoice	24/06/2015 18:23
889	Pay	10/07/2015 12:03
889	Prepare delivery	21/07/2015 11:22
889	Make delivery	22/07/2015 13:00
889	Confirm payment	23/07/2015 10:53

Each of the 1266 cases in Table 1 can be described as such a *trace*. There are 503 cases that correspond to the trace ⟨*place order, send invoice, pay, prepare delivery, make delivery, confirm payment*⟩ (including order 889). Table 3 shows the frequency of each observed trace. As we will see later, the control-flow aspect of an event log Table 3 can be formally represented as a multi-set of activity sequences.

Clearly, Table 3 contains only a fraction of the information in the original event log (compare with the fragment in Table 1). However, this information is sufficient to construct a control-flow model, e.g., a Petri net. To further simplify things, assume that we abstract from activity *sr = send reminder*, i.e., we create a new log without this activity and feed it to various discovery algorithms. Figure 1 shows a few discovered process models that will be discussed later.

Control-flow models such as the ones shown in Fig. 1 are just the starting point for process mining. By replaying the event log on the discovered model, one can show bottlenecks, etc. Actually, the control-flow model may be extended with additional perspectives: the organizational perspective ("What are the

Table 3. Event log represented as a multi-set of traces using the short names *po* = *place order*, *si* = *send invoice*, *py* = *pay*, *pd* = *prepare delivery*, *md* = *make delivery*, *cp* = *confirm payment*, *co* = *cancel order*, *sr* = *send reminder*.

Trace	Frequency
⟨po, si, py, pd, md, cp⟩	503
⟨po, si, sr, py, pd, md, cp⟩	247
⟨po, si, sr, sr, co⟩	141
⟨po, si, sr, sr, py, pd, md, cp⟩	139
⟨po, si, py, pd, cp, md⟩	135
⟨po, si, sr, py, pd, cp, md⟩	57
⟨po, si, sr, sr, py, pd, cp, md⟩	36
⟨po, py, si, pd, md, cp⟩	6
⟨po, py, si, pd, cp, md⟩	2
Total	1266

organizational roles and which resources are performing particular activities?"), the case perspective ("Which characteristics of a case influence a particular decision?"), and the time perspective ("Where are the bottlenecks in my process?"). These additional perspectives are very important from a practical point of view. Analysts applying process mining are interested in questions such as:

- What is the main bottleneck in the process and is this caused by particular resources?
- What do the incompliant cases have in common? Do they lead to higher costs?
- Will we be able to handle 95% of the cases in two hours in the coming days?
- Does the workload have an effect on the durations of activities?

For all of these questions, we first need to have a control-flow model. Moreover, we need to be able to link events in the log to activities in the process model in order to discuss bottlenecks, deviations, decisions, etc. Transitional machine learning, data mining, and optimization techniques can be used once the control-flow model is in place and the event log is aligned with the model. However, these techniques cannot be used to handle the control-flow perspective. Therefore, this paper focuses on control-flow.

1.3 Process Discovery ≠ Synthesis

Data science techniques need to deal with uncertainty and incompleteness [3]. Suppose that we learn a decision tree that adequately predicts that one is less likely to claim insurance when being middle-aged and female and that one is more likely to claim insurance when being young and male. However, even a properly constructed decision tree will not be able to classify things correctly

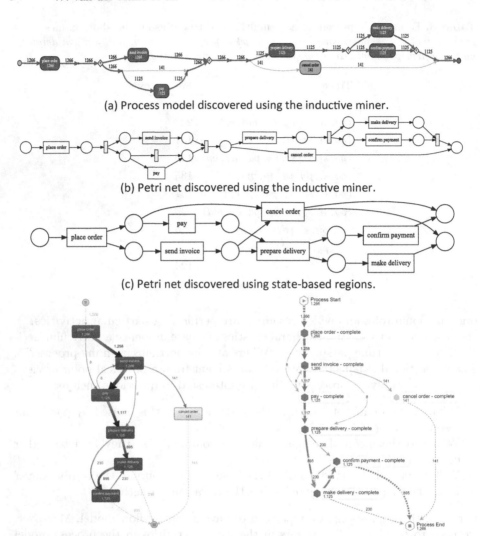

(a) Process model discovered using the inductive miner.

(b) Petri net discovered using the inductive miner.

(c) Petri net discovered using state-based regions.

(d) Process model discovered using Disco. (e) Process model discovered using Celonis.

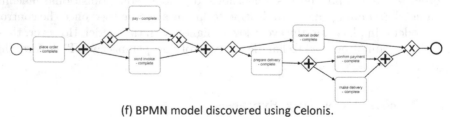

(f) BPMN model discovered using Celonis.

Fig. 1. Different process models discovered using plug-ins of ProM (models (a), (b), and (c)), Disco (model (d)), and Celonis (models (e) and (f)). Process models (d) and (e) are so-called directly follows graphs. Process models (a), (b), and (f) all describe the same process, but use different notations. Process model (c) describes the process best, because *pay* and *cancel order* are mutually exclusive.

and we accept that there will be young males not claiming insurance or middle-aged females that do claim insurance. Moreover, we do not expect to see all combinations in our input data. For example, when customers are described by 10 different attributes age, gender, income, region, brand, etc., we cannot expect to see all combinations in our training data. Despite these limitations, the decision tree still provides valuable insights. When performing process discovery, we are facing similar challenges. The event log is just a sample and the fact that something happens in the logs once does not imply that the model should allow for it. Consider for example the 1266 cases in Table 3. Some traces are very frequent, e.g., $\langle po, si, py, pd, md, cp \rangle$ occurred 503 times. Other traces are very infrequent, e.g., $\langle po, py, si, pd, cp, md \rangle$ happened only twice. Traces in an event log typically follow a Pareto distribution, i.e., some are frequent, but many may be infrequent (for example, 80% of the log is described by 20% of the trace variants). When using a larger process model with many activities, one often witnesses that many traces in the event log are unique. Taking a different sample of the same process will lead to different unique variants. For example, if we take only the first 500 cases from the log shown in Table 3, we may not see $\langle po, py, si, pd, cp, md \rangle$. Moreover, if we observe the same process for a longer time, we will see cases that are not in Table 3 (e.g., new trace variants with three or more reminders). We may also be just interested in the dominant behavior. For example, to resolve a bottleneck, we may want to focus on the most frequent behavior. However, when investigating fraud we may be most interested in the least frequent behavior.

Researchers working on formal methods and concurrency theory often have problems dealing with uncertainty and incompleteness. Most synthesis approaches assume that the input provides a *full* and *unambiguous* description of *all* possible behavior [4]. When performing process mining in reality such assumptions are very unrealistic. To reason about the correctness of process mining results one needs to use a data science approach (e.g., cross-validation using test and training data) and apply notions such as precision and recall to assess the quality.

1.4 Industrial Uptake of Process Mining and Commercial Software

Around 2002, only simple stand-alone process mining tools were available (e.g., *MiMo*, *EMiT*, *Little Thumb*, and *InWolvE*) [1]. This triggered the development of the *ProM framework*. In 2004, the first fully functional version of the ProM framework was released. Since then, ProM has been the de facto standard for process mining research. ProM provides a "plug-able" open-source platform where developers can contribute new analysis techniques in the form of plug-ins. Currently, there are over 1500 plug-ins. ProM served as an example for a "wave" of commercial process mining tools.

Reports by Gartner [5] and Forrester [6] report on the uptake of process mining in industry. There are more than 25 commercial tools supporting process mining. Examples include: *Disco* (Fluxicon), *Celonis Process Mining* (Celonis), *ProcessGold Enterprise Platform* (ProcessGold), *QPR ProcessAnalyzer* (QPR),

SNP Business Process Analysis (SNP AG), *minit* (Gradient ECM), *myInvenio* (Cognitive Technology), *Everflow* (Accelera Labs), *ProDiscovery* (Puzzeldata), *PAFnow* (Process Analytics Factory), *Stereologic* (Stereologic), *ARIS Process Mining* (Software AG), *Mehrwerk Process Mining* (Mehrwerk), *Logpickr Process Explorer* (Logpickr), and *Lana Process Mining* (Lanalabs) [1]. See the website of the IEEE Task Force on Process Mining for examples of successful case studies [7]. For example, within Siemens there are currently over 2,500 active users of *Celonis Process Mining*. Siemens reported savings of "double-digit millions euros" as a result of the worldwide application of process mining [8]. Celonis is a process mining start-up founded in 2011 that is now valued over one billion dollar. The many tools available and the widespread adoption illustrate the relevance of process mining. However, at the same time, there are many challenges, as demonstrated in this paper.

As described in Gartner [5], there is an increasing need for process mining capabilities that go beyond process discovery. Initially, commercial tools focused on process discovery and often discovery capabilities were limited to the so-called *directly-follows graph*. In such a graph, nodes represent activities and arcs represent a simplistic view on causality. Activities a and b are connected if a is frequently followed by b. In the event log used before, $po = place\ order$ is directly followed by $si = send\ invoice$ 1258 times, and $po = place\ order$ is directly followed by $py = pay$ only 8 times. Depending on the threshold set, an arc connecting *place order* to *pay* is added to the directly-follows graph.

Figure 1 shows six process models discovered based on the event log described in Tables 1 and 3 (without activity *send reminder*). Figure 1(a) shows a screenshot of ProM's visual inductive miner [9]. This model can be converted into a Petri net as shown in Fig. 1(b). Note that the model allows for the trace where *pay* is followed by *cancel order*, i.e., the model is "underfitting" and allowing for unseen behavior (in the event log there are no orders for which both *pay* and *cancel order* occurred). Figure 1(c) shows a model generated by ProM based on state-based regions. This Petri net does not allow for the unseen behavior mentioned before (there is an exclusive choice between *pay* and *cancel order*). Figure 1(d) shows a directly-follows graph generated using Disco, the software from Fluxicon. The directly-follows graph is annotated with the frequencies. Using two sliders, it is possible to remove infrequent arcs and activities. Figure 1(e) shows a similar directly-follows graph generated using Celonis. Celonis also supports the basic inductive mining algorithm [1,10]. The result is presented as a BPMN model as shown in Fig. 1(f). Note that models (a), (b), and (f) are semantically equivalent, but use different notations. Each of the models in Fig. 1 defines a language, i.e., the (possibly infinite) set of traces accepted by the model. Ideally, the model accepts most of the traces in the event log (but not necessarily all, e.g., outlier behavior may not be accepted) and does not accept behavior that is unlikely given the event log. Later, these notions will be defined more precisely.

Although the initial focus of commercial tools was on discovery, vendors started to add functionality related to conformance checking (i.e., monitoring deviations by comparing model and log), social network/organizational mining, root-cause analysis using machine learning, and case prediction. For example, it is possible to find possible causes for bottlenecks and compliance problems and generate statements like "The extreme waiting times are caused by unnecessary rework involving an external party and under-staffing of the backoffice on Fridays" and "Most of the compliance violations were caused by this new manager that approved requests without the necessary checks". Note that these more advanced questions have been researched for over more than a decade [1]. All of the analysis features that can be found in today's commercial tools, had been implemented in ProM years before. Note that, currently, ProM provides over 1500 plug-ins providing a wide range of analysis techniques.

In this paper, we focus on conformance checking and limit the scope to control-flow [1,2]. Given a trace $\sigma_1 = \langle po, si, pd, py, md, cp \rangle$ and a process model like Fig. 1(c), we would like to decide if the trace fits the model and if not, diagnose the difference(s). In σ_1, $pd = prepare\ delivery$ is performed before the $py = pay$ which is impossible according to the Petri net in Fig. 1(c). In $\sigma_2 = \langle po, py, si, py, pd, md, cp \rangle$, the order is paid twice. In $\sigma_3 = \langle po, py, si, md, cp \rangle$, the order was delivered without the mandatory preparation step. Given an event log with possibly millions of events, conformance checking techniques need to provide aggregate diagnostics (e.g., the activity was skipped 1500 times).

1.5 Killer App for Petri Nets

The process mining discipline is highly relevant (see the uptake of commercial process mining tools and activities) and provide interesting scientific challenges. Progress in this young scientific discipline has been remarkable and this resulted in powerful techniques that are highly scalable. However, many of the problems identified (process discovery, conformance checking, etc.) are notoriously difficult and have not been solved satisfactorily. Process mining provides a great opportunity for Petri net researchers. When studying real-life processes, concurrency should be a starting point for analysis and not added as an afterthought (locality of actions). Many other modeling approaches start from a sequential view on the world and then add special operators to introduce concurrency. Petri nets are inherently concurrent. Although Petri nets are often seen as a procedural language, in essence, Petri nets are declarative. A Petri net without any places and a non-empty set of transitions allows for any behavior involving the activities represented by these transitions. Adding a place is like introducing a constraint. The idea that transitions (modeling activities or actions) are independent (i.e., concurrent) unless specified otherwise is foundational [11].

Most data science approaches do not consider behavioral aspects and when they do, they typically use sequential models (e.g., sequential patterns, Markov models, etc.). As the evolution of the process mining discipline shows, the combination of event data and Petri nets is very powerful. Therefore, we encourage Petri-net researchers to tackle the open problems identified in this paper. However, as Sect. 1.3 describes this is not "business as usual": One needs to approach the problem from a data-science perspective.

1.6 Outline

The remainder of this paper is organized as follows. Section 2 introduces preliminaries, including basic notations, event logs, and Petri nets. Section 3 discusses the challenges related to process discovery and presents inductive and region-based discovery techniques. Section 4 is devoted to conformance checking. The focus here is on alignment-based techniques. Alignments relate seen behavior to modeled behavior even when there are deviations. Such alignments are used for a range of analyses (bottleneck analysis, predictions, etc.). Section 5 focuses on the quality of a process model in the context of a given event log. Notions such as fitness, precision, generalization, and simplicity are discussed. Section 6 concludes the paper.

2 Preliminaries

2.1 Mathematical Notation

A multi-set is like a set in which each element may occur multiple times. For example, $[a, b^2, c^3, d^2, e]$ is the multi-set with nine elements: one a, two b's, three c's, two d's, and one e. The following three multi-sets are identical: $[a, b, b, c^3, d, d, e]$, $[e, d^2, c^3, b^2, a]$, and $[a, b^2, c^3, d^2, e]$. Formally, $\mathbb{B}(A) = A \to \mathbb{N}$ is the set of multi-sets (bags) over a finite domain A, i.e., $X \in \mathbb{B}(A)$ is a multi-set, where for each $a \in A$, $X(a)$ denotes the number of times a is included in the multi-set. For example, if $X = [a, b^2, c^3]$, then $X(b) = 2$ and $X(e) = 0$.

The sum of two multi-sets $(X \uplus Y)$, the difference $(X \setminus Y)$, the presence of an element in a multi-set $(x \in X)$, and the notion of subset $(X \leq Y)$ are defined as usual.

For a given set A, A^* is the set of all finite sequences over A. A finite sequence over A of length n is a mapping $\sigma \in \{1, \ldots, n\} \to A$. Such a sequence is represented by a string, i.e., $\sigma = \langle a_1, a_2, \ldots, a_n \rangle$ where $a_i = \sigma(i)$ for $1 \leq i \leq n$. $|\sigma|$ denotes the length of the sequence, i.e. $|\sigma| = n$. $\sigma \oplus a' = \langle a_1, \ldots, a_n, a' \rangle$ is the sequence with element a' appended at the end. Similarly, $\sigma_1 \oplus \sigma_2$ appends sequence σ_2 to σ_1 resulting a sequence of length $|\sigma_1| + |\sigma_2|$.

$hd^k(\sigma) = \langle a_1, a_2, \ldots, a_{k \min n} \rangle$, i.e., the "head" of the sequence consisting of the first k elements (if possible). Note that $hd^0(\sigma)$ is the empty sequence and for $k \geq n$: $hd^k(\sigma) = \sigma$. $pref(\sigma) = \{hd^k(\sigma) \mid 0 \leq k \leq n\}$ is the set of prefixes of σ.

$tl^k(\sigma) = \langle a_{(n-k+1) \max 1}, a_{k+2}, \ldots, a_n \rangle$, i.e., the "tail" of the sequence composed of the last k elements (if possible). Note that $tl^0(\sigma)$ is the empty sequence and for $k \geq n$: $tl^k(\sigma) = \sigma$.

$\sigma \uparrow X$ is the projection of σ onto some subset $X \subseteq A$, e.g., $\langle a, b, c, a, b, c, d \rangle \uparrow \{a, b\} = \langle a, b, a, b \rangle$ and $\langle d, a, a, a, a, a, a, d \rangle \uparrow \{d\} = \langle d, d \rangle$.

For any sequence $\sigma = \langle a_1, a_2, \ldots, a_n \rangle$ over A, $\{a \in \sigma\} = \{a_1, a_2, \ldots, a_n\}$ and $[a \in \sigma] = [a_1, a_2, \ldots, a_n]$, e.g., if $\sigma = \langle d, a, a, a, a, a, a, d \rangle$, then $\{a \in \sigma\} = \{a, d\}$ and $[a \in \sigma] = [a^6, d^2]$.

2.2 Process Models as Petri Nets

Definition 1 ((Labeled) Petri net). *A (labeled) Petri Net [12] is a tuple* $N = \langle P, T, \mathcal{F}, M_0, M_f, \Sigma, \lambda \rangle$, *where P is the set of places, T is the set of transitions (with $P \cap T = \emptyset$), $\mathcal{F} : (P \times T) \cup (T \times P) \to \{0, 1\}$ is the flow relation, $M_0 \in \mathbb{B}(P)$ is the initial marking, $M_f \in \mathbb{B}(P)$ is the final marking, Σ is an alphabet of actions and $\lambda : T \to \Sigma \cup \{\tau\}$ labels every transition by an action, or as a silent action denoted by the symbol τ.*

A marking $M \in \mathbb{B}(P)$ is an assignment of a non-negative integer to each place. If k is assigned to place p by marking M (denoted $M(p) = k$), we say that p is marked with k tokens. Given a node $x \in P \cup T$, its pre-set and post-set are denoted by $\bullet x$ and $x \bullet$ respectively.

A transition t is *enabled* in a marking M when all places in $\bullet t$ are marked. When a transition t is enabled, it can *fire* by removing a token from each place in $\bullet t$ and putting a token to each place in $t \bullet$. A marking M' is *reachable* from M if there is a sequence of firings $t_1 \ldots t_n$ that transforms M into M', denoted by $M[t_1 \ldots t_n\rangle M'$. We define the *language* of N as the set of *full runs* defined by $\mathcal{L}(N) := \{\lambda(t_1) \ldots \lambda(t_n) \mid M_0[t_1 \ldots t_n\rangle M_f\}$. A Petri net is *$k$-bounded* if no reachable marking assigns more than k tokens to any place. A Petri net is bounded if there exist a k for which it is k-bounded. A Petri net is *safe* if it is 1-bounded. A bounded Petri net has an *executable loop* if it has a reachable marking M and sequences of transitions $u_1 \in T^*$, $u_2 \in T^+$, $u_3 \in T^*$ such that $M_0[u_1\rangle M[u_2\rangle M[u_3\rangle M_f$.

2.3 Process Event Data

Processes are crucial to manage the operations in organizations. Often, processes are complex, due to many reasons: they can comprise a large number of activities, can have many decision points, there can be many participants working jointly on a case, etc. Also, there can be many work streams running in parallel, possibly with the support of multiple information systems and also external suppliers. Regardless of their complexity, processes in organizations have a common element: They generate data. We call the event data generated in a process *event logs*.

The collection and analysis of event logs opens various opportunities for improving the underlying process. Table 4 shows a simple event log, which contains information about the underlying process, including *data* in the form of *event attributes*. The event log underlying traces are $\{\langle a, b, c, d \rangle, \langle c, d, a, b \rangle, \langle b, a, d, c \rangle, \langle d, c, b, a \rangle\}$ and they correspond to 'case IDs' 1, 2, 3, and 4, respectively. We assume that the set of attributes is fixed and the function *attr* maps pairs of events and attributes to the corresponding values. For each *event* e the log contains the case ID $case(e)$, the activity name $act(e)$, and the set of attributes defined for e, e.g., $attr(e, timestamp)$. For instance, for the event log in Table 4, $case(e_7) = 2$, $act(e_7) = a$, $attr(e_7, timestamp) = $ "10-04-2015 10:28pm", and $attr(e_7, cost) = 19$.

Table 4. An example event log

Event	Case ID	Activity	Timestamp	Temperature	Resource	Cost	Risk
1	1	a	10-04-2015 9:08am	25.0	Martin	17	Low
2	2	c	10-04-2015 10:03am	28.7	Mike	29	Low
3	2	d	10-04-2015 11:32am	29.8	Mylos	16	Medium
4	1	b	10-04-2015 2:01pm	25.5	Silvia	15	Low
5	1	c	10-04-2015 7:06pm	25.7	George	14	Low
6	1	d	10-04-2015 9:08pm	25.3	Peter	17	Medium
7	2	a	10-04-2015 10:28pm	30.0	George	19	Low
8	2	b	10-04-2015 10:40pm	29.5	Peter	22	Low
9	3	b	11-04-2015 9:08am	22.5	Mike	31	High
10	4	d	11-04-2015 10:03am	22.0	Mylos	33	High
11	4	c	11-04-2015 11:32am	23.2	Martin	35	High
12	3	a	11-04-2015 2:01pm	23.5	Silvia	40	Medium
13	3	d	11-04-2015 7:06pm	28.8	Mike	43	High
14	3	c	11-04-2015 9:08pm	22.9	Silvia	45	Medium
15	4	b	11-04-2015 10:28pm	23.0	Silvia	50	High
16	4	a	11-04-2015 10:40pm	23.1	Peter	35	Medium

For the sake of understandability, in this paper we will focus on the data corresponding to the temporal relation of activities, i.e., the *control-flow* perspective of a process. The reader can find in the literature references to also explore process mining techniques that go beyond this perspective.

Formally, an event log is a collection of traces, where a trace may appear more than once. Formally:

Definition 2 (Event Log). *An event log $L \in I\!B(\Sigma^*)$ (over an alphabet of actions Σ) is a multiset of traces. $\sigma \in \Sigma^*$ is a possible trace.*

The event log in Fig. 2 can be compactly represented as $L = [\langle po, si, py, pd, md, cp \rangle^{503}, \langle po, si, sr, py, pd, md, cp \rangle^{247}, \langle po, si, sr, sr, co \rangle^{141}, \langle po, si, sr, sr, py, pd, md, cp \rangle^{139}, \langle po, si, py, pd, cp, md \rangle^{135}, \langle po, si, sr, py, pd, cp, md \rangle^{57}, \langle po, si, sr, sr, py, pd, cp, md \rangle^{36}, \langle po, py, si, pd, md, cp \rangle^{6}, \langle po, py, si, pd, cp, md \rangle^{2}]$.

3   Discovering Process Models

This section focuses on process discovery and introduces various techniques. The goal is not to be complete, but to provide valuable insights into the different solution approaches.

3.1   Definition of Discovery

The input for process discovery is an event log $L \in I\!B(\Sigma^*)$. As mentioned before, we focus on control flow and leave out other perspectives (time, costs, data, etc.).

Process discovery algorithms take an event log as input and aim to output a process model that satisfies certain properties. To judge the quality of the discovered model the following four quality dimensions of process mining [1] are used:

- *fitness* (also called *recall*): the discovered model should allow for the behavior seen in the event log (avoiding "non-fitting" behavior),
- *precision*: the discovered model should not allow for behavior completely unrelated to what was seen in the event log (avoiding "underfitting"),
- *generalization*: the discovered model should generalize the example behavior seen in the event log (avoiding "overfitting"), and
- *simplicity*: the discovered model should not be unnecessarily complex.

The simplicity dimension refers to Occam's Razor: "one should not increase, beyond what is necessary, the number of entities required to explain anything". In the context of process mining, this is often operationalized by quantifying the complexity of the model (number of nodes, number of arcs, understandability, etc.). The other three dimensions typically abstract from the representation. This means that an event log $L \in \mathbb{B}(\Sigma^*)$ is a multiset of traces and a model $\mathcal{M} \subseteq \Sigma^*$ is simply seen as a set of traces (a language). Is this paper $\mathcal{M}$ corresponds to the language of a labeled accepting Petri net $N = \langle P, T, \mathcal{F}, M_0, M_f, \Sigma, \lambda \rangle$, i.e., $\mathcal{M} = \mathcal{L}(N)$.

Many conformance measures have been proposed throughout the years. There seems to be consensus on the four quality dimensions, but these can be operationalized in many different ways. In [13], 21 so-called conformance propositions were defined to discuss desirable properties of existing measures for recall, precision, and generalization. It has been shown that seemingly obvious conformance propositions are violated by existing approaches. Furthermore, [13] also shows the importance of probabilistic conformance measures that also take into account trace likelihoods in process models. However, to date, very few conformance measures exist that can actually support probabilistic process models.

Another approach to get handle on the problem of evaluating a process model in the context of an event log is to assume the existence of an underlying system $\mathcal{S} \subseteq \Sigma^*$ that generated the event log L. (Note that we assume $\mathcal{S}$ to be a language just like $\mathcal{M}$.). Process mining techniques aim at extracting a process model $\mathcal{L}(N) = \mathcal{M}$ from a log L with the goal to elicit the process underlying system $\mathcal{S}$. By relating the behaviors of L, $\mathcal{M}$ and $\mathcal{S}$, particular concepts can be defined [14]. A log is *incomplete* if $\mathcal{S} \setminus \{\sigma \in L\} \neq \emptyset$. A model $\mathcal{M}$ *fits* log L perfectly if $\{\sigma \in L\} \subseteq \mathcal{M}$. A model is *precise* in describing a log L if $\mathcal{M} \setminus \{\sigma \in L\}$ is small. A model $\mathcal{M} = \mathcal{L}(N)$ represents a *generalization* of log L with respect to system $\mathcal{S}$ if some behavior in $\mathcal{S} \setminus \{\sigma \in L\}$ exists in $\mathcal{M}$.

As shown in [1,13], the above line of thinking is rather naive when it comes to real-life problems. The following properties make process discovery and the evaluation of process discovery results particularly challenging.

- In reality one will never know the underlying system $\mathcal{S}$. This is only possible in a simulation setting.

- One cannot witness negative examples, i.e., the event log does not show what could not happen.
- The event log only contains a tiny fraction of the set of all possible traces.
- If the model has loops, then the set $\mathcal{M} \setminus \{\sigma \in L\}$ is infinitely large and expressions such as $\mathcal{M} \setminus \{\sigma \in L\}$ do not make any sense.
- If one observes a system for a longer time, one can see new behaviors. This may be due to the large number of possible variants, the low probability of some variants, or concept drift (the underlying system changes over time). Hence, Murphy's law for process mining states that "anything is possible if one waits long enough". Therefore, one cannot assume the existence of a system oracle that acts as a binary classifier, and therefore it is vital to take frequencies, incompleteness, and probabilities into account to get the full picture [13].

The above challenges explain why a range of discovery approaches have been developed over time. In the remainder of this section we aim to provide insights into a few representative examples.

3.2   Process Discovery Using Inductive Mining

Process models discovered from event logs describe possible life-cycles of cases (i.e., process instances). Such models have a start and end. In terms of Petri nets, we are interested in models that have an initial marking $M_0 \in \mathbb{B}(P)$ and a final marking $M_f \in \mathbb{B}(P)$. All accepted traces correspond to paths from M_0 to M_f. However, Petri nets may be deadlocking or livelocking. It may even be the case that there is no accepting trace (i.e., M_f is not reachable from M_0). The notion of soundness was first introduced in the context of workflow nets [15], but can be relaxed and generalized to accepting Petri nets. Soundness cannot be decided locally, yet process discovery techniques may need to make local decisions. Therefore, most process discovery techniques may generate unsound models. One way to ensure soundness is to only consider block-structured process models. Inductive process discovery, as presented in [9,10,16,17], uses *process trees* to ensure soundness. Moreover, the hierarchical nature of a process tree also enables a divide-and-conquer approach. Rather than using a single pass through the event log, it is also possible to try and break the problem into smaller problems. Inductive process discovery approaches split the event log recursively into smaller sublogs. For example, if one group of activities is preceded by another group of activities, but never the other way around, then we may deduce that these two groups are in a sequence relation. Subsequently, the event log is decomposed based on the two groups of activities. Next to the sequence relation, it is also possible to detect choices, concurrency, and loops, and split the log accordingly. This divide-and-conquer approach is repeated until each sublog refers to single activity. Leemans et al. developed a family of inductive mining techniques [9,10,16,17]. Some of these techniques are tailored to dealing with huge event logs and process models, other techniques address challenges such as infrequent behavior and incompleteness of logs. In this section, we only consider the basic algorithm also described in [1].

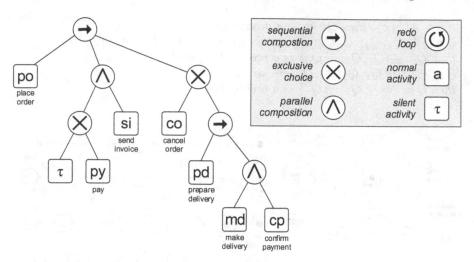

Fig. 2. Process tree $\rightarrow(po, \wedge(\times(\tau, py), si), \times(co, \rightarrow(pd, \wedge(md, cp))))$ discovered for the event log described in Tables 1 and 3 (without activity $sr = send\ reminder$). The process tree corresponds to models (a), (b), and (f) in Fig. 1.

Figure 2 shows a process tree. This process tree was discovered using the event log introduced in Sect. 1. Next to the visual tree representation, process trees also have a textual description $\rightarrow(po, \wedge(\times(\tau, py), si), \times(co, \rightarrow(pd, \wedge(md, cp))))$. The root node is of type $\rightarrow$ meaning that the three subtrees po, $\wedge(\times(\tau, py), si)$, and $\times(co, \rightarrow(pd, \wedge(md, cp)))$ are executed in sequence. The subtree in the middle $\wedge(\times(\tau, py), si)$ is of type $\wedge$ meaning that its two parts are executed concurrently. The subtree on the left $\times(co, \rightarrow(pd, \wedge(md, cp)))$ is of type $\times$ meaning that there is an exclusive choice between its two children.

Next to $\rightarrow$ (sequential composition), $\times$ (exclusive choice), and $\wedge$ (parallel composition), there is also the $\circlearrowleft$ (redo loop) operator. The redo loop operator $\circlearrowleft$ has at least two children. The first child is the "do" part and the other children are alternative "redo" parts. Process tree $\circlearrowleft(a, b, c)$ allows for traces $\{\langle a\rangle, \langle a, b, a\rangle, \langle a, c, a\rangle, \langle a, b, a, b, a\rangle, \langle a, c, a, c, a\rangle, \langle a, c, a, b, a\rangle, \langle a, b, a, c, a\rangle, \ldots\}$. Activity a is executed at least once and the process always starts and ends with a. The "do" part alternates with the "redo" parts b or c. When looping back either b or c is executed. The redo loop operator $\circlearrowleft$ is often used in conjunction with silent activity τ. For example, $\circlearrowleft(\tau, a, b, c, \ldots, z)$ allows for any trace (including the empty one) involving activities $a, b, c, \ldots, z$.

The same activity may appear multiple times in the same process tree. For example, process tree $\rightarrow(a, a, a)$ models a sequence of three a activities. From a behavioral point of view $\rightarrow(a, a, a)$ and $\wedge(a, a, a)$ are indistinguishable. Both allow for one possible trace: $\langle a, a, a\rangle$.

Definition 3 (Process tree). *Let $A \subseteq \Sigma$ be a finite set of activities with $\tau \notin \Sigma$. $\bigoplus = \{\rightarrow, \times, \wedge, \circlearrowleft\}$ is the set of process tree operators. Process trees are defined inductively:*

- *if* $a \in A \cup \{\tau\}$, *then* $Q = a$ *is a process tree,*
- *if* $n \geq 1$, $Q_1, Q_2, \ldots, Q_n$ *are process trees, and* $\oplus \in \{\rightarrow, \times, \wedge\}$,
  then $Q = \oplus(Q_1, Q_2, \ldots Q_n)$ *is a process tree, and*
- *if* $n \geq 2$ *and* $Q_1, Q_2, \ldots, Q_n$ *are process trees,*
  then $Q = \circlearrowleft(Q_1, Q_2, \ldots Q_n)$ *is a process tree.*

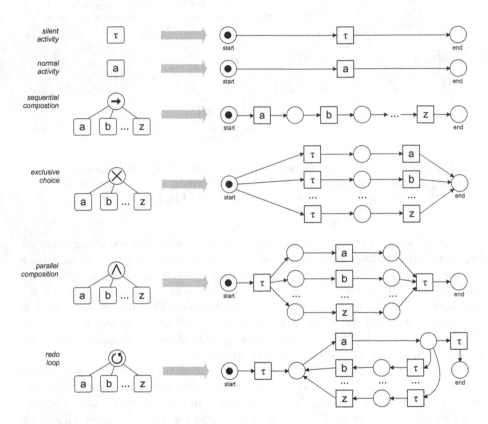

Fig. 3. Mapping process trees onto Petri to illustrate their semantics.

$\mathcal{L}(Q)$ is the *language* of a process tree Q. Formal definitions are provided in [1,9,10,16,17]. Here we only provide a mapping to Petri nets and some examples. Figure 3 shows the semantics of each operator in terms of Petri nets. The following examples further illustrate the process tree operators and their semantics: $\mathcal{L}(\tau) = \{\langle \ \rangle\}$, $\mathcal{L}(a) = \{\langle a \rangle\}$, $\mathcal{L}(\rightarrow(a,b,c)) = \{\langle a, b, c \rangle\}$, $\mathcal{L}(\times(a,b,c)) = \{\langle a \rangle, \langle b \rangle, \langle c \rangle\}$, $\mathcal{L}(\wedge(a,b,c)) = \{\langle a, b, c \rangle, \langle a, c, b \rangle, \langle b, a, c \rangle, \langle b, c, a \rangle, \langle c, a, b \rangle, \langle c, b, a \rangle\}$, $\mathcal{L}(\circlearrowleft(a,b)) = \{\langle a \rangle, \langle a, b, a \rangle, \langle a, b, a, b, a \rangle, \ldots\}$, $\mathcal{L}(\rightarrow(a, \times(b,c), \wedge(a,a))) = \{\langle a, b, a, a \rangle, \langle a, c, a, a \rangle\}$, $\mathcal{L}(\times(\tau, a, \tau, \rightarrow(\tau, b), \wedge(c, \tau))) = \{\langle \ \rangle, \langle a \rangle, \langle b \rangle, \langle c \rangle\}$, and $\mathcal{L}(\circlearrowleft(a, \tau, c)) = \{\langle a \rangle, \langle a, a \rangle, \langle a, a, a \rangle, \langle a, c, a \rangle, \langle a, a, c, a \rangle, \langle a, c, a, c, a \rangle, \ldots\}$.

Consider process tree $Q = \rightarrow(po, \wedge(\times(\tau, py), si), \times(co, \rightarrow(pd, \wedge(md, cp))))$, i.e., the process tree shown in Fig. 2. $\mathcal{L}(Q) = \{\langle po, si, py, co \rangle, \langle po, si, py, pd, md,$

$cp\rangle, \langle po, si, py, pd, cp, md\rangle, \langle po, py, si, co\rangle, \ \langle po, py, si, pd, md, cp\rangle, \langle po, py, si, pd, cp, md\rangle, \langle po, si, co\rangle, \langle po, si, pd, md, cp\rangle, \langle po, si, pd, cp, md\rangle\}$.

Given an event log $L \in \mathbb{B}(\Sigma^*)$ we would like to discover a process tree Q_L. For example, for event log $L = [\langle a, b, c, e\rangle^{85}, \langle a, c, b, e\rangle^{56}, \langle a, d, e\rangle^{34}]$ we would like to discover $Q = \rightarrow(a, \times(\wedge(b, c), d), e)$. For event log $L = [\langle a, b, d\rangle^{33}, \langle a, b, c, b, d\rangle^{25}, \langle a, b, c, b, c, b, d\rangle^{12}, \langle a, b, c, b, c, b, c, b, d\rangle^{6}, \langle a, b, c, b, c, b, c, b, c, b, d\rangle^{2}]$ we would like to discover $Q = \rightarrow(a, \circlearrowleft(b, c), d)$. The general idea of the inductive mining approach is to build a *directly-follows graph* and decompose the event log based on a particular *cut* of the directly-follows graph. The decomposition partitions the set of activities, i.e., sublogs are created in such a way that each event appears in precisely one of the sublogs. This is repeated until each sublog refers to only one activity. Note that each of the four operators, i.e., $\rightarrow$ (sequential composition), $\times$ (exclusive choice), $\wedge$ (parallel composition), and $\circlearrowleft$ (redo loops), corresponds to a particular type of cut and decomposes the event log accordingly.

To be able to define cuts, we first formalize the notion of a directly-follows graph.

Definition 4 (Directly-Follows Graph). *Let L be an event log, i.e., $L \in \mathbb{B}(\Sigma^*)$. The directly-follows graph of L is $G(L) = (A_L, \mapsto_L, A_L^{start}, A_L^{end})$ with:*

- *$A_L = \{a \in \sigma \mid \sigma \in L\}$ is the set of activities in L,*
- *$\mapsto_L = \{(a, b) \in A \times A \mid a >_L b\}$ is the directly-follows relation,[1]*
- *$A_L^{start} = \{a \in A \mid \exists_{\sigma \in L} a = first(\sigma)\}$ is the set of start activities, and*
- *$A_L^{end} = \{a \in A \mid \exists_{\sigma \in L} a = last(\sigma)\}$ is the set of end activities.*

The *Inductive Mining* (*IM*) algorithm iteratively splits the initial event log into smaller *sublogs*. For any sublog L we can create a directly-follows graph $G(L)$. $a \mapsto_L b$ if a was directly followed by b somewhere in L. $a \not\mapsto_L b$ if a was never directly followed by b. $\mapsto_L^+$ is the transitive closure of $\mapsto_L$. $a \mapsto_L^+ b$ if there is a non-empty *path* from a to b in $G(L)$, i.e., there exists a sequence of activities $a_1, a_2, \ldots, a_k$ such that $k \geq 2$, $a_1 = a$ and $a_k = b$ and $a_i \mapsto_L a_{i+1}$ for $i \in \{1, \ldots, k-1\}$. $a \not\mapsto_L^+ b$ if there is no path from a to b in the directly-follows graph.

Note that $a \mapsto_L b$ if a was directly followed by b only once in L. It is also possible to set thresholds to filter out infrequent behavior [16].

Definition 5 (Cut). *Let L be an event log with corresponding directly-follows graph $G(L) = (A_L, \mapsto_L, A_L^{start}, A_L^{end})$. Let $n \geq 1$. An n-ary cut of $G(L)$ is a partition of A_L into pairwise disjoint sets $A_1, A_2, \ldots A_n$: $A_L = \bigcup_{i \in \{1, \ldots, n\}} A_i$ and $A_i \cap A_j = \emptyset$ for $i \neq j$. Notation: $(\oplus, A_1, A_2, \ldots A_n)$ with $\oplus \in \{\rightarrow, \times, \wedge, \circlearrowleft\}$. For each type of operator ($\rightarrow$, $\times$, $\wedge$, and $\circlearrowleft$) specific conditions apply:*

- *An* exclusive-choice cut *of $G(L)$ is a cut $(\times, A_1, A_2, \ldots A_n)$ such that*
  - *$\forall_{i,j \in \{1, \ldots n\}} \forall_{a \in A_i} \forall_{b \in A_j} \ i \neq j \implies a \not\mapsto_L b$.*

---

[1] $a >_L b$ if and only if there is a trace $\sigma = \langle t_1, t_2, t_3, \ldots t_n\rangle$ and $i \in \{1, \ldots, n-1\}$ such that $\sigma \in L$ and $t_i = a$ and $t_{i+1} = b$.

- A sequence cut *of $G(L)$ is a cut* $(\rightarrow, A_1, A_2, \ldots A_n)$ *such that*
  - $\forall_{i,j \in \{1,\ldots n\}} \forall_{a \in A_i} \forall_{b \in A_j} \; i < j \implies (a \mapsto_L^+ b \land b \not\mapsto_L^+ a)$.
- A parallel cut *of $G(L)$ is a cut* $(\land, A_1, A_2, \ldots A_n)$ *such that*
  - $\forall_{i \in \{1,\ldots n\}} \; A_i \cap A_L^{start} \neq \emptyset \land A_i \cap A_L^{end} \neq \emptyset$ *and*
  - $\forall_{i,j \in \{1,\ldots n\}} \forall_{a \in A_i} \forall_{b \in A_j} \; i \neq j \implies a \mapsto_L b$.
- A redo-loop cut *of $G(L)$ is a cut* $(\circlearrowleft, A_1, A_2, \ldots A_n)$ *such that*
  - $n \geq 2$,
  - $A_L^{start} \cup A_L^{end} \subseteq A_1$,
  - $\{a \in A_1 \mid \exists_{i \in \{2,\ldots n\}} \exists_{b \in A_i} \; a \mapsto_L b\} \subseteq A_L^{end}$,
  - $\{a \in A_1 \mid \exists_{i \in \{2,\ldots n\}} \exists_{b \in A_i} \; b \mapsto_L a\} \subseteq A_L^{start}$,
  - $\forall_{i,j \in \{2,\ldots n\}} \forall_{a \in A_i} \forall_{b \in A_j} \; i \neq j \implies a \not\mapsto_L b$,
  - $\forall_{i \in \{2,\ldots n\}} \forall_{b \in A_i} \exists_{a \in A_L^{end}} \; a \mapsto_L b \implies \forall_{a' \in A_L^{end}} \; a' \mapsto_L b$, *and*
  - $\forall_{i \in \{2,\ldots n\}} \forall_{b \in A_i} \exists_{a \in A_L^{start}} \; b \mapsto_L a \implies \forall_{a' \in A_L^{start}} \; b \mapsto_L a'$.

A cut $(\oplus, A_1, A_2, \ldots A_n)$ *with* $\oplus \in \{\rightarrow, \times, \land, \circlearrowleft\}$ *of directly-follows graph* $G(L)$ *is* maximal *if there is no cut* $(\oplus, A_1', A_2', \ldots A_m')$ *with* $m > n$. *Cut* $(\oplus, A_1, A_2, \ldots A_n)$ *is called* trivial *if $n = 1$.*

Definition 6 (Projection). *Let L be an event log and $(\oplus, A_1, A_2, \ldots A_n)$ a cut with $\oplus \in \{\rightarrow, \times, \land, \circlearrowleft\}$ based on the directly-follows graph $G(L)$ ($n \geq 2$). L is split into sublogs $L_1, L_2, \ldots, L_n$ such that each event ends up in precisely one log and $\cup_{\sigma \in L_i} \{a \in \sigma\} = A_i$ for any $1 \leq i \leq n$.*

The precise way in which the event log is split depends on the operator [9,10,16]. Consider cut $(\rightarrow, \{a\}, \{b, c, d\}, \{e\})$ in the context of $L = [\langle a, b, c, e \rangle^{85}, \langle a, c, b, e \rangle^{56}, \langle a, d, e \rangle^{34}]$. L will be split into $L_1 = [\langle a \rangle^{175}]$, $L_2 = [\langle b, c \rangle^{85}, \langle c, b \rangle^{56}, \langle d \rangle^{34}]$, and $L_3 = [\langle e \rangle^{175}]$. Consider cut $(\times, \{a, b\}, \{c, d\})$ in the context of $L = [\langle a, b \rangle^{10}, \langle b, a \rangle^{10}, \langle c, d \rangle^{20}]$. L will be split into $L_1 = [\langle a, b \rangle^{10}, \langle b, a \rangle^{10}]$ and $L_2 = [\langle c, d \rangle^{20}]$. Consider cut $(\land, \{a, b\}, \{c\})$ in the context of $L = [\langle a, b, c \rangle^{10}, \langle b, a, c \rangle^{10}, \langle a, c, b \rangle^{10}, \langle b, c, a \rangle^{10}, \langle c, a, b \rangle^{10}, \langle c, b, a \rangle^{10}]$. L will be split into $L_1 = [\langle a, b \rangle^{30}, \langle b, a \rangle^{30}]$ and $L_2 = [\langle c \rangle^{60}]$. Consider cut $(\circlearrowleft, \{a, b\}, \{c, d\})$ in the context of $L = [\langle a, b \rangle^{10}, \langle a, b, c, d, a, b \rangle^{4}, \langle a, b, c, d, a, b, c, d, a, b \rangle^{2}]$. L will be split into $L_1 = [\langle a, b \rangle^{24}]$ and $L_2 = [\langle c, d \rangle^{8}]$. Note that each iteration creates a new case, e.g., case $\langle a, b, c, d, a, b, c, d, a, b \rangle$ is split into three $\langle a, b \rangle$ cases and two $\langle c, d \rangle$ cases.

The IM *algorithm works as follows* [1]. *IM* is a function that converts an event log into a process tree. Given a log or sublog L, $Q = IM(L)$ is the corresponding (sub)tree. Given an event log, the directly-follows graph is constructed. If there is a non-trivial exclusive-choice cut, then a maximal exclusive-choice cut is applied, splitting the event log into smaller event logs. If there is no non-trivial exclusive-choice cut, but there is a non-trivial sequence cut, then a maximal sequence cut is applied splitting the event log into smaller event logs. If there are no non-trivial exclusive-choice and sequence cuts, but there is a non-trivial parallel cut, then a maximal parallel cut is applied splitting the event log into smaller event logs. If there are no non-trivial exclusive-choice, sequence and parallel cuts, but there is a redo-loop cut, then a maximal redo-loop cut is applied splitting the event log into smaller event logs. After splitting the event log into sublogs the procedure is repeated until a *base case* (sublog with only one activity) is reached.

How the event log is split into sublogs, depends on the operator (see before). Empty traces are handled in a dedicated manner (based on the operator) and may result in the insertion of τ activities. If there are no non-trivial cuts meeting the requirements in Definition 5, a *fall-through* is selected. The part that cannot be split is presented by a so-called *flower model* ("anything can happen"). Note that such a model can be easily represented as process tree $\circlearrowright(\tau, a, b, \ldots)$ allowing for any trace involving the activities in the different redo parts. The fall-through serves as a last resort ensuring fitness, but possibly resulting in lower precision.

In the base case, the sublog contains only events corresponding to a particular activity, say a. If the sublog is of the form $L = [\langle a \rangle^k]$ with $k \geq 1$ (i.e., a occurs once in each trace), then the subtree a is returned. If the sublog is of the form $L = [\langle \ \rangle^k, \langle a \rangle^l]$ with $k, l \geq 1$, then the subtree $\times(a, \tau)$ is returned because a is sometimes skipped. If a is executed at least once in each trace in the sublog and sometimes multiple times (e.g., $L = [\langle a \rangle^9, \langle a, a \rangle^2, \langle a, a, a \rangle]$), then the subtree $\circlearrowright(a, \tau)$ is returned. In all other cases (e.g., $L = [\langle \ \rangle^3, \langle a \rangle^4, \langle a, a, a \rangle]$), the subtree $\circlearrowright(\tau, a)$ is returned because a is executed zero or more times in the traces of sublog L.

First, we show a larger worked out example showing the iterative process of splitting the event logs into sublogs based on cuts.

- Let $L_{abcdef} = [\langle a, b, c, d \rangle^3, \langle a, c, b, d \rangle^4, \langle a, b, c, e, f, b, c, d \rangle^2, \langle a, c, b, e, f, b, c, d \rangle^2,$ $\langle a, b, c, e, f, c, b, d \rangle, \langle a, c, b, e, f, b, c, e, f, c, b, d \rangle]$ be an event log. Based on the directly-follows graph we identify a maximal sequence cut $(\rightarrow, \{a\}, \{b, c, e, f\}, \{d\})$ splitting the event log into L_a, L_{bcef}, L_d.
  - $L_a = [\langle a \rangle^{13}]$ is a base case. Hence, $IM(L_a) = a$.
  - $L_{bcef} = [\langle b, c \rangle^3, \langle c, b \rangle^4, \langle b, c, e, f, b, c \rangle^2, \langle c, b, e, f, b, c \rangle^2, \langle b, c, e, f, c, b \rangle, \langle c, b, e, f, b, c, e, f, c, b \rangle]$. There are no non-trivial exclusive-choice, sequence or parallel cuts. Therefore, we apply the maximal redo-loop cut $(\circlearrowright, \{b, c\}, \{e, f\})$ splitting the event log into L_{bc} and L_{ef}.
    * $L_{bc} = [\langle b, c \rangle^{11}, \langle c, b \rangle^9]$. Based on the maximal parallel cut $(\wedge, \{b\}, \{c\})$, we obtain L_b and L_c.
      . $L_b = [\langle b \rangle^{20}]$ is a base case. Hence, $IM(L_b) = b$.
      . $L_c = [\langle c \rangle^{20}]$ is a base case. Hence, $IM(L_c) = c$.
    * $L_{ef} = [\langle e, f \rangle^7]$. Based on the maximal sequence cut $(\rightarrow, \{e\}, \{f\})$, we obtain L_e and L_f.
      . $L_e = [\langle e \rangle^7]$ is a base case. Hence, $IM(L_e) = e$.
      . $L_f = [\langle f \rangle^7]$ is a base case. Hence, $IM(L_f) = f$.
  - $L_d = [\langle d \rangle^{13}]$ is a base case. Hence, $IM(L_d) = d$.
- After splitting the event log to reach the base cases, we can construct the overall tree:
  - $IM(L_{b,c}) = \wedge(IM(L_b), IM(L_c)) = \wedge(b, c)$.
  - $IM(L_{e,f}) = \rightarrow(IM(L_e), IM(L_f)) = \rightarrow(e, f)$.
  - $IM(L_{bcef}) = \circlearrowright(IM(L_{b,c}), IM(L_{e,f})) = \circlearrowright(\wedge(b, c), \rightarrow(e, f))$.
  - $IM(L_{abcdef}) = \rightarrow(IM(L_a), IM(L_{bcef}), IM(L_d)) = \rightarrow(a, \circlearrowright(\wedge(b, c), \rightarrow(e, f)), d)$ represents the overall process tree for the whole event log.

To illustrate the handling of base cases we show a few smaller examples:

- If $L = [\langle a, a \rangle^{12}, \langle a, a, a \rangle^6]$, then $IM(L) = \circlearrowleft(a, \tau)$.
- If $L = [\langle a, b, c \rangle^{20}, \langle a, c \rangle^{30}]$, then $IM(L) = \rightarrow(a, \times(b, \tau), c)$.
- If $L = [\langle b \rangle^{12}, \langle a, b \rangle^6, \langle b, c \rangle^5, \langle a, b, c \rangle^4]$, then $IM(L) = \rightarrow(\times(a, \tau), b, \times(c, \tau))$.
- If $L = [\langle a, c \rangle^2, \langle a, b, c \rangle^3, \langle a, b, b, c \rangle^2, \langle a, b, b, b, c \rangle^2, \langle a, b, b, b, b, b, b, c \rangle]$, then $IM(L) = \rightarrow(a, \circlearrowleft(\tau, b), c)$.

In this section, we only described the basic *IM* algorithm [1,10]. This algorithm provides many guarantees. *The discovered models are always sound and are guaranteed to be able to replay the whole event log.* Moreover, for large subclasses of process trees, rediscoverability is guaranteed (i.e., a sufficiently large event log obtained by simulating a model is sufficient to reconstruct a behaviorally equivalent model). However, the basic algorithm presented here cannot abstract from infrequent behavior and does not handle incompleteness well. The log is assumed to be directly-follows complete and frequencies are not taken into account. Fortunately, the inductive mining framework is quite flexible. Using the basic ideas presented in this section, a family of inductive mining techniques has been developed [1,9,10,16,17]. All use a *divide-and-conquer* approach in combination with process trees that are *sound by construction.* As demonstrated in [9,17] the approach can be made highly scalable and can be applied to huge event logs.

3.3   Process Discovery Using Region-Based Approaches

In contrast to inductive mining, which is able to guarantee a sound workflow model, the existing approaches that rely on the notion of *region theory* [18] search for a process model that is both fitting and precise [19]. This section shows two branches of region-based approaches for process discovery: state and language-based approaches.

State-Based Region Approach for Process Discovery. State-based region approaches for process discovery need to convert the event log into a state-based representation, that will be used to discover the Petri net. The techniques described in [20] present many variants for solving this first step. The basic idea to incorporate state information is to look at the pre/post history of a subtrace in the event log. Figure 4(a)–(b) show an example, where states are decided by looking at the set of common prefixes.

A *transition system* (TS) is a tuple (S, Σ, A, s_{in}), where S is a set of *states*, Σ is an alphabet of *activities*, $A \subseteq S \times \Sigma \times S$ is a set of *(labeled) arcs*, and $s_{in} \in S$ is the *initial state*. We will use $s \xrightarrow{e} s'$ as a shortcut for $(s, e, s') \in A$, and the transitive closure of this relation will be denoted by $\xrightarrow{*}$. Figure 4(b) presents an example of a transition system.

1	r,s,sb,p,ac,ap,c
2	r,sb,em,p,ac,ap,c
3	r,sb,p,em,ac,rj,rs,c
4	r,em,sb,p,ac,ap,c
5	r,sb,s,p,ac,rj,rs,c
6	r,sb,p,s,ac,ap,c
7	r,sb,p,em,ac,ap,c

(a)

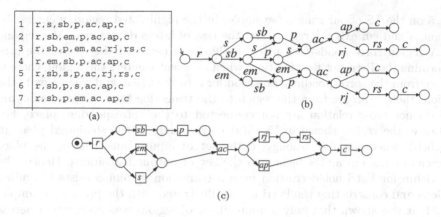

(b)

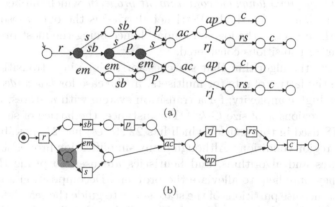

(c)

Fig. 4. State-based region discovery: (a) log L, (b) a transition system corresponding to L, (c) derived Petri net.

(a)

(b)

Fig. 5. (a) Example of region (three shadowed states). The predicates are r enters, s and em exits, and the rest of events do not cross, (b) Corresponding place shadowed in the Petri net.

A *region*[2] in a transition system is a set of states that satisfy an homogeneous relation with respect to the set of arcs. In the simplest case, this relation can be described by a predicate on the set of states considered. Formally, let S' be a subset of the states of a TS, $S' \subseteq S$. If $s \notin S'$ and $s' \in S'$, then we say that transition $s \xrightarrow{a} s'$ *enters* S'. If $s \in S'$ and $s' \notin S'$, then transition $s \xrightarrow{a} s'$ *exits* S'. Otherwise, transition $s \xrightarrow{a} s'$ *does not cross* S': it is completely inside ($s \in S'$ and $s' \in S'$) or completely outside ($s \notin S'$ and $s' \notin S'$). A set of states $r \subseteq S$ is a region if for each event $e \in E$, exactly one of the three predicates (*enters*, *exits* or *does not cross*) holds for each of its arcs. An example of region is presented in

---

[2] In this paper we will use region to denote a 1-bounded region. However, when needed we will use k-bounded region to extend the notion, necessary to account for k-bounded Petri nets.

Fig. 5 on the TS of our running example. In the highlighted region, r enters the region, s and em exit the region, and the rest of labels do not cross the region.

A region corresponds to a place in the Petri net, and the role of the arcs determine the Petri net flow relation: when an event e enters the region, there is an arc from the corresponding transition for e to the place, and when e exits the region, there is an arc from the region to the transition for e. Events satisfying the do not cross relation are not connected to the corresponding place. For instance, the region shown in Fig. 5(a) corresponds to the shadowed place in Fig. 5(b), where event r belongs to the set of input transitions of the place whereas events em and s belong to the set of output transitions. Hence, the algorithm for Petri net derivation from a transition system consists in finding regions and constructing the Petri net as illustrated with the previous example. In [21] it was shown that only a minimal set of regions was necessary, whereas further relaxations to this restriction can be found in [19]. The Petri net obtained by this method is guaranteed to accept the language of the transition system, and satisfy the *minimal language containment property*, which implies that if all the minimal regions are used, the Petri net derived is the one whose language difference with respect to the log is minimal, hence being the most precise Petri net for the set of transitions considered.

In any case, the algorithm that searches for regions in a transition system must explore the lattice of sets (or multisets, in the case for *k-bounded* regions), thus having a high complexity: for a transition system with n states, the lattice for k-bounded regions is of size $\mathcal{O}(k^n)$. For instance, the lattice of sets of states for the toy TS used in this article (which has 22 states) has 2^{22} possibles sets to check for the region conditions. Although many simplification properties, efficient data structures and algorithms, and heuristics are used to prune this search space [19], they only help to alleviate the problem. Decomposition alternatives, which for instance use partitions of the state space to guide the search for regions, significantly alleviate the complexity of the state-based region algorithm, at the expense of not guaranteeing the derivation of precise models [22]. Other state-based region approaches for discovery have been proposed, which complement the approach described in this section [23–25].

Language-Based Region Approach for Process Discovery. In Language-based region theory [26–31] the goal is to construct the smallest Petri net such that the behavior of the net is equal to the given input language (or minimally larger). [32] provides an overview for language-based region theory for different classes of languages: step languages, regular languages, and (infinite) partial languages.

More formally, let $L \in \mathbb{B}(\Sigma^*)$ be an event log, then language based region theory constructs a Petri net with the set of transitions equals to Σ and in which all traces of L are a firing sequence. The Petri net should have only minimal firing sequences not in the language L (and all prefixes in L). This is achieved by adding places to the Petri net that restrict unobserved behavior, while allowing for observed behavior. The theory of regions provides a method to identify these places, using *language regions*.

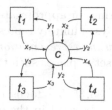

Fig. 6. Region for a language over four activities [33].

Definition 7 (Prefix Closure). *Let $L \in \mathbb{B}(\Sigma^*)$ be an event log. The prefix closed language $\mathcal{L} \subseteq \Sigma^*$ of L is defined as: $\mathcal{L} = \{\sigma \in \Sigma^* \mid \exists_{\sigma' \in \Sigma^*} \sigma \circ \sigma' \in L\}$.*

The prefix closure of a log is simply the set of all prefixes in the log (including the empty prefix).

Definition 8 (Language Region). *Let Σ be a set of activities. A region of a prefix-closed language $\mathcal{L} \in \Sigma^*$ is a triple $(\boldsymbol{x}, \boldsymbol{y}, c)$ with $\boldsymbol{x}, \boldsymbol{y} \in \{0,1\}^\Sigma$ and $c \in \{0,1\}$, such that for each non-empty sequence $w = w' \circ a \in \mathcal{L}$, $w' \in \mathcal{L}$, $a \in \Sigma$:*

$$c + \sum_{t \in \Sigma} (\boldsymbol{w'}(t) \cdot \boldsymbol{x}(t) - \boldsymbol{w}(t) \cdot \boldsymbol{y}(t)) \geq 0$$

This can be rewritten into the inequation system:

$$c \cdot 1 + M' \cdot \boldsymbol{x} - M \cdot \boldsymbol{y} \geq 0$$

where M and M' are two $|\mathcal{L}| \times |\Sigma|$ matrices with $M(w, t) = \boldsymbol{w}(t)$, and $M'(w, t) = \boldsymbol{w'}(t)$, with $w = w' \circ a$. The set of all regions of a language is denoted by $\mathfrak{R}(\mathcal{L})$ and the region $(\boldsymbol{0}, \boldsymbol{0}, 0)$ is called the trivial region.

Intuitively, vectors $\boldsymbol{x}$, $\boldsymbol{y}$ denote the set of incoming and outgoing arcs of the place corresponding to the region, respectively, and c sets if it is initially marked. Figure 6 shows a region for a language over four activities, i.e. each solution $(\boldsymbol{x}, \boldsymbol{y}, c)$ of the inequation system can be regarded in the context of a Petri net, where the region corresponds to a feasible place with preset $\{t | t \in T, \boldsymbol{x}(t) = 1\}$ and postset $\{t | t \in T, \boldsymbol{y}(t) = 1\}$, and initially marked with c tokens. Note that we do not assume arc-weights here, while the authors of [26–28, 34] do.

Since the place represented by a region is a place which can be added to a Petri net, without disturbing the fact that the net can reproduce the language under consideration, such a place is called a *feasible* place.

Definition 9 (Feasible place). *Let $\mathcal{L}$ be a prefix-closed language over Σ and let $N = ((P, \Sigma, F), m)$ be a marked Petri net. A place $p \in P$ is called* feasible *if and only if there exists a corresponding region $(\boldsymbol{x}, \boldsymbol{y}, c) \in \mathfrak{R}(\mathcal{L})$ such that $m(p) = c$, and $\boldsymbol{x}(t) = 1$ if and only if $t \in {}^\bullet p$, and $\boldsymbol{y}(t) = 1$ if and only if $t \in p^\bullet$.*

In general, there are many feasible places for any given event log (when considering arc-weights in the discovered Petri net, there are even infinitely many).

Several methods exist for selecting an appropriate subset of these places. The authors of [28,34] present two ways of finitely representing these places, namely a *basis representation* and a *separating representation*. Both representations maximize precision, i.e. they select a set of places such that the behavior of the model outside of the log is minimal.

In contrast, the authors of [33,35–37] focus on those feasible places that express some causal dependency observed in the event log, and/or ensure that the entire model is a connected workflow net. They do so by introducing various cost functions favouring one solution of the equation system over another and then selecting the top candidates.

Process Discovery vs. Region Theory. The goal of region theory is to find a Petri net that perfectly describes the observed behavior (where this behavior is specified in terms of a language or a statespace). As a result the Petri nets are perfectly fitting and maximally precise.

As a consequence, the assumption on the input is that it provides a *full behavioral specification*, i.e. that the input is *complete* and *noise free*. Furthermore, the assumption on the output is that it is a *compact, exact representation* of the input behavior.

When applying region theory in the context of process mining, it is therefore very important to perform any generalization *before* calling region theory algorithms. For state-based regions, the challenges are in the construction of the statespace from the event log and in language based regions in the selection of the appropriate prefixes to include in the final prefix-closed language in order to ensure some level of generalization.

4 Conformance Checking and the Challenge of Alignments

Conformance checking is a crucial dimension in process mining: by relating modelled and observed behavior, process models that have either been discovered or manually created, can be confronted with event data [2]. On its core, conformance checking relies on the fundamental problem of identifying, among the set of runs of a process model (which can be infinite), the run that mostly resembles an observed trace. In this section we overview the problem of computing alignments, and provide applications to be build on top of alignments.

4.1 Formal Definition of Alignments

An *alignment* of an observed trace and a process model relates events of the observed trace to elements of the model and vice versa. Such an alignment reveals how the given trace can be replayed on the process model. The classical notion of aligning an event log and process model was introduced by [38]. To achieve an alignment, we need to relate *moves* in the observed trace to *moves* in the

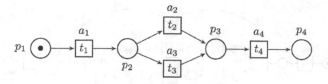

Fig. 7. Process model N_1.

model. It may be the case that some of the moves in the observed trace can not be mimicked by the model and vice versa. For instance, consider the model N_1 in Fig. 7, with the following labels, $\lambda(t_1) = a_1, \lambda(t_2) = a_2, \lambda(t_3) = a_3$ and $\lambda(t_4) = a_4$, and the trace $\sigma = \langle a_1, a_1, a_4, a_2 \rangle$; four possible alignments are:

$$\alpha_1 = \begin{array}{|c|c|c|c|c|} a_1 & a_1 & \bot & a_4 & a_2 \\ \hline t_1 & \bot & t_3 & t_4 & \bot \end{array} \quad \alpha_2 = \begin{array}{|c|c|c|c|c|} a_1 & a_1 & \bot & a_4 & a_2 \\ \hline \bot & t_1 & t_2 & t_4 & \bot \end{array}$$

$$\alpha_3 = \begin{array}{|c|c|c|c|c|} a_1 & a_1 & a_4 & a_2 & \bot \\ \hline t_1 & \bot & \bot & t_2 & t_4 \end{array} \quad \alpha_4 = \begin{array}{|c|c|c|c|c|} a_1 & a_1 & a_4 & a_2 & \bot \\ \hline \bot & t_1 & \bot & t_2 & t_4 \end{array}$$

The moves are represented in tabular form, where moves by the trace are at the top, and moves by the model are at the bottom of the table. For example the first move in α_2 is $(a_1, \bot)$ and it means that the observed trace moves a_1, while the model does not make any move. Formally, an alignment is defined as follows:

Definition 10 (Alignment). *Given a labeled Petri net N and an alphabet of events Σ, Let A_M and A_L be the alphabet of transitions in the model and events in the log, respectively, and $\bot$ denote the empty set, then:*

- *(X, Y) is a synchronous move if $X \in A_L$, $Y \in A_M$ and $X = \lambda(Y)$*
- *(X, Y) is a move in log if $X \in A_L$ and $Y = \bot$.*
- *(X, Y) is a move in model if $X = \bot$ and $Y \in A_M$.*
- *(X, Y) is an illegal move, otherwise.*

The set of all legal moves is denoted as A_{LM} and given an alignment $\alpha \in A^*_{LM}$, the projection of the first element (ignoring $\bot$), $\alpha \upharpoonright_{A_L}$, results in the observed trace σ, and projecting the second element (ignoring $\bot$), $\alpha \upharpoonright_{A_M}$, results in the model trace.

For the previous example, $\alpha_1 \upharpoonright_{A_M} = t_1 t_3 t_4$ and $\alpha_1 \upharpoonright_{A_L} = a_1 a_1 a_4 a_2$.

Costs can be associated to the different types of moves in Definition 10. Traditionally, the approaches in the literature use a cost function that assigns higher costs to asynchronous moves (move in model/log) than to synchronous moves, and the model trace that minimizes the cost (hence, minimizing the number of asynchronous moves) is computed. When a cost function is in place, then one can consider optimality: an *optimal alignment* is an alignment with minimal cost. The most simple cost function that satisfies this requirement is the *standard cost function*, which assigns cost 1 to asynchronous moves, and cost

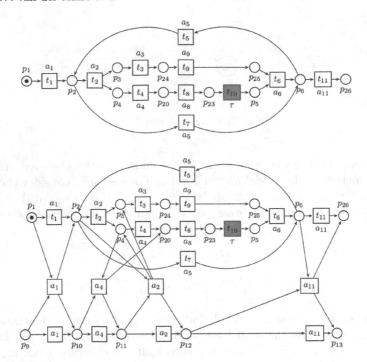

Fig. 8. (Top) Process model, (Bottom) Synchronous product net over the trace $\langle a_1, a_4, a_2, a_{11} \rangle$.

0 to synchronous moves. In this paper we will assume the standard cost function, which will for instance assign cost 3 to the alignments $\alpha_1 - \alpha_4$ shown before. According to the standard cost function, all four alignments are optimal.

4.2   Techniques for the Computation of Alignments

In this section we report some of the existing alternatives to compute alignments. First we describe the reference technique nowadays for alignment computation. Then, we provide pointers to other techniques so that the reader gets an overall impression on the vivid field of alignment computation.

A* Technique over the Synchronous Product Net. The reference technique for alignment computation was presented in the context of Arya Adriansyah's PhD thesis [38]. It is based on the notion of *synchronous product net* (SPN, for short), which we define informally now with the help of an example.

Figure 8 (Top) describes a process model. Now let us assume that it should be aligned with the trace $\sigma = \langle a_1, a_4, a_2, a_{11} \rangle$. The idea underlying the method from [38] is to first create an SPN, that encompasses the joint behavior between the process model and σ. This SPN is described in Fig. 8 (Bottom). Transitions in the SPN can be partitioned into three sets

- Transitions at the top (i.e., $t_1, \ldots t_{11}$) correspond to original model transitions.
- Transitions at the bottom correspond to the Petri net representation of σ.
- Transitions in the middle represent the joint synchronization of the model and the trace states.

Accordingly, transitions in the SPN will be assigned a cost[3]: transitions in the top of the SPN, and in the bottom of the SPN, will receive cost one. Transitions in the middle, will receive cost 0. Informally, this cost assignment penalizes the process model or the log independent executions, and in contrast favours synchronous executions, that are executed without cost. Then, a search for the cost-minimal path between initial and final marking of the state space of the created SPN is computed.

The cost-minimal path search can be done either once the full state-space is computed, or more intelligently by applying a A^* strategy to avoid, whenever possible, the full exploration of the state-space. This can be achieved by using heuristics that at each state reached, estimate the minimal cost to reach the final marking, and prune the exploration for successor states that are not promising. Several heuristics can be applied, that include the use of the *marking equation of Petri nets* [12] to estimate the remaining distance by solving an (integer) linear program that provides a lower bound to the real distance.

Other Techniques. Alternatives to A^* have appeared very recently: in the approach presented in [39], the alignment problem is mapped as an *automated planning* instance. Unlike the A^*, the aforementioned work is only able to produce one optimal alignment (not all optimal), but it is expected to consume considerably less memory. Automata-based techniques have also appeared [40, 41]. In particular, the technique in [40] can compute all optimal alignments. The technique in [40] relies on state space exploration and determinization of automata, whilst the technique in [41] is based on computing several subsets of activities and projecting the alignment instances accordingly.

The work in [42] presented the notion of *approximate* alignment to alleviate the computational demands of the current challenge by proposing a recursive paradigm on the basis of structural theory of Petri nets. In spite of resource efficiency, the solution is not guaranteed to be executable. A follow-up work of [42] is presented in [43], which proposes a trade-off between complexity and optimality of solutions, and guarantees executable properties of results. The technique in [44] presents a framework to reduce a process model and the event log accordingly, with the goal to alleviate the computation of alignments. The obtained alignment, called *macro-alignment* since some of the positions are high-level elements, is expanded based on the information gathered during the initial reduction. Decompositional techniques have been presented [45, 46] that, instead of computing optimal alignments, they focus on the *decisional problem* of whereas a given trace fits or not a process model.

---

[3] Remember that we are assuming the standard cost function that assigns cost 1 to synchronous moves and cost 0 to asynchronous moves.

Recently, two different approaches by the same authors have appeared: the work in [47] proposes using binary decision diagrams to alleviate the computation of alignments. The work in [48], which has the goal of maximizing the synchronous moves of the computed alignments, uses a pre-processing step on the model.

4.3   Alignments Applications

In this section we overview different use cases of alignments. The reader can find a complete and detailed presentation in [2].

Model Enhancement. Alignments open the door to incorporate information from the event log to the process model. More concretely, from an alignment α, one can transfer the information from the log trace $\alpha \uparrow_{A_L}$ to the model activities in $\alpha \uparrow_{A_M}$ corresponding to the synchronous moves. This information is contained in the events attributes. For instance, after aligning the complete log, one can realize that in reality a given model transition is only performed by a limited set of persons or roles, or the cost of executing it is within certain margins.

Alignments can also be used to *animate* the process model. This has already illustrated early in the paper (see for instance the annotations in Fig. 1(a)). This helps to understand better the visualization of real traces through the process model, and is one of the most interesting features of several existing process mining tools.

Furthermore, more elaborated information can be projected on top of the process model, with the help of a tailored analysis. One possibility is *performance analysis*, in order to display performance information like *activity durations*, *waiting times* and *routing probabilities*. For instance, in the left-most decision of Fig. 1 (a), one can see that 11.1% (141 out of 1266) of the cases the activity *pay* is skipped.

Alternatively, the event log attributes can be used to explain decision in process models, once alignments are obtained. This is known as *decision point analysis*. Decision point analysis is based on building prediction models next to each decision point, using the data available. Often, the data used to feed these models can be obtained from the events corresponding to the prefixes of traces that lead to the decision point. Examples of data attributes that can be used for the previous example are product, resource, prod-price, quantity and address in Table 1. Continuing with the previous example, one can for instance infer that the explanation on why the activity *pay* is sometimes skipped is due to products orders placed which both contain more than one item and the item price is less than 500.

Model and Log Repair. Another interesting application of using alignments is the possibility to repair the model or the log, so that they are more aligned. Model repair can be performed by selecting and resolving particular asynchronous moves in an alignment [49]. Intuitively, to resolve model moves one need to insert routing logic that allows to skip certain activities to be executed. Symmetrically, to resolve log moves one should extend the process model so that new behavior is possible in particular situations.

On the other hand, log repair considers that models contain the truth and whenever deviations are found, they should be corrected at the log level. Correcting deviations at log level is straightforward: model moves amounts to insert events in the observed trace, whilst log moves imply to remove events in the observed trace. For instance, the log repair of the alignment α_1 for trace $\sigma = \langle a_1, a_1, a_4, a_2 \rangle$, and the model shown in Fig. 7, derives the new trace $\sigma' = \langle a_1, a_3, a_4 \rangle$.

5   Evidence-Based Quality Metrics for Process Models

With the aim of quantifying the relation between observed and modeled behavior, conformance checking techniques consider four quality dimensions: fitness, precision, generalization and simplicity [50]. In Sect. 3.1 we already described the intuition behind these dimensions. In this section we do a step further, and present ways to measure them.

When alignments are available, most of the quality dimensions can be defined on top [2]. In a way, alignments are optimistic: although observed behavior may deviate significantly from modeled behavior, it is always assumed that the least deviations are the best explanation (from the model's perspective) for the observed behavior. For the first three dimensions, the alignment between a process model and an event log is of paramount importance, since it allows relating modelled and observed behavior.

5.1   Fitness

Fitness evaluates to which extent the observed behavior is possible according to the modelled behavior. Intuitively (and abusing a bit the notation), if L is also considered the language of the log, then the fitness of L with respect to a process model N can be computed by the following formula:

$$fitness = \frac{|L \cap \mathcal{L}(N)|}{|L|}$$

Alternatively, a refined metric for fitness can be measured through alignments. The way alignments are constructed, i.e. by looking for a shortest path through the state space of the SPN, is not necessarily deterministic. There may be more than one shortest path. However, the final cost of the alignment is minimal and therefore deterministic. On the basis of this cost, alignment-based fitness is defined as:

$$fitness_a = 1 - \frac{\text{cost of the optimal alignment}}{\text{cost of worst-case alignment}}$$

Again, for any log L we have:

$$fitness_a^l = 1 - \frac{\sum_{\sigma \in L} \text{cost of the optimal alignment for } \sigma}{\sum_{\sigma \in L} \text{cost of worst-case alignment for } \sigma}$$

For alignment-based fitness, two costs are of interest, namely the cost of the optimal alignment and the cost of the worst-case alignment. The former is obtained by the alignment algorithm defined in Sect. 4.2. The latter is simply defined as the cost of aligning the empty trace in the model plus the cost of treating all events as log moves.

5.2 Precision

One important metric in conformance checking is to assess the precision of the model with respect to the observed executions, i.e., characterize the ability of the model to produce behavior unrelated to the one observed. Therefore, precision can be measured by the following formula:

$$precision = \frac{|L \cap \mathcal{L}(N)|}{|\mathcal{L}(N)|}$$

The formula above poses a problem for measuring precision in practice: since in general a model can have an infinite language, the formula on its limit tends to 0. Below we provide a couple of alternatives to fight this problem, thus coining metrics for estimating the precision dimension.

Escaping Arcs. Precision can be approximated by exploring the behavior of the process model using as a reference the traces of the log, and stopping the exploration each time modelled behavior deviates from recorded behavior [51]. The following formula would then be used to estimate precision:

$$precision_{ea}(L, N) = \frac{\sum_{\sigma \in L, e \in \sigma} enabled_L(e)}{\sum_{\sigma \in L, e \in \sigma} enabled_N(e)}$$

where $enabled_L(e)$ provides the activities that are possible in L after the same prefix that contains the event e has been observed, and $enabled_N(e)$ denotes the number of tasks that can be executed in the state right before executing the task corresponding to e. In the formula above, we assume fitting models, i.e., $enabled_L(e) \subseteq enabled_N(e)$. This assumption can be lifted with the help of the alignments [52].

Anti-alignments. The idea of *anti-alignments* [53] is to seek in the language of a model N, what are the runs which differ considerably with all the observed traces. Hence, this is the opposite of the notion of alignment. Anti-alignments can be used to measure precision [53,54]. The intuition behind the metric based on anti-alignments is as follows. A very precise process model allows for exactly the traces to be executed and not more. Hence, if one trace σ is removed from the log, this trace becomes the anti-alignment for the remaining log, as it is the only execution of the model that is not in the log. This idea would lead to a *trace-based* precision metric grounded in anti-alignments, that penalizes the model precision for traces σ in the log that are very deviating from the anti-alignment obtained when removing σ in the log. Alternatively, a *log-based* precision metric can also be defined, if not per trace but instead a single anti-alignment with respect to the whole log is used.

5.3 Generalization

In process mining, the challenge is not to discover *the correct* process model for a given log, but to discover the process model that provides the most insights into the process from which the log originated. This is comparable to the area of process modelling, where a model is made of a process in order to describe that process in a meaningful way [55].

In process modelling, there are essentially two fundamental concepts that describe a lot of behavior using only a limited number of modelling elements, namely parallelism and loops. Both constructs have in common that they allow for many traces to be executed, while at the same time the number of states a process model can be in remains limited. In order to generalize from a set of observed sequences, process mining techniques make use of the inference of concurrency and loops. All process discovery techniques presented in Sect. 3 do so, however the way they decide is different.

Another fundamental property of correctly modelling processes is *abstraction*, i.e. one should only include parts of a model if they are relevant enough. By making models too detailed, an end-user can no longer see the important parts of the process.

Generalization is a quality dimension that tries to measure whether the inference of parallelism and loops and abstraction are done properly. For generalization, only few metrics exist [14,53,56]. The approach in [14] considers frequency of use, where models are assumed to generalize if all parts of the model are used equally frequently when reproducing the event log, i.e. this metric focusses entirely on proper abstraction of the observed data. The metric in [56] uses artificial negative events, i.e. events that were not observed at a particular point in the trace and uses a confidence in these events to measure generalization. Finally, in [53] a metric based on anti-alignments is presented which focusses entirely on the relation between the number of states and the number of traces in the model.

5.4 Simplicity

When a process model is discovered, one crucial metric for its evaluation is simplicity: is the derived process model the most simple explanation of the underlying process? This metric refers to the *Occam's Razor* principle. One method for measuring simplicity is by analysing the complexity of the underlying graph. In [57] some examples of such complexity metrics (e.g., *size, diameter, connectivity*) can be found. Alternatively, process model related metrics can also be defined, such as *sequentiality, structuredness*, among others.

6   Concluding Remarks

Process mining is a discipline which already impacts organizations in the present, but promises to impact them even more in the near future. In this paper we have overviewed the area, focusing on the fundamental algorithmic challenges that need to be confronted, to make these promises to become realities soon. Behind these challenges lie traditional theory that has been there several decades already: Petri net theory, data science methods, optimization, business process management, to cite the most important ones. We believe research in these areas would contribute to the development of the process mining field as well, a phenomena that we have already observed in the last decade.

Acknowledgments. This work has been supported by MINECO and FEDER funds under grant TIN2017-86727-C2-1-R.

References

1. van der Aalst, W.M.P.: Process Mining - Data Science in Action, 2nd edn. Springer, Heidelberg (2016). https://doi.org/10.1007/978-3-662-49851-4
2. Carmona, J., van Dongen, B.F., Solti, A., Weidlich, M.: Conformance Checking - Relating Processes and Models. Springer, Heidelberg (2018). https://doi.org/10.1007/978-3-319-99414-7
3. Pearl, J.: Reasoning under uncertainty. Ann. Rev. Comput. Sci. 4(1), 37–72 (1990)
4. Badouel, E., Bernardinello, L., Darondeau, P.: Petri Net Synthesis. Texts in Theoretical Computer Science. An EATCS Series. Springer, Heidelberg (2015). https://doi.org/10.1007/978-3-662-47967-4
5. Kerremans, M.: Gartner Market Guide for Process Mining, Research Note G00353970 (2018). www.gartner.com
6. Koplowitz, R., Mines, C., Vizgaitis, A., Reese, A.: Process Mining: Your Compass for Digital Transformation: The Customer Journey Is The Destination (2019). www.forrester.com
7. TFPM: Process Mining Case Studies (2017). http://www.win.tue.nl/ieeetfpm/doku.php?id=shared:process_mining_case_studies
8. Celonis: Process Mining Success Story: Innovation is an Alliance with the Future (2017). http://www.win.tue.nl/ieeetfpm/lib/exe/fetch.php?media=:casestudies:siemens_celonis_story_english.pdf
9. Leemans, S., Fahland, D., van der Aalst, W.: Scalable process discovery and conformance checking. Softw. Syst. Modeling **17**, 599–631 (2016)
10. Leemans, S.J.J., Fahland, D., van der Aalst, W.M.P.: Discovering block-structured process models from event logs - a constructive approach. In: Colom, J.-M., Desel, J. (eds.) PETRI NETS 2013. LNCS, vol. 7927, pp. 311–329. Springer, Heidelberg (2013). https://doi.org/10.1007/978-3-642-38697-8_17
11. Aalst, W.: Discovering the "glue" connecting activities - exploiting monotonicity to learn places faster. In: Boer, F., Bonsangue, M., Rutten, J. (eds.) It's All About Coordination, pp. 1–20 (2018)

12. Murata, T.: Petri nets: properties, analysis and applications. Proc. IEEE **77**(4), 541–574 (1989)
13. Aalst, W.: Relating process models and event logs: 21 conformance propositions. In: Proceedings of the International Workshop on Algorithms and Theories for the Analysis of Event Data (ATAED 2018), vol. 2115, pp. 56–74. CEUR Workshop Proceedings, CEUR-WS.org (2018)
14. Buijs, J.C.A.M., van Dongen, B.F., van der Aalst, W.M.P.: Quality dimensions in process discovery: the importance of fitness, precision, generalization and simplicity. Int. J. Coop. Inf. Syst. **23**(1), 1440001 (2014)
15. van der Aalst, W.M.P., et al.: Soundness of workflow nets: classification, decidability, and analysis. Formal Asp. Comput. **23**(3), 333–363 (2011)
16. Leemans, S.J.J., Fahland, D., van der Aalst, W.M.P.: Discovering block-structured process models from event logs containing infrequent behaviour. In: Lohmann, N., Song, M., Wohed, P. (eds.) BPM 2013. LNBIP, vol. 171, pp. 66–78. Springer, Cham (2014). https://doi.org/10.1007/978-3-319-06257-0_6
17. Leemans, S.: Robust process mining with guarantees. Ph.D thesis, Eindhoven University of Technology (2017)
18. Ehrenfeucht, A., Rozenberg, G.: Partial (set) 2-structures. Part I, II. Acta Informatica **27**, 315–368 (1990)
19. Carmona, J., Cortadella, J., Kishinevsky, M.: New region-based algorithms for deriving bounded Petri nets. IEEE Trans. Comput. **59**(3), 371–384 (2009)
20. van der Aalst, W.M.P., Rubin, V., Verbeek, H.M.W.E., van Dongen, B.F., Kindler, E., Günther, C.W.: Process mining: a two-step approach to balance between underfitting and overfitting. Softw. Syst. Modeling **9**, 87 (2009)
21. Desel, J., Reisig, W.: The synthesis problem of Petri nets. Acta Inf. **33**(4), 297–315 (1996)
22. Carmona, J.: Projection approaches to process mining using region-based techniques. Data Min. Knowl. Discov. **24**(1), 218–246 (2012)
23. Solé, M., Carmona, J.: Light region-based techniques for process discovery. Fundam. Inform. **113**(3–4), 343–376 (2011)
24. Solé, M., Carmona, J.: Incremental process discovery. In: Jensen, K., Donatelli, S., Kleijn, J. (eds.) Transactions on Petri Nets and Other Models of Concurrency V. LNCS, vol. 6900, pp. 221–242. Springer, Heidelberg (2012). https://doi.org/10.1007/978-3-642-29072-5_10
25. Solé, M., Carmona, J.: Region-based foldings in process discovery. IEEE Trans. Knowl. Data Eng. **25**(1), 192–205 (2013)
26. Darondeau, P.: Deriving unbounded Petri nets from formal languages. In: Sangiorgi, D., de Simone, R. (eds.) CONCUR 1998. LNCS, vol. 1466, pp. 533–548. Springer, Heidelberg (1998). https://doi.org/10.1007/BFb0055646
27. Badouel, E., Bernardinello, L., Darondeau, P.: Polynomial algorithms for the synthesis of bounded nets. In: Mosses, P.D., Nielsen, M., Schwartzbach, M.I. (eds.) CAAP 1995. LNCS, vol. 915, pp. 364–378. Springer, Heidelberg (1995). https://doi.org/10.1007/3-540-59293-8_207
28. Lorenz, R., Juhás, R.: How to synthesize nets from languages - a survey. In: Proceedings of the Winter Simulation Conference, WSC 2007 (2007)
29. Bergenthum, R., Desel, J., Lorenz, R., Mauser, S.: Synthesis of petri nets from infinite partial languages. In: ACSD, pp. 170–179 (2008)
30. Lorenz, R.: Towards synthesis of petri nets from general partial languages. In: AWPN, pp. 55–62 (2008)

31. Bergenthum, R., Desel, J., Mauser, S., Lorenz, R.: Synthesis of petri nets from term based representations of infinite partial languages. Fundam. Inform. **95**(1), 187–217 (2009)

32. Mauser, S., Lorenz, R.: Variants of the language based synthesis problem for petri nets. In: ACSD, pp. 89–98 (2009)

33. van der Aalst, W.M.P., van Dongen, B.F.: Discovering petri nets from event logs. In: Jensen, K., van der Aalst, W.M.P., Balbo, G., Koutny, M., Wolf, K. (eds.) Transactions on Petri Nets and Other Models of Concurrency VII. LNCS, vol. 7480, pp. 372–422. Springer, Heidelberg (2013). https://doi.org/10.1007/978-3-642-38143-0_10

34. Bergenthum, R., Desel, J., Lorenz, R., Mauser, S.: Process mining based on regions of languages. In: Alonso, G., Dadam, P., Rosemann, M. (eds.) BPM 2007. LNCS, vol. 4714, pp. 375–383. Springer, Heidelberg (2007). https://doi.org/10.1007/978-3-540-75183-0_27

35. van der Werf, J.M.E.M., van Dongen, B.F., Hurkens, C.A.J., Serebrenik, A.: Process discovery using integer linear programming. Fundam. Inform. **94**(3–4), 387–412 (2009)

36. van Zelst, S.J., van Dongen, B.F., van der Aalst, W.M.P., Verbeek, H.M.W.: Discovering workflow nets using integer linear programming. Computing **100**(5), 529–556 (2018)

37. van Zelst, S.J., van Dongen, B.F., van der Aalst, W.M.P.: ILP-based process discovery using hybrid regions. In: van der Aalst, W.M.P., Bergenthum, R., Carmona, J. (eds.) Proceedings of the International Workshop on Algorithms & Theories for the Analysis of Event Data, ATAED 2015, Satellite event of the conferences: 36th International Conference on Application and Theory of Petri Nets and Concurrency Petri Nets 2015 and 15th International Conference on Application of Concurrency to System Design, ACSD 2015, 22–23 June 2015, Brussels, Belgium, vol. 1371, pp. 47–61. CEUR Workshop Proceedings. CEUR-WS.org (2015)

38. Adriansyah, A.: Aligning observed and modeled behavior. Ph.D. thesis, Technische Universiteit Eindhoven (2014)

39. de Leoni, M., Marrella, A.: Aligning real process executions and prescriptive process models through automated planning. Expert Syst. Appl. **82**, 162–183 (2017)

40. Reißner, D., Conforti, R., Dumas, M., Rosa, M.L., Armas-Cervantes, A.: Scalable conformance checking of business processes. In: Panetto, H., et al. (eds.) OTM 2017. LNCS, vol. 10573, pp. 607–627. Springer, Heidelberg (2017). https://doi.org/10.1007/978-3-319-69462-7_38

41. Leemans, S.J.J., Fahland, D., van der Aalst, W.M.P.: Scalable process discovery and conformance checking. Softw. Syst. Modeling **17**(2), 599–631 (2018)

42. Taymouri, F., Carmona, J.: A recursive paradigm for aligning observed behavior of large structured process models. In: La Rosa, M., Loos, P., Pastor, O. (eds.) BPM 2016. LNCS, vol. 9850, pp. 197–214. Springer, Cham (2016). https://doi.org/10.1007/978-3-319-45348-4_12

43. van Dongen, B., Carmona, J., Chatain, T., Taymouri, F.: Aligning modeled and observed behavior: a compromise between computation complexity and quality. In: Dubois, E., Pohl, K. (eds.) CAiSE 2017. LNCS, vol. 10253, pp. 94–109. Springer, Cham (2017). https://doi.org/10.1007/978-3-319-59536-8_7

44. Taymouri, F., Carmona, J.: Model and event log reductions to boost the computation of alignments. In: Ceravolo, P., Guetl, C., Rinderle-Ma, S. (eds.) SIMPDA 2016. LNBIP, vol. 307, pp. 1–21. Springer, Cham (2018). https://doi.org/10.1007/978-3-319-74161-1_1

45. Munoz-Gama, J., Carmona, J., Van Der Aalst, W.M.P.: Single-entry single-exit decomposed conformance checking. Inf. Syst. **46**, 102–122 (2014)
46. van der Aalst, W.M.P.: Decomposing petri nets for process mining: a generic approach. Distrib. Parallel Databases **31**(4), 471–507 (2013)
47. Bloemen, V., van de Pol, J., van der Aalst, W.M.P.: Symbolically aligning observed and modelled behaviour. In: 18th International Conference on Application of Concurrency to System Design, ACSD, Bratislava, Slovakia, 25–29 June, pp. 50–59 (2018)
48. Bloemen, V., van Zelst, S.J., van der Aalst, W.M.P., van Dongen, B.F., van de Pol, J.: Maximizing synchronization for aligning observed and modelled behaviour. In: Weske, M., Montali, M., Weber, I., vom Brocke, J. (eds.) BPM 2018. LNCS, vol. 11080, pp. 233–249. Springer, Cham (2018). https://doi.org/10.1007/978-3-319-98648-7_14
49. Fahland, D., van der Aalst, W.M.P.: Model repair - aligning process models to reality. Inf. Syst. **47**, 220–243 (2015)
50. Rozinat, A., van der Aalst, W.M.P.: Conformance checking of processes based on monitoring real behavior. Inf. Syst. **33**(1), 64–95 (2008)
51. Munoz-Gama, J.: Conformance Checking and Diagnosis in Process Mining - Comparing Observed and Modeled Processes. LNBIP. Springer, Heidelberg (2016). https://doi.org/10.1007/978-3-319-49451-7
52. Adriansyah, A., Munoz-Gama, J., Carmona, J., van Dongen, B.F., van der Aalst, W.M.P.: Measuring precision of modeled behavior. Inf. Syst. E-Business Manag. **13**(1), 37–67 (2015)
53. Chatain, T., Carmona, J.: Anti-alignments in conformance checking – the dark side of process models. In: Kordon, F., Moldt, D. (eds.) PETRI NETS 2016. LNCS, vol. 9698, pp. 240–258. Springer, Cham (2016). https://doi.org/10.1007/978-3-319-39086-4_15
54. van Dongen, B.F., Carmona, J., Chatain, T.: A unified approach for measuring precision and generalization based on anti-alignments. In: La Rosa, M., Loos, P., Pastor, O. (eds.) BPM 2016. LNCS, vol. 9850, pp. 39–56. Springer, Cham (2016). https://doi.org/10.1007/978-3-319-45348-4_3
55. Dumas, M., Rosa, M.L., Mendling, J., Reijers, H.A.: Fundamentals of Business Process Management, 2nd edn. Springer, Heidelberg (2018). https://doi.org/10.1007/978-3-662-56509-4
56. vanden Broucke, S.K.L.M., Weerdt, J.D., Vanthienen, J., Baesens, B.: Determining process model precision and generalization with weighted artificial negative events. IEEE Trans. Knowl. Data Eng. **26**(8), 1877–1889 (2014)
57. Mendling, J., Neumann, G., van der Aalst, W.: Understanding the occurrence of errors in process models based on metrics. In: Meersman, R., Tari, Z. (eds.) OTM 2007. LNCS, vol. 4803, pp. 113–130. Springer, Heidelberg (2007). https://doi.org/10.1007/978-3-540-76848-7_9

How Petri Net Theory Serves Petri Net Model Checking: A Survey

Karsten Wolf[✉]

Institut für Informatik, Universität Rostock, Rostock, Germany
`karsten.wolf@uni-rostock.de`

Abstract. Structure theory is a unique treasure of the Petri net community. It was originally studied as a set of stand-alone techniques for exploring Petri net properties such as liveness, boundedness, reachability, and deadlock freedom. Today, methods based on the exploration of the reachability graph (state space methods) dominate Petri net verification. Thanks to the concept of model checking, these methods can deal with a much larger range of verification problems, and thanks to state space reduction methods (symmetries, partial order reduction, and other abstraction techniques), they became tractable for many practical applications. However, in the course of pushing model checking technology to its limits, several elements of Petri net structure theory celebrate a resurrection, being viewed from a different angle. This time, they are used for acceleration of the state space methods. In this article, we give an overview on the use of structural methods in Petri net model checking. We further report on our experience with combining state space and structural methods.

Keywords: Model checking · State equation · Petri net invariant · Siphons and traps · Conflict cluster

1 Introduction

Petri net structure theory had a golden age in the 1970s. At that time, decidability of the boundedness problem was known [24], but for many other problems [18], including reachability, decidability was open until the mid 1980s [26,32]. Consequently, researchers tried to approach these problems from various angles. They found relations between the incidence matrix of the net and its behaviour (the state equation and invariant calculi [28,34]), they discovered patterns in the net topology (siphons, traps, components [17]), and identified net classes where verification problems can be solved more efficiently than in the general case (state machines, marked graphs, free-choice nets [9,17,21]). Since then, results were gradually generalised and extended. Today, the body of structural results for Petri net verification collectively forms a unique theory and it is widely agreed to be an outstanding selling point for the Petri net formalism as such. It

© Springer-Verlag GmbH Germany, part of Springer Nature 2019
M. Koutny et al. (Eds.): ToPNoC XIV, LNCS 11790, pp. 36–63, 2019.
https://doi.org/10.1007/978-3-662-60651-3_2

is extremely useful for the verification of infinite-state systems since most structural results do not require that the net under consideration has only finitely many reachable markings.

With the first tools appearing, the year 1986 is often reported as the starting point of model checking [8]. This technology is not bound to a particular formalism as its main ingredients are defined on transitions systems. In difference to the Petri net theory of the 1970s, model checking permits the specification of a property using temporal logic [11,31]. Consequently, a large variety of verification problems can be approached. In the 1990s, model checking technology made significant progress. Thanks to the concept of symbolic model checking [5] using binary decision diagrams (BDD [3]), and thanks to various state space reduction methods (the symmetry method [22,43,46], partial order reduction [16,39,51]), model checking became tractable for more and more practically relevant applications. Petri net tools started to embrace model checking. Explicit (non-symbolic) model checking was not very difficult to adopt since several Petri net verification tools used state space methods anyway for deciding Petri net standard properties such as deadlock freedom, liveness, or reachability [41,57]. The Petri net community also took an active part in the development of the symmetry method, the partial order reduction and other state space reduction techniques that made explicit model checking tractable. BDDs were explored in tools for stochastic Petri nets [35], and the first BDD based model checkers for Petri nets appeared [38]. With a model checker based on branching processes [12,33], Petri nets contributed a first completely unique piece of technology to model checking.

After the year 2000, several researchers focussed on pushing model checking technology to its limits. Technology differentiated in order to exploit the specific features of various application domains. For Petri nets, explicit model checking preserved higher attention than elsewhere. This is due to the focus of Petri nets on concurrent systems. Here, partial order reduction can substantially reduce the state space considered while symbolic tools sometimes struggle with the nondeterminism that is inherent in concurrent systems. At the same time, traditional Petri net theory was revisited, this time as a tool for accelerating the otherwise state-space based model checking technology. The yearly model checking contest (MCC [25]) that took place in 2011 for the first time, put a substantial incentive for further development and implementation of the techniques. Today, leading Petri net model checking tools exhibit an extensive integration of state space methods with structural methods. It would be difficult for these tools to switch to another modeling formalism without substantial performance loss. In this sense, Petri net model checking has become its own branch of model checking technology.

In this article, we highlight the interaction between model checking and Petri net structure theory. We collect known results and present also some new results. We first introduce the basic notions for Petri nets, temporal logics, and model checking. Then we introduce various items of Petri net structure theory and discuss their benefits for Petri net model checking: the state equation, place invariants, transition invariants, siphons and traps, and finally conflict clusters. We do not cover the area of net reduction since this is a technique for pre-processing that works for every subsequent verification, not only for state space methods. We also do not cover the verification of infinite-state systems, i.e. Petri nets with infinitely many reachable markings. The survey shall contain some personal judgements of the author, based on more than 25 years of experience with implementing competitive Petri net verification tools.

2   Basic Terminology

Definition 1 (Place/transition net). *A place/transition net consists of a finite set P of places, a finite set T of transitions (disjoint to P), a set $F \subseteq (P \times T) \cup (T \times P)$ of arcs, a weight function $W : (P \times T) \cup (T \times P) \to \mathbb{N}$ where $[x, y] \notin F$ if and only if $W(x, y) = 0$, and a marking m_0, the initial marking. A marking is a mapping $m : P \to \mathbb{N}$.*

For a marking m, the fact $m(p) = k$ is interpreted as "p carries k tokens". A place carrying at least one token is called "marked".

Definition 2 (Behaviour of a place/transition net). *Transition t is enabled in marking m if, for all $p \in P$, $W(p, t) \leq m(p)$. If t is enabled in m, t can fire, producing a new marking m' where, for all $p \in P$, $m'(p) = m(p) - W(p, t) + W(t, p)$. This firing relation is denoted as $m \xrightarrow{t} m'$. It can be extended to firing sequences by the following inductive scheme: $m \xrightarrow{\varepsilon} m$ (for the empty sequence ε), and $m \xrightarrow{w} m' \wedge m' \xrightarrow{t} m'' \implies m \xrightarrow{wt} m''$ (for a sequence w and a transition t). The reachability graph of N has a set of vertices that is the set of all markings that are reachable by any sequence from the initial marking of N. Every element $m \xrightarrow{t} m'$ of the firing relation defines an edge from m to m' annotated with t.*

In the sequel, we shall only consider Petri nets with finitely many reachable markings.

Definition 3 (Navigation in a Petri net). *Consider a place/transition net $N = [P, T, F, W, m_0]$. Elements of $P \cup T$ are collectively called* nodes. *For a node x, $\bullet x = \{y \mid [y, x] \in F\}$ is its* preset *while $x \bullet = \{y \mid [x, y] \in F\}$ is its* postset. *These notions are lifted to sets of nodes: for $X \subseteq P \cup T$, $\bullet X = \bigcup_{x \in X} \bullet x$ and $X \bullet = \bigcup_{x \in X} x \bullet$.*

Definition 4 (Syntax of CTL). *Let $N = [P, T, F, W, m_0]$ be a place/transition net.* TRUE, FALSE, FIREABLE(t) *(for $t \in T$),* DEADLOCK, *and $k_1 p_1 + \cdots + k_n p_n \leq k$ ($k_i, k \in \mathbb{Z}, p_i \in P$) are atomic propositions. $\mathcal{P} = \{A, E\}$ is called the set of path quantifiers, $\mathcal{U} = \{X, F, G\}$ the set of unary temporal operators, and $\mathcal{B} = \{U, R\}$ the set of binary temporal operators.*
Every atomic proposition is a CTL formula. If ϕ and ψ are CTL formulas, so are $\neg\phi$, $(\phi \wedge \psi)$, $(\phi \vee \psi)$, $QY\phi$ (with $Q \in \mathcal{P}$ and $Y \in \mathcal{U}$), and $Q(\phi Z \psi)$ (with $Q \in \mathcal{P}$ and $Z \in \mathcal{B}$).

Atomic propositions are properties of markings that can be evaluated by investigation of just the marking (and its enabled transitions). The unary temporal operators are called *nextstate* (X), *finally* or *future* (F), and *globally* (G). The binary temporal operators are called *until* (U) and *release* (R). They are interpreted on paths (sequences of markings). The path quantifiers range on all paths that start in a given markings and otherwise behave as the universal and existential path quantifiers in first order logic.

The logic LTL is defined similarly. The only difference is that the path quantifiers are not used in LTL.

Definition 5 (Semantics of CTL). *Marking m satisfies CTL formula ϕ ($m \models \phi$) according to the following inductive scheme:*

- $m \models$ TRUE, $m \not\models$ FALSE;
- $m \models$ FIREABLE(t) *if t is enabled in m;*
- $m \models$ DEADLOCK *if there are no enabled transitions in m;*
- $m \models k_1 p_1 + \cdots + k_n p_n \leq k$ *if $k_1 m(p_1) + \cdots + k_n m(p_n) \leq k$;*
- $m \models \neg\phi$ *if $m \not\models \phi$;*
- $m \models (\phi \wedge \psi)$ *if $m \models \phi$ and $m \models \psi$;*
- $m \models EX\phi$ *if there is a t and an m' with $m \xrightarrow{t} m'$ and $m' \models \phi$;*
- $m \models E(\phi U \psi)$ *if there is a path $m_1 m_2 \ldots m_k$ ($m_1 = m, k \geq 1$) in the reachability graph where $m_k \models \psi$ and, for all i with $1 \leq k_i < k$, $m_i \models \phi$;*
- $m \models A(\phi U \psi)$ *if, for each maximal path (i.e. infinite or ending in a deadlock) $m_1 m_2 \ldots$ in the reachability graph with $m_1 = m$, there is a k ($k \geq 1$) where $m_k \models \psi$ and, for all i with $1 \leq k_i < k$, $m_i \models \phi$.*

The semantics of the remaining CTL operators is defined using the tautologies $(\phi \vee \psi) \iff \neg(\neg\phi \wedge \neg\psi)$, $AX\phi \iff \neg EX\neg\phi$, $EF\phi \iff E(\text{TRUE } U\phi)$, $AF\phi \iff A(\text{TRUE } U\phi)$, $AG\phi \iff \neg EF\neg\phi$, $EG\phi \iff \neg AF\neg\phi$, $E(\phi R\psi) \iff \neg A(\neg\phi U\neg\psi)$, and $A(\phi R\psi) \iff \neg E(\neg\phi U\neg\psi)$. A place/transition net N satisfies a CTL formula if its initial marking m_0 does.

For LTL, all temporal modalities concern the same single path. Moreover, only infinite paths are considered. A maximal finite path is transformed into an infinite path by infinitely repeating the last (deadlock) marking. Otherwise, evaluation accords with CTL. A place/transition net N satisfies an LTL formula if all paths starting from m_0 do.

A CTL or LTL formula not containing temporal operators is called *state predicate*.

3   Explicit CTL Model Checking

We consider *local* model checking, that is, we want to evaluate a given CTL formula just for the initial marking m_0. Other markings are only considered as far as necessary for determining the value at m_0. In global model checking, one would be interested in the value of the given formula in all reachable markings.

As a reference for our work, we use the algorithm of [55]. In the sequel, we briefly sketch this algorithm.

We assume that, attached to every marking, there is a vector that has an entry for every subformula of the given CTL query. The value of a single entry can be true, false, or unknown. Whenever we want to access the value of a subformula ϕ in a marking m, we inspect the corresponding value. If it is unknown, we recursively launch a procedure to evaluate ϕ in m.

If ϕ is an atomic proposition or a Boolean combination of subformulas, evaluation is trivial. For evaluating a formula of shape $EX\ \phi'$ or $AX\ \phi'$, we proceed to the immediate successor states and evaluate ϕ' in those states. If the successor marking has not been visited yet, we add it to the set of visited markings. A marking is always added with all its vector entries set to unknown.

If ϕ has the shape $A(\psi U\chi)$, we launch a depth-first search from m, aiming at the detection of a counterexample. The search proceeds through markings that satisfy ψ, violate χ, and for which $A(\psi U\chi)$ is recorded as unknown. Whenever we leave the space of states satisfying these assumptions, there is a reaction that does not require continuation of the search beyond that marking, as follows.

If χ is satisfied, or $A(\psi U\chi)$ is recorded as true in any marking m', we backtrack since there cannot be a counterexample path containing m'. If χ and ψ are violated, or $A(\psi U\chi)$ is recorded as false, we exit the search since the search stack forms a counterexample for $A(\psi U\chi)$ in m. If we hit a marking m' that is already on the search stack, we have found a counterexample, too (a path where ψ and not χ hold forever). The depth-first search assigns a value different from unknown to all states visited during the search: For markings on the search stack (i.e. participating in the counterexample), $A(\psi U\chi)$ is false, while for states that have been visited but already removed from the search stack, $A(\psi U\chi)$ is actually true.

If ϕ has the shape $E(\psi U\chi)$, we launch a similar depth-first search, aiming at the detection of a witness path. This time, we integrate Tarjan's algorithm [48] for detecting the strongly connected components (SCC) during the search. It proceeds through markings that satisfy ψ, violate χ, and for which $E(\psi U\chi)$ is recorded as unknown. If we hit a marking m' where χ is satisfied, or $E(\psi U\chi)$ is true, we have found our witness. In states where ψ and χ are violated, or $E(\psi U\chi)$ is known to be false, we backtrack since there cannot be a witness path containing such a marking. Again, we assign a value different from unknown to every marking visited during the search. Markings that are on the search stack as well as markings that are not on the search stack but appear in SCC that have not yet been completely explored, is given true. An SCC is not yet fully explored if it contains states that are still on the search stack. In this case, however, a path to the search stack extended by the remaining portion of the search stack forms a witness. For markings appearing in an SCC that has been completely explored, $E(\psi U\chi)$ is false.

We can see that existentially and universally quantified until operators are not fully symmetric. This is due to the fact that a cycle of markings that satisfy

ψ and violate χ, forms a counterexample for universal until, but no witness for existential until.

The remaining CTL operators can be traced back to the two until operators using tautologies. Since every search assigns values to all visited markings, the overall run time of the algorithm is $O(|\phi||R|)$ where $|\phi|$ is the length of ϕ (the number of subformulas), and $|R|$ is the number of markings reachable from m_0. Due to the state explosion problem, $|R|$ is the dominating factor for complexity.

4   Explicit LTL Model Checking

Explicit LTL model checking [54] typically uses automata that are able to accept infinite sequences, e.g. Büchi automata [4]. In a Büchi automaton, some states are defined to be *accepting states*. A Büchi automaton accepts an infinite path if and only if a corresponding run of the automaton visits an accepting state infinitely often.

Every LTL formula ϕ can be transformed into a finite Büchi automaton that accepts precisely the paths that satisfy ϕ (the "language" L_ϕ). The reachability graph of a Petri net N can be interpreted as a Büchi automaton that accepts all infinite paths of N (the "language" L_N). The term *language* is set in quotes since the term "formal languages" usually refers to a set of *finite* sequences.

N satisfies ϕ if and only if $L_N \subseteq L_\phi$. This is the case if and only if $L_N \cap L_{\neg\phi} = \emptyset$. Given Büchi automata for L_N and $L_{\neg\phi}$, a construction similar to the well-known product automaton construction for finite automata, yields a Büchi automaton that accepts the intersection of both languages. The resulting product automaton does not accept the empty language if and only if it has a reachable SCC that has more than one element (or a single element with a self-loop) and contains an accepting state. If such an SCC is found, a path from the initial marking to the accepting SCC, followed by a round-trip through that SCC, forms a counterexample that proves the property to be false.

For an LTL formula ϕ, the accepting Büchi automaton may have up to $2^{|\phi|}$ states. With the complexity of building the product automaton, and for the detection of SCC, we obtain a complexity $O(2^{|\phi|}|R|)$ for explicit model checking. Formulas are often quite small, and the resulting Büchi automaton is often much smaller than suggested by the exponential worst-case complexity. That is why the size of the reachability graph is generally the limiting parameter for the performance of LTL model checking.

5   Symbolic Model Checking

There are several approaches to symbolic model checking. In this paper, we refer mainly to BDD-based model checking [5] and SAT-based model checking [7]. Both approaches rely on coding markings as Boolean vectors of some fixed length n and a representation of the transition relation $TR = \{[m, m'] \mid \exists t : m \xrightarrow{t} m'\}$ as a Boolean function with arity $2n$. TR can be derived from the given place/transition net.

A binary decision diagram [3] is a data structure that represents a Boolean function of some arity n. This function can be viewed as the characteristic function of a set of bit vectors of length n. Given such a set M of markings, coded as bit vectors represented as a BDD, and a BDD representing TR, there exist algorithms [5] that can produce a BDD representing the set $Pred(M)$ of predecessors $\{m \mid \exists m' \in M \exists t \in T : m \xrightarrow{t} m'\}$. Other BDD based operations can be used for implementing intersection, union, and complement of sets [3]. This way, given a BDD for the set M_ϕ of markings satisfying ϕ and a BDD for the set M_ψ of markings satisfying ψ, a BDD for the set $M_{E(\phi U \psi)}$ can be obtained by the following procedure:

– $Z := M_\psi$;
– Repeat until Z is not changed:
  • $Z := Z \cup (Pred(Z) \cap M_\phi)$;

Other CTL operators can be explored similarly. The BDD data structure was generalised such that integer vectors instead of Boolean vectors are used [23, 40, 47]. This way, the step of Boolean coding of markings can be dropped. Otherwise, model checking is similar to BDD based model checking.

In SAT based LTL model checking, we start with the coding of markings as bit vectors of some length n. A path $m_0 m_1 \dots m_k$ is coded as a Boolean formula using variables $x_i^j (1 \leq i \leq n, 0 \leq j \leq k)$. x_i^j represents the i-th bit in the coding of m_j. Starting with a formula ϕ_0 over $x_1^0 \dots x_n^0$ that is true just for the initial marking (and can be constructed with little effort), the path $m_0 \dots m_k$ can be coded as

$$\phi_0 \wedge TR(x_1^0, \dots x_n^0, x_1^1 \dots x_n^1) \wedge TR(x_1^1, \dots x_n^1, x_1^2 \dots x_n^2) \wedge \dots$$

$$\dots \wedge TR(x_1^{k-1}, \dots x_n^{k-1}, x_1^k \dots x_n^k)$$

To this formula, additional Boolean constraints can be added such that, given LTL formula ϕ, the formula is satisfied if and only if there is a counterexample for ϕ of length k. A SAT solver is used for solving this question. Thus, SAT based model checking relies on the quite impressive performance of present-day SAT solvers. This approach is repeated for rising values of k until a counterexample is found, or k is large enough to have enough trust in the validity of ϕ. SAT based model checking can prove ϕ to be true only if there is an estimation for the maximal length of a counterexample.

Recent developments try to replace SAT by SMT (satisfiability modulo theory), that is, first order logic generalisations of SAT solving.

6   The Petri Net State Equation

The state equation is a linear algebraic approximation of the Petri net reachability relation.

6.1   Theory

A marking can be interpreted as an integer vector with index set P. We assume that the vector is written as a column. For every transition $t \in T$, we can assign the *effect vector* Δt with same index set P by setting $\Delta t[p] = W(t,p) - W(p,t)$. From Definition 2, it is easy to see that $m \xrightarrow{t} m'$ implies $m' = m + \Delta t$.

The *incidence matrix* C of Petri net N is a matrix with rows indexed by P and columns by T such that the column for transition t is just Δt. The Petri net state equation generalises the observation above to transition sequences. For a transition sequence w, let $\Psi(w)$ be a vector with index set T such that $\Psi(w)[t]$ is the number of occurrences of t in w. $\Psi(w)$ is called the *Parikh vector* of sequence w. With these ingredients, the desired generalisation spells as follows:

Proposition 1. *Let m, m' be markings and w a transition sequence of some place/transition net N with incidence matrix C. If $m \xrightarrow{w} m'$ then*

$$m' = m + C\Psi(t).$$

We can deduce that m' is reachable from m only if the equation $m' = m + C \cdot \underline{x}$, the *Petri net state equation*, has a solution in the natural numbers. $\underline{x}$ is a vector of $|T|$ unknowns. Furthermore, the set $\{m_0 + C\underline{x} \mid \underline{x} \in \mathbb{N}^{|P|}\}$ is a superset of the set of reachable markings of N.

6.2   Invariance of Atomic Propositions

An atomic proposition is invariant if it has the same value for all reachable markings. If an atomic proposition can be proven to be true or false, it can be replaced by the respective Boolean constant. A formula simplified by such a replacement typically permits the application of various rewrite rules that can make subformulas or even the whole formula collapse into a constant. In either case, the resulting formula is simpler than the original formula and the model checker (whether symbolic or explicit) has an easier job when solving the problem.

Let $P = \{p_1, \cdots, p_n\}$. Using the fact that the linear space $\{m_0 + C\underline{x} \mid \underline{x} \in \mathbb{N}^{|P|}\}$ over-approximates the set of reachable markings, we can prove an atomic proposition $k_1 p_1 + \cdots + k_n p_n \leq k$ is invariantly true if the system of inequations

$$m_0 + C\underline{x} = (x_{p_1} \ldots x_{p_n})^T$$
$$\underline{x} \geq \underline{0}$$
$$k_1 x_{p_1} + \cdots + k_n x_{p_n} > k$$

has no solution in the integer numbers where the vectors $\underline{x}$ and $(x_{p_1} \ldots x_{p_n})$ serve as unknowns. Correspondingly, the proposition is invariantly false if the system of inequations

$$m_0 + C\underline{x} = (x_{p_1} \ldots x_{p_n})^T$$
$$\underline{x} \geq \underline{0}$$
$$k_1 x_{p_1} + \cdots + k_n x_{p_n} \leq k$$

has no solution [2].

We can equivalently treat the expression $k_1 x_{p_1} + \cdots + k_n x_{p_n}$ as target function in a linear program where $m_0 + C\underline{x} = (x_{p_1} \ldots x_{p_n})^T$ and $\underline{x} \geq \underline{0}$ serve as side conditions. The desired invariance results can be derived from maximising (resp. minimising) the target function.

In recent MCCs, between 25% and 33% of the model checking problems can be reduced to formulas that do not contain any temporal operators and can thus be verified by just inspecting the initial marking. For a substantial number of further problems, the formula left for the model checker is much simpler (has less temporal operators) than the original formula. It often contains less visible transitions (transitions that effect any atomic proposition in the formula). Having not too many visible transitions is known to be valuable for applying the stubborn set method which is the most effective state space reduction method for Petri nets.

The approach covers atomic propositions FIREABLE(t) since such a proposition can be rewritten to

$$\bigwedge_{p \in \bullet t} p \geq W(p, t).$$

DEADLOCK is covered as well since this is equivalent to

$$\bigwedge_{t \in T} \neg \text{FIREABLE}(t).$$

However, we found that approach not to perform very well in practice since the resulting formulas tend to be too complicated.

6.3   Enabling On-The-Fly Determination of Precise Bounds

A *bound* for a place p is a natural number b such that, for all reachable markings m, $m(p) \leq b$. The smallest number of that kind is called the *precise* bound of p. Knowing good or even precise bounds is quite valuable for model checking, as discussed later in this paper. In general, precise bounds are determined by exploring the set of all reachable markings. There is a stubborn set method[1] [42] that preserves the precise bound but the (reduced) state space needs to be explored completely for computing the precise bound. Using the considerations above, the linear program

$$\text{maximise } x_p \text{ where}$$
$$m_0 + C\underline{x} = (x_{p_1} \ldots x_{p_n})^T$$
$$\underline{x} \geq \underline{0}$$

yields a value b that is a valid bound b for p. Since the state equation over-approximates the set of reachable markings, b is not necessarily precise. If, however, the search in the reachability graph reveals a marking m with $m(p) = b$,

---

[1] Stubborn set methods are a class of state space reduction methods. In every marking m, they explore only a subset ("stubborn set") of the enabled transitions. The stubborn sets are selected such that a given property is preserved (holds in the reduced state space if and only it holds in the full state space). The surveys [52,53] present stubborn sets in full breadth.

we known that b is precise and we can abort state space exploration. The same approach is applicable for determining a precise upper bound of formal sums of places which is a separate category in the MCC. Application of the state equation as just described, increased the success rate of LoLA [57] in the MCC 2018 from 40% to 80% of the problem instances.

6.4   Portfolio Approach to Reachability

A portfolio is a collection of tools that run in parallel and approach the same problem. The first approach that terminates with a definite answer determines the run-time of the portfolio. The participating methods may compete for both memory and processor time (single core), for memory only (multicore), or for no resource at all (network of workstations). Participating methods may be complete or incomplete (that is, have "unknown" as potential return value). As long as there is at least one complete method in the portfolio, the portfolio as a whole is a complete verification method. Participating methods must, however, be sound, that is, they should never return true or false if that is not the correct output for the verification problem. If methods compete for memory, all but one method should have a modest memory footprint.

The state equation can be used as an incomplete method in a portfolio for verifying CTL queries $EF\phi$, $AG\phi$ and LTL query $G\phi$ where ϕ is a state predicate. As these formulas can be reduced to each other by negation, we only consider $EF\phi$.

We assume that all atomic propositions have the shape $k_1 p_1 + \cdots + k_n p_n \leq k$. This does not restrict generality since other atomic propositions can be transformed into the desired shape.

Assume first that ϕ is a conjunction $\phi_1 \wedge \cdots \wedge \phi_k$ of atomic propositions. In this case, we can reuse the argument of Subsect. 6.2 to see that ϕ must be false if the system

$$m_0 + C\underline{x} = (x_{p_1} \ldots x_{p_n})^T$$
$$\underline{x} \geq \underline{0}$$
$$\phi_1$$
$$\ldots$$
$$\phi_k$$

has no solution. If it has a solution, a minimal one can be found by transforming the system into a linear program, adding the target function

$$\text{minimise } \underline{x}[1] + \cdots + \underline{x}[|T|].$$

In a next step, we may check whether there is a firing sequence w that corresponds to the solution for $\underline{x}$. This is a reachability problem with a severe depth restriction that still permits the application of a stubborn set method [42] for further state space reduction. If a realisable sequence can be found, we can answer the query $EF\phi$ with true. This way, both negative and positive answers can be obtained from the state equation. If it turns out that the Parikh vector given by the solution of the linear program does not correspond to any realisable firing sequence, we may add constraints to the system that exclude the spurious solution and repeat the approach. This way, the "unknown" area of the approach can be further reduced [19,56]. The method stays, however, incomplete. This can be easily seen by observing that the state equation approach works in polynomial space. Integer linear programming is known to be NP-complete [10] while the reachability problem for place/transition nets is EXPSPACE-hard [30].

If ϕ is not a simple conjunction, we may transform it into disjunctive normal form and apply the approach to each conjunction separately. Unfortunately, construction of a disjunctive normal form may explode. That is why it is recommendable to alleviate this kind of explosion, for instance by taking care of duplicate subformulas, as suggested in [58].

In the MCC, about 30% of solutions produced by the tool LoLA in the reachability category of recent MCCs are delivered by the state equation approach.

Since $EF\phi$ is a necessary condition for $AGEF\phi$, $EFAGEF\phi$, $EFAG\phi$, $AGEFAG\phi$, $E(\chi U\phi)$, $A(\chi U\psi)$ and others, the sketched approach can be used as a (less effective but correct) portfolio member in the verification of the mentioned formulas. Likewise, $AG\phi$ is a sufficient criterion for $AGEF\phi$, $EFAGEF\phi$, $EFAG\phi$, and $AGEFAG\phi$ and yields another portfolio member.

7   Place Invariants

Place invariants [28,34] are perhaps the most popular structural verification method. Many extensions of the Petri net formalism permit the formulation of place invariants as a structural verification technique.

7.1   Theory

A *place invariant* is a place-indexed vector i with values in the integer numbers such that $i^T \cdot C = \underline{0}$. A place-indexed vector v assigns an integer *weight* $v^T \cdot m = \sum_{p \in P} v[p] \cdot m(p)$ to every marking m. For a place invariant i, all reachable markings have the same weight: $m \xrightarrow{w} m'$ implies $i^T \cdot m = i^T \cdot m'$.

An obvious necessary criterion for reachability can be derived but the state equation mentioned above is stronger in this regard. However, some other interesting accelerations in model checking are derivable from place invariants.

The *support* $supp(i)$ of a place invariant i is the set of all places p with $i(p) \neq 0$. A place invariant is *semi-positive* if, for all places p, $i(p) \geq 0$.

7.2   Coding of Markings I: Significant Places

If i is a place invariant, the equation

$$i[p_1]m(p_1) + \ldots i[p_{j-1}]m(p_{j-1}) + i[p_j]m(p_j) + i[p_{j+1}]m(p_{j+1}) + \cdots + i[p_n]m(p_n)$$
$$= i^T \cdot m_0$$

is valid for all reachable markings m. Assuming that p_j is in the support of i, this equation can be rewritten to

$$m(p_j) = i^T \cdot m_0 - \frac{i[p_1]m(p_1)+\cdots+i[p_{j-1}]m(p_{j-1})+i[p_{j+1}]m(p_{j+1})+\cdots+i[p_n]m(p_n)}{i[p_j]} \quad (1)$$

In other words, the marking on p_j functionally depends on the marking on the other places. If we have k linear independent place invariants, k different places can be identified as being functionally dependent on the remaining $n - k$ places.

When storing markings in explicit model checking, places that are functionally dependent on other places can be disregarded: the remaining places uniquely identify the marking. This way, marking vectors to be stored get 30% to 70% shorter. Run time for searching and inserting markings during state space exploration is also reduced by approximately the same ratio. This is due to the fact that searching and inserting markings consume most of the run time of an explicit model checker.

If the reachability graph is explored using depth-first search, the actual invariants that prove functional dependency do not need to be stored. As long as the marking currently explored is present for all places, including the functionally dependent ones, we can proceed forward by adding the effect vector of the executed transition, and we can backtrack later on by subtracting this effect vector. There is no need to recover the marking of functionally dependent places from a stored (shortened) marking. This means that the approach causes virtually no overhead during run-time [44].

For computing functional dependencies, we do not need to compute the actual place invariants either. We can derive the required information from an intermediate situation of that computation, as follows. An equation of the system $i^T \cdot C = \underline{0}$ has the shape $i[p_1]C[p_1, t] + \cdots + i[p_n]C[p_n, t] = 0$, for some $t \in T$. We can chose some place p where $C(p, t) \neq 0$ and mark it as functionally dependent on the remaining places. Then we can eliminate p from all other equations in the sense of the Gaussian elimination algorithm. We can then repeat the procedure for the resulting system, excluding the selected equation. This way, we can mark the maximal number of functionally dependent places. This can be executed in a fraction of a second even for large nets in the MCC benchmark.

In symbolic model checking, we have two options for exploiting functional dependencies. First, we may eliminate the functionally dependent places from the bit vector representation of a marking. This requires, however, that the effect of an eliminated place is rewritten in the representation of the transition relation TR, exploiting Eq. 1.

A second option, especially for BDD based model checkers, is to exploit functional dependencies just in the calculation of a variable ordering. It is well known that the size of a BDD is quite sensitive to the order in which variables are considered in the decision diagram. Knowing functional dependencies helps in placing them in the variable order in a position where they cost no additional blow-up in the BDD size [35,37,38].

7.3   Coding of Markings II: 0/1-Invariants

A place invariant i is a 0/1 invariant if, for all $p \in P$, $i[p] \in \{0,1\}$, and $i^T \cdot m_0 = 1$. If place/transition net N has a 0/1 invariant i, then, for every reachable marking m, exactly one place in $supp(i)$ is marked, and it carries exactly one token. This observation can be used for a compressed bit vector representation of markings. For the places in $supp(i)$ it is sufficient to store the *index* of the place that carries the unique token. This requires $\log |supp(i)|$ bits compared to $|supp(i)|$ bits in the conservative representation [20,37]. The approach of the previous section can reduce the number of bits only from $|supp(i)|$ to $|supp(i)| - 1$.

The sketched idea has been reported for BDD based model checking. Coding and decoding the compressed bit vector representation must be implemented in the Boolean function TR representing the transition relation. Since SAT based model checking uses TR as well, the method is applicable to SAT based model checking as well.

To the best knowledge of the author, 0/1 invariants have not yet been exploited in explicit model checking. This may be due to the fact that programmers shy away from the effort for coding bit vectors.

A good 0/1 invariant can be easily detected using the following integer linear program:

$$\text{maximise } i[p_1] + \ldots i[p_n] \text{ where}$$
$$i^T \cdot C = \underline{0}$$
$$\underline{0} \leq i \leq \underline{1}$$
$$i^T \cdot m_0 = 1.$$

Similar logarithmic coding is possible if the net has state machine components, or nested units [15]. The easiest way of detecting such structures is to monitor the process of creation of the net, e.g. by translation from some other modeling formalism.

7.4   Estimation of Place Bounds

If place p is in the support of a semi-positive place invariant i, the following bound for p can be derived. For all reachable m,

$$m(p) \leq \frac{i^T \cdot m_0}{i[p]}$$

A known bound b for a place p, if not derivable from other structural information, is valuable as it permits a bitvector representation for the marking of p with just $\log(b+1)$ bits. Semi-positive place invariants can coded as a linear system

$$i^T \cdot C = \underline{0}$$
$$i \geq \underline{0}$$

In practice, one could try to add a target function turning the system into a linear program. However, we have to deal with the following trade-off. If we minimise the support of the resulting invariant, we obtain tight bounds but have to solve many linear programs. If we maximise the support (which may be tricky since we do not want to maximise the entries as such), we have to solve less linear programs but get bounds that are less tight. In many use cases, time is not a decisive factor for model checking and we can go with the tight bounds, i.e. invariants with minimal support. For repeated calculation, we may force the system of inequations to cover new places. If bounds for all places in some subset P^* of P are already known, the additional constraint

$$\sum_{p \notin P^*} p \geq 1$$

makes sure that another solution of the proposed system adds information.

If run time is a critical resource (for instance in the MCC), we propose another (previously unpublished) approximation that drastically reduces the number of systems that need to be solved for covering all places. We construct a partition $\mathcal{P} = \{P_1, \ldots, P_k\}$ of P and a partition $\mathcal{T} = \{T_1, \ldots, T_l\}$ of T such that, for all $i \leq k$ and $j \leq l$, $p, q \in P_i$ and $t, u \in T_j$ implies $W(p, t) = W(q, u)$ and $W(t, p) = W(u, q)$. Such a partition induces a Petri net $\mathcal{N} = [\mathcal{P}, \mathcal{T}, \mathcal{F}, \mathcal{W}, \mathcal{M}_0]$ where $\mathcal{W}(X, Y) = W(x, y)$ for any $x \in X$ and $y \in Y$, $[X, Y] \in \mathcal{F}$ if and only if $\mathcal{W}(X, Y) \neq 0$, and $\mathcal{M}_0(P_i) = \sum_{p \in P_i} m_0(p)$. It can be shown that, for all m reachable in N, the marking $\mathcal{M}$ with $\mathcal{M}(P_i) = \sum_{p \in P_i} m(p)$ is reachable in $\mathcal{N}$. Consequently, a bound b for a place P_i of $\mathcal{N}$ is a valid bound for all $p \in P_i$ in net N. As $\mathcal{N}$ tends to be much smaller than N, less linear problems need to be solved for deriving bounds for $\mathcal{N}$.

There are two ways to obtain partitions of P and T as required. If N is a place/transition net that is translated from a high-level net where all firing modes of a transition consume (resp. produce) the same *number* of tokens from every place, the partition can be directly derived from the high-level net. A class is comprised of all nodes that result from the same node in the high-level net. In this case, $\mathcal{N}$ is referred to as the *skeleton* of the high-level net.

If no high-level net is available for directly deriving a partition, we can start with an initial partition where nodes with same in/out-degree from a single class. Then classes can be split according to the required conditions: Given a pair X and Y of classes, X can be separated into $X_0, X_1, X_2, \ldots$ where, for all i, $X_i = \{x \in X \mid \exists y \in Y : W(x, y) = i\}$ while Y can be separated accordingly with $Y_i = \{y \in Y \mid \exists x \in X : W(x, y) = i\}$. This process has to be repeated until no splitting of classes results. The resulting partition is the unique coarsest partition that satisfies the required conditions.

Calculation of the partition takes less than a second for even the largest place/transition nets in the MCC benchmark. Of course, resulting bounds are not very tight but still useful. Consider for example the 10,000 philosophers net in the MCC. Its places are 1-bounded. The sketched approach unifies all places where different philosophers are in the same local states, to a single place in the skeleton. We obtain a bound of 10,000 for every place. This information permits the use of just 14 bits for representing the marking of a place, compared to 32 bits that are spent in LoLA if no bound is known for a place. As discussed earlier, having a bit vector of less than half of the original size doubles the performance and the memory capacity of explicit model checking. Thanks to the skeleton approach, the bound of 10,000 can be determined in less than a second while the calculation of tighter bounds with the place invariant approach would take more than half an hour.

The effect is similar for most Petri nets that are derived from high-level Petri nets or from another high-level formalism. The advantage of the proposed method is that we do not depend on the presence of that high-level model itself.

7.5   Simplification of Atomic Propositions

A place invariant i induces an equation $i^T \cdot m = i^T \cdot m_0$ that holds for every reachable m. In this equation, m_0 is constant and given. That is, such an equation may look like: $p_1 + p_2 + p_3 = 1$, a completely arbitrary example. For a net with that place invariant, we may want to check a formula where $p_1 + 2p_2 + 2p_3 + p_4 \leq 3$ appears as an atomic proposition. Exploiting the invariant, we may rewrite the proposition to $p_2 + p_3 + p_4 \leq 2$. The proposition is still non-constant and state space exploration may be required to evaluate the formula. However, the formal sum is smaller and can thus be evaluated more quickly. Since a typical state space exploration processes millions of markings, already little run-time improvements matter if they need to be executed for every marking. Moreover, p_1 is no longer mentioned in the proposition. Consequently, transitions in the vicinity of p_1 may become invisible. A transition is invisible if its occurrence does not influence the value of any atomic proposition of a formula under investigation. A small number of invisible transitions is desirable for the application of the stubborn set method or other partial order reduction techniques.

Given an atomic proposition $\sum_{p \in P_+} k_p p - \sum_{p \in P_-} k_p p \sim k$, for natural numbers k_p and k, two disjoint subsets P_+ and P_- of P, and $\sim \in \{=, \neq, >, \geq, <\leq\}$, a place invariant suitable for simplification can be calculated as the solution to the following linear program:

$$\text{maximise} \sum_{p \in P_+} i[p] - \sum_{p \in P_-} i[p] \text{ where}$$
$$i^T \cdot C = \underline{0}$$
$$0 \leq i[p] \leq k_p, \text{ for } p \in P_+$$
$$k_p \leq i[p] \leq 0, \text{ for } p \in P_-$$
$$i[p] = 0, \text{ for } p \in P \setminus (P_+ \cup P_-)$$

The approach has not been published before, but is implemented in the LoLA tool and yields considerable speed-up.

8   Transition Invariants

Structurally, this concept is dual to place invariants. It is still one of the biggest mysteries of Petri net theory that this structural duality does not extend to the number and significance of known results.

8.1 Theory

A transition invariant i is defined as a T-indexed vector of integers that solves the equation $C \cdot i = \underline{0}$. If, for some transition sequence w and markings m and m' we have $m \xrightarrow{w} m'$ then $m = m'$ if and only if the Parikh vector $\Psi(w)$ is a transition invariant. If $m \xrightarrow{w_1} m'$ and $m \xrightarrow{w_2} m'$ then $\Psi(w_1) - \Psi(w_2)$ is a transition invariant. That is, transition invariants give hints to cycles and diamond structures in the reachability graph.

A nontrivial transition invariant i proves that the effect vector of one of the transitions in $supp(i)$ is a linear combination of the others. Let U be a linearly independent set of transitions. From the observations made so far we may conclude

(1) A firing sequence consisting of transitions in U never forms a cycle, and
(2) Two firing sequences consisting of transitions in U can only form a diamond (i.e. start and end in the same marking) if they have precisely the same Parikh vector.

In the sequel, let U^* be a maximal set (w.r.t. set inclusion) of transitions with linearly independent effect vectors.

8.2 Selective Storage of Markings

When searching the state space, visited markings are stored. This way, we avoid multiple exploration of the same marking and force termination (for nets with a finite reachability graph). Since memory is the limiting resource in several use cases of model checking, we may think of not storing at least some of the explored markings thus trading better memory efficiency for the additional run time needed for multiple exploration of markings. It is, however, desirable to make sure that the state space exploration still terminates. This is the case if, for every cycle in the reachability, at least one marking is stored [1].

This is where transition invariants, or more precisely the information on linear dependency of effect vectors, come into play. A valid strategy is to store a marking m if a transition t from $T \backslash U^*$ is fired in m. Since every cycle in the state space must include an occurrence of a transition in $T \setminus U^*$, the required storage of at least one marking in every cycle is indeed guaranteed [44].

We deliberately used the term "is fired" instead of "is enabled". This way, we take care of reduced reachability graph construction where, for instance, the stubborn set method does not necessarily explore all enabled transitions. In fact, use of stubborn sets is recommended for using selective storage, for two reasons. First, the smaller sets of transitions to be explored increase the likelihood that a marking does not explore a transition in $T \setminus U^*$ and does not need to be stored. Hence, selective storage yields better space reduction. Second, markings that are not stored and need to be explored more than once, have a smaller branching degree thus requiring less time for exploration. Hence, the run time penalty of the method is smaller than with unreduced state space generation.

8.3   A Progress Measure for the Sweep-Line Method

The sweep-line method [6] is another strategy for selective storage. It requires a progress measure π that assigns an integer number $\pi(m)$ to every marking m (the *progress value* of m). Markings are explored in the order of their progress value (with small progress values first). A progress measure is perfect if $m \xrightarrow{t} m'$ implies $\pi(m) < \pi(m')$, for all markings m, m' and transitions t. In this case, the question "has m' already been visited?" always concerns the set of markings with progress values greater than $\pi(m)$. That is, markings with progress values that are smaller than the one of the currently explored marking can be removed from storage without any impact on the correctness of the search algorithm. This way, memory is released that can be used for storing other markings, and we can ultimately explore larger state spaces.

If the reachability graph of the net contains a cycle, it is impossible to find a perfect progress measure. The best we can do is to make $\pi(m) = \pi(m')$ for transitions t participating in the cycle. In a reversible net (a net with a strongly connected reachability graph), this would, however, mean that all markings have the same progress value and the method is ineffective. In fact, another strategy has been proposed for nets with cycles [27]. We use an imperfect progress measure and still remove markings that have a progress value smaller than the currently explored one. That is, we may have transitions in the reachability graph where $m \xrightarrow{t} m'$ but $\pi(m) > \pi(m')$. Such transitions are called *regressing*. For m', we have no information whether or not we have visited it before. Hence, we mark m' as persistent (that is, we do not remove it any time subsequently). After completion of the search procedure, we launch a new search using the recently added persistent markings as starting point. This procedure is iterated until no fresh markings will be marked as persistent. It can be proven that this strategy will eventually visit all reachable markings. An extension of the method to LTL model checking has been proposed [14].

We call a progress measure π incremental if we can assign an integer number $\delta(t)$ to every transition t such that, for all markings m, m', $m \xrightarrow{t} m'$ implies $\pi(m') = \pi(m) + \delta(t)$. An incremental measure is perfect if and only if, for all t, $\delta(t) > 0$. For the sweep-line method being effective, it is desirable to have as few as possible transitions with $\delta(t) < 0$.

An incremental progress measure has several advantages. First, it permits the calculation of progress values in constant time (just add $\delta(t)$ to $\pi(m)$ when firing t in m). Second, it permits a delayed strategy for removing markings. When exploring a marking m, we remove only the markings with progress value smaller than $\pi(m) + \min_{t \in T} \delta(t)$. That is, we remove marking later than originally proposed. This way, we increase the likelihood that markings obtained by regressing transitions are still in memory and thus do not need to be marked persistent.

A third advantage of incremental progress measures is that we can automatically compute one, based on transition invariants (or again: linear dependencies of transitions) [45]. In fact, an arbitrary assignment δ does not necessarily establish a progress measure. To see this, we first lift an incremental progress measure to transition sequences: if, for a transition sequence w, $m \xrightarrow{w} m'$, we have $\pi(m') = \pi(m) + \sum_{t \in T} \Psi(w)[t]\delta(t)$ with $\Psi(w)$ being the already mentioned Parikh vector of w. Now, every m can have only one progress value $\pi(m)$, regardless of the transition sequence reaching it. In other words, the assignment δ must be such that $m \xrightarrow{w_1} m'$ and $m \xrightarrow{w_2} m'$ implies

$$\sum_{t \in T} \Psi(w_1)[t]\delta(t) = \sum_{t \in T} \Psi(w_2)[t]\delta(t).$$

This condition can be satisfied using the following strategy that involves a maximal set U^* of transitions with linear independent effect vectors: For transitions in U^*, set $\delta(t) = 1$. For a transition u in $T \setminus U^*$, there is a unique linear combination $\Delta u = \sum_{t \in U^*} k_t \Delta t$. Then set $\delta(u) = \sum_t \in U^* k_t \delta(t)$. This setting obviously satisfies the consistency requirement for incremental progress measures: for all paths w, w' leading from the initial marking m_0 to a fixed marking m,

$$\sum_{t \in T} \Psi(w)[t]\delta(t) = \sum_{t \in T} \Psi(w')[t]\delta(t).$$

Only if this requirement is satisfied, the assignment δ yields a well-defined progress measure.

The measure obtained this way is definitely increasing for all transitions in U^*. Transitions in $T \setminus U^*$ may be progressing, regressing, or have a δ of 0. It is not necessarily the best possible measure (in the sense that the number of regressing transitions is minimal). Its quality can be influenced by two degrees of freedom. First, a set of transitions usually has more than one maximal set of linearly independent effect vectors. Different U^* may yield different δ values for the transitions in $T \setminus U^*$. Second, the setting $\delta(t) = 1$ for transitions in U^* is arbitrary. Every setting of δ for the transitions in U^* actually yields a consistent measure. Different values for transitions in U^* cause different values for the transitions in $T \setminus U^*$ and may turn a regressing transition into a progressing one. Some attempts have been made to optimise a progress measure using these degrees of freedom. However, resulting progress measures have so far not been substantially better than those obtained by arbitrary choice of U^* and simple setting of $\delta(t) = 1$ for $t \in U^*$.

8.4 Portfolio Member for LTL Model Checking

In explicit LTL model checking, we perform a search in the product of the reachability graph and the Büchi automaton of the negated formula, aiming at a cycle that contains accepting states. Usually, the product is constructed for the Büchi automaton and the reachability graph. We can, however, turn the Büchi automaton into a Petri net and synchronise it with the net under investigation such that the synchronised net produces the same product state space. The result is a Petri net in which we want to find a realisable accepting cycle. A necessary condition for such a cycle to exist is a transition invariant. This invariant must contain transitions that are connected to the places representing the acceptance condition of the Büchi automaton. If such a transition invariant does not exist, we may conclude that the net satisfies the given formula [13].

9   Siphons and Traps

Siphons and traps [17] are topological patterns in a Petri net that preserve absence resp. presence of tokens. Their interplay is related to properties such as deadlock freedom and liveness.

For simplicity, we generally assume in this section that, for all nodes x and y, $W(x, y) \in \{0, 1\}$. Most results can be lifted to more general assumptions on arc multiplicities. But then results require additional, sometimes tricky additional assumptions that lead beyond the scope of this survey.

9.1   Theory

A *siphon* is a nonempty set of places S such that $\bullet S \subseteq S \bullet$. Once a siphon does not contain any token, it cannot be marked subsequently since all transitions that could put tokens into S have an unmarked pre-place in S. If a marking m is a deadlock (i.e. does not enable any transition), the set $\{p \in P \mid m(p) = 0\}$ forms an unmarked siphon.

A *trap* is a (not necessarily nonempty) set of places Q such that $Q \bullet \subseteq \bullet Q$. Once a trap contains a token, it contains at least one token in any subsequent marking. From the facts collected so far, we may conclude that a net where every siphon includes an initially marked trap [17], is deadlock free.

The relation "every siphon contains a marked trap" is called the *siphon/trap property*. Since the union of traps is a trap again, every siphon includes a unique maximal trap (w.r.t. set inclusion). That is, for checking the siphon/trap property it is sufficient to check: for every minimal siphon (w.r.t. set inclusion), its maximal trap is marked. This way, the complexity of evaluating the siphon/trap property is substantially reduced but remains exponential since a Petri net may have an exponential number of minimal siphons. To date, the best way to evaluate the siphon/trap property appears to be its translation to a satisfiability problem in propositional logic [36]. This approach benefits from the impressive progress in the area of SAT checking and appears to be feasible for even large nets.

The propositional formula ϕ is built such that it is satisfiable if and only if there exists a siphon S where the maximal trap is initially unmarked. It uses $|P|(|P| + 1)$ variables called $p_1^0, \ldots, p_n^0, p_1^1, \ldots, p_n^1, \ldots, p_1^n, \ldots, p_n^n$ assuming that $P = \{p_1, \ldots, p_n\}$. The formula is a conjunction of subformulas $\phi = \phi_0 \wedge \phi_1 \wedge \cdots \wedge \phi_n \wedge \phi_f$.

Formula ϕ_0 is satisfied by assignment β if and only if $\{p \mid \beta(p^0)\}$ is a siphon. To achieve this,

$$\phi_0 = \bigvee_{p \in P} p^0 \wedge \bigwedge_{t \in T} \left(\bigvee_{p \in t\bullet} p^0 \implies \bigvee_{q \in \bullet t} q^0 \right)$$

Subformulas ϕ_i, for $i > 0$ mimick the process of computing a maximal trap in this siphon. To this end,

$$\phi_i = \bigwedge_{p \in P} (p^i \iff p^{i-1} \wedge \bigwedge_{t \in T, p \in \bullet t} \bigvee_{q \in t \bullet} q^{i-1})$$

represents the removal of all places where some transition removes a token without putting a token back into the remaining set of places. This process must be repeated since the removal of places can change the situation for other places. It can, however, not be iterated more than $n = |P|$ times since we either remove a place or converge. That is, $\{p \mid \beta(p^n)\}$ is the maximal trap in the above mentioned siphon. The final subformula ϕ_f specifies emptiness of the trap:

$$\bigwedge_{p : m_0(p) > 0} \neg p^n$$

In this shape, the SAT checker sometimes requires a prohibitive run-time for large nets. That is why we typically use a smaller formula in practice. It relies on the fact that, for computing the maximal trap, all n iterations are necessary only in extreme cases. That is, we fix some number k and use only $\phi_0 \wedge \phi_1 \wedge \cdots \wedge \phi_k \wedge \phi_f$ (with variables in ϕ_f adjusted to k). If that formula is satisfiable, there is some unmarked superset of the maximal trap which means that the maximal trap as such is unmarked, too. If the calculation of the maximal trap requires less than k iterations, the formula $\phi^* = \bigwedge_{p \in P} p^k \iff p^{k-1}$ is satisfied. If this is the case and ϕ is otherwise not satisfiable, then the siphon/trap property actually holds. In the remaining cases, the formula does not permit any conclusion concerning the validity of the siphon/trap property. The SAT problem can be modified such that all mentioned situations are covered. With k being small compared to n, the siphon/trap property can be decided for even large nets.

9.2 Using the Siphon/Trap Property

If the model checking problem is EF DEADLOCK, or AG ¬ DEADLOCK, an evaluation of the siphon/trap property can be added to the portfolio. Since the problem is passed to a SAT checker that requires only polynomial space, there is not too much competition for the memory resource.

This seems to be only a rather singular support for model checking. However, deadlock checking is one of the most fundamental problems and occurs far above average in practice. In the MCC, there is a dedicated subcategory for checking reachability of deadlocks.

If enough run time is available, the check for the siphon/trap property can be executed prior to actual model checking. If the siphon/trap property happens to hold, every single occurrence of the atomic proposition DEADLOCK in the formula under investigation can be replaced by FALSE. This modification then may enable a large number of rewrite rules making the formula or at least a subformula of the verification problem collapse.

9.3   Using Individual Siphons and Traps

The defining property of traps (once marked - always marked) and siphons (once unmarked, always unmarked) can be used to prove atomic propositions to be invariantly true or false. They can be replaced by the respective Boolean constant and enable further simplification of the problem under investigation through rewrite rules. For an atomic proposition $\alpha = k_1 p_1 + \ldots k_n p_n \sim k$ with $\sim \in \{=, \neq, <, \leq, >, \geq\}$, let the support $supp(\alpha)$ be the set $\{p_i \mid k_i \neq 0\}$. Assume further that all k_i are greater or equal to zero. We claim

- if the support of $k_1 p_1 + \ldots k_n p_n \geq 1$ includes an initially marked trap, it is invariantly true.
- if the support of $k_1 p_1 + \ldots k_n p_n \leq 0$ includes an initially marked trap, it is invariantly false.
- if the support of $k_1 p_1 + \ldots k_n p_n \geq 1$ is included in an initially unmarked siphon, it is invariantly false.
- if the support of $k_1 p_1 + \ldots k_n p_n \leq 0$ is included in an initially unmarked siphon, it is invariantly true.

In rare cases, we may be able to prove stronger results such as: If the support of $k_1 p_1 + \ldots k_n p_n \geq 2$ includes two disjoint initially marked traps, it is invariantly true.

Finding a suitable trap is rather cheap. We start with the support of the proposition and iteratively remove all places that violate the trap property. This results in the maximal trap contained in the support of the proposition. If any included trap is marked, this one is marked as well. Similarly, we compute a wrapping siphon by starting with the set of all unmarked places and iteratively remove all places that violate the siphon property. If any unmarked siphon is a superset of the support of the given proposition, this siphon is. The calculation of suitable siphons and traps is much cheaper than the state equation approach to atomic propositions mentioned in Sect. 6.2. Furthermore, siphons and traps may prove propositions to be invariant where the state equation is not able to prove this fact. We have sufficient evidence for this claim from our experience in the MCC.

A valid question in this context is: is an initially unmarked siphon a pattern that we may expect in a Petri net? Indeed, for an initially unmarked siphon S, all surrounding transitions ($\bullet S \cup S \bullet$) are dead already in the initial marking. Of course, nobody would model such a broken system. However, we would like to turn the reader's attention to the fact that many place/transition nets are not the result of a carefully thought modeling process using a graphical editor. They can be the result of a long chain of automated translation processes from other formalisms. Such nets may very well include dead parts, as artifacts of the translation process. Unmarked siphons may also occur as the result of modeling errors. A verification tool should therefore be able to cope with broken net parts as efficiently as possible.

10   Conflict Clusters

Looking at conflict clusters, we study the way transitions compete for tokens. Several net classes with strong structural results (free-choice nets, asymmetric choice nets) are defined by the shape of their conflict clusters.

10.1   Theory

For a net $N = [P, T, F, W, m_0]$, define $F_{pre} = F \cap (P \times T)$. Two nodes x and y are said to be *conflict equivalent* if $[x, y]$ occurs in the reflexive, symmetric, and transitive closure of F_{pre}. The resulting equivalence classes are called *conflict clusters*. Using Tarjan's union/find algorithm [49], conflict clusters can be computed in little more than $O(|F_{pre}|)$ time.

A conflict cluster C is called *free choice* if there is a number k such that, for all places $p \in C$ and all transitions $t \in C$, $W(p, t) = k$. If a transition in a free choice cluster is enabled at a marking, all transitions in the cluster are enabled.

10.2   Quick Stubborn Set Computation

With the stubborn set method [51], we compute a reduced reachability graph. Starting exploration with the initial marking, for every explored marking m, a stubborn set $stub(m)$ of transitions is computed and only enabled transitions in $stub(m)$ are explored in m. There are numerous results that specify how stubborn sets must be chosen for preserving various classes of properties [52, 53]. Many of them include the following rules for calculating a stubborn set U:

- Add an enabled transition to U;
- If an enabled transition t is in U, add all transitions in $(\bullet t)\bullet$ to U;
- If a disabled transition t is in U, add all transitions in $\bullet p$ to U, for some insufficiently marked pre-place p of t.

If a transition is enabled at a marking and its conflict cluster is free choice, this procedure will always result in $C \cap T$ being a very small stubborn set and hence desirable. Hence, we give transitions in free choice clusters higher priority in stubborn set calculation thus speeding up model checking. Among the free choice clusters, the ones with few contained transitions get priority over the ones with many contained transitions. This way, many calculations of stubborn sets use less time and still produce good results.

Unfortunately, this strategy may, in rare cases, yield worse reduction than other strategies for computing stubborn sets. However, every known procedure for computing stubborn sets appears to suffer from that problem. Hence, the observation is no argument not to use the information on free choice clusters.

10.3   Preprocessing

If we explore a marking m' during state space exploration, we need to calculate the set of transitions enabled in m'. Doing that in brute force fashion amounts to evaluating the enabling condition of $|T|$ transitions. Given the large number of markings to be explored, the number of transitions to be checked deserves to be reduced. To this end, we exploit the fact that, for all markings other than the initial marking, m' is the result of firing some transition t in a marking m. t removes a token only from places in $\bullet t$, and adds tokens only to places in $t\bullet$. If a transition is enabled in m, it can be disabled in m' only if is in $(\bullet t)\bullet$. If a transition is disabled in m, it can be enabled in m' only if it is in $(t\bullet)\bullet$. Hence, when exploring m', only for enabled transitions in m appearing in $(\bullet t)\bullet$ and for disabled transitions in m appearing in $(t\bullet)\bullet$, enabledness needs to be checked in m' while for all other transitions, enabledness did not change w.r.t. m. For implementing this speed-up, it is recommended to store sets $(\bullet t)\bullet$ and $(t\bullet)\bullet$ explicitly for all transitions t, since these sets are traversed millions of times.

The set $(\bullet t)\bullet$, called conflict set of t, is interesting for stubborn set generation as well (see above). According to our experience, storing the conflict sets may require a lot of space, in worst case $O(|T|^2)$. We observed, however, that quite frequently, we have transitions t and t' where $(\bullet t)\bullet = (\bullet t')\bullet$. Such transitions can share the memory for storing their conflict sets.

Two transitions can have the same conflict set only if they appear in the same conflict cluster. Consequently, when searching for transitions with sharable conflict sets, we may process the conflict clusters separately, even in parallel. This way, preprocessing (that can take several minutes just for computation of the sets $(\bullet t)\bullet$) can be accelerated significantly.

11   Summary and Open Issues

There is a large variety of approaches where elements of Petri net structure theory are used for accelerating model checking. They target several entities in model checking that we shall briefly discuss in the sequel.

Elimination or Simplification of Atomic Propositions. The state equation, siphons, traps, and the siphon/trap property can be used for replacing atomic propositions by Boolean constants. Place invariants may simplify propositions. The replacement permits the application of logical tautologies for rewriting the property under investigation. The result may be a much simpler verification problem. In some cases (about 20% to 30% of the problems in the MCC), the whole verification problem collapses into a constant. This means that the problem could be verified completely by structure theory.

The applied approaches have in common that they establish *inductive invariants*. An inductive invariant is a property of markings that is true for the initial marking and preserved by transition occurrences. Consequently it is true for some superset of the set of all reachable markings. In future work, we may look more systematically into other kinds of inductive invariants and their capabilities for eliminating atomic propositions (or larger subformulas). It is known that, for every unreachable marking m, there is an inductive invariant expressible in Presburger arithmetic violated by m [29]. However, there is no apparent characterisation of all inductive invariants that can be calculated efficiently, although progress is being made in this direction [50].

Coding Markings. With place invariants and other structural entities (state machine components, nested units [15]), we are able to reduce the length of bit vector representations of markings. This saves space but also accelerates the procedures for accessing the bit vectors. In symbolic model checking, we get additional information concerning good variable orderings. This connection between structure theory and model checking appears to be well explored.

Detecting Cycles and Diamonds. For several methods that use selective storage of markings, we need to care about cycles and diamonds. Transition invariants give us valuable information. Cycles in the reachability graph have outstanding importance for model checking. They are involved in accepting criteria for LTL model checking and for several calculations necessary for advanced stubborn set methods. As a consequence, future research may come up with additional ideas for exploiting transition invariants.

In general, structural methods concerning transitions have still less applications than structural methods involving places. Given the formal duality of places and transitions, this gap is still surprising. One could be tempted to conclude that we are still missing the right angle to look at transition related structures in Petri nets.

Dealing with Conflicts. Another almost blank spot in structure theory appears to be the study of conflict clusters. Intuitively, the way in which transitions compete for tokens seems to have decisive influence on behaviour. Yet, conflict clusters occur in only very few structural results and have some, but not too impressive value for current model checking procedures. There are structural results for nets where *all* conflict clusters have a certain structure (free choice nets, asymmetric choice nets). But what if *some* clusters have this structure? We may not be able to find purely structural results for such nets. But can model checking benefit from nets that have many free choice conflicts, but some other conflict clusters, too? Questions like this deserve further investigation.

Summary. At first glance, model checking is a technology that is independent of the formalism used for modeling systems. As soon as the behaviour can be expressed in terms of a transition system, operators of temporal logic are well defined and the general algorithms are applicable. However, model checking suffers from the state explosion problem. There are some general state space reduction methods such as partial order reduction and the symmetry method. But then, there exist many domain specific complementary methods such as the model checking algorithms based on finite branching prefixes, and there are domain specific accelerations of the general methods. For Petri nets, its well developed structure theory is a unique selling point. We may finally conclude that model checking Petri nets has developed into a specific branch of model checking technology, with unique contributions to the technology and its own domain specific approaches.

References

1. Behrmann, G., Larsen, K.G., Pelánek, R.: To store or not to store. In: Hunt, W.A., Somenzi, F. (eds.) CAV 2003. LNCS, vol. 2725, pp. 433–445. Springer, Heidelberg (2003). https://doi.org/10.1007/978-3-540-45069-6_40
2. Bønneland, F., Dyhr, J., Jensen, P.G., Johannsen, M., Srba, J.: Simplification of CTL formulae for efficient model checking of Petri nets. In: Khomenko, V., Roux, O.H. (eds.) PETRI NETS 2018. LNCS, vol. 10877, pp. 143–163. Springer, Cham (2018). https://doi.org/10.1007/978-3-319-91268-4_8
3. Bryant, R.E.: Symbolic Boolean manipulation with ordered binary-decision diagrams. ACM Comput. Surv. **24**(3), 293–318 (1992)
4. Büchi, J.R.: On a decision method in restricted second order arithmetic. In: Proceedings of the International Congress on Logic, Method, and Philosophy of Science, pp. 1–12 (1962)
5. Burch, J.R., Clarke, E.M., McMillan, K.L., Dill, D.L., Hwang, L.J.: Symbolic model checking: 10^{20} states and beyond. In: Proceedings of the LICS, pp. 428–439. IEEE (1990)
6. Christensen, S., Kristensen, L.M., Mailund, T.: A sweep-line method for state space exploration. In: Margaria, T., Yi, W. (eds.) TACAS 2001. LNCS, vol. 2031, pp. 450–464. Springer, Heidelberg (2001). https://doi.org/10.1007/3-540-45319-9_31
7. Clarke, E.M., Biere, A., Raimi, R., Zhu, Y.: Bounded model checking using satisfiability solving. Formal Methods Syst. Des. **19**(1), 7–34 (2001)
8. Clarke, E.M., Emerson, E.A., Sistla, A.P.: Automatic verification of finite-state concurrent systems using temporal logic specifications. ACM Trans. Program. Lang. Syst. **8**(2), 244–263 (1986)
9. Commoner, F., Holt, A.W., Even, S., Pnueli, A.: Marked directed graphs. J. Comput. Syst. Sci. **5**(5), 511–523 (1971)
10. Cook, S.A.: The complexity of theorem-proving procedures. In: Proceedings of the 3rd Annual ACM Symposium on Theory of Computing, pp. 151–158 (1971)
11. Emerson, E.A., Clarke, E.M.: Using branching time temporal logic to synthesize synchronization skeletons. Sci. Comput. Program. **2**(3), 241–266 (1982)
12. Esparza, J.: Model checking using net unfoldings. In: Gaudel, M.-C., Jouannaud, J.-P. (eds.) CAAP 1993. LNCS, vol. 668, pp. 613–628. Springer, Heidelberg (1993). https://doi.org/10.1007/3-540-56610-4_93
13. Esparza, J., Melzer, S.: Model checking LTL using constraint programming. In: Azéma, P., Balbo, G. (eds.) ICATPN 1997. LNCS, vol. 1248, pp. 1–20. Springer, Heidelberg (1997). https://doi.org/10.1007/3-540-63139-9_26
14. Evangelista, S., Kristensen, L.M.: A sweep-line method for Büchi automata-based model checking. Fundam. Inform. **131**(1), 27–53 (2014)
15. Garavel, H.: Nested-unit Petri nets: a structural means to increase efficiency and scalability of verification on elementary nets. In: Devillers, R., Valmari, A. (eds.) PETRI NETS 2015. LNCS, vol. 9115, pp. 179–199. Springer, Cham (2015). https://doi.org/10.1007/978-3-319-19488-2_9

16. Godefroid, P., Wolper, P.: A partial approach to model checking. Inf. Comput. **110**(2), 305–326 (1994)
17. Hack, M.: Analysis of production schemata by Petri nets. Technical report, MS thesis, Department of Electrical Engineering, MIT, Cambridge, Massachusetts (1972)
18. Hack, M.: Decidability questions for Petri nets. Outstanding Dissertations in the Computer Sciences. Garland Publishing, New York (1975)
19. Hajdu, Á., Vörös, A., Bartha, T.: New search strategies for the Petri net CEGAR approach. In: Devillers, R., Valmari, A. (eds.) PETRI NETS 2015. LNCS, vol. 9115, pp. 309–328. Springer, Cham (2015). https://doi.org/10.1007/978-3-319-19488-2_16
20. Heiner, M., Rohr, C., Schwarick, M., Tovchigrechko, A.A.: MARCIE's secrets of efficient model checking. In: Koutny, M., Desel, J., Kleijn, J. (eds.) Transactions on Petri Nets and Other Models of Concurrency XI. LNCS, vol. 9930, pp. 286–296. Springer, Heidelberg (2016). https://doi.org/10.1007/978-3-662-53401-4_14
21. Holt, A.W., Commoner, F.: Events and conditions. In: MAC Conference on Concurrent Systems and Parallel Computation, pp. 3–52 (1970)
22. Jensen, K.: Condensed state spaces for symmetrical coloured Petri nets. Formal Methods Syst. Des. **9**(1/2), 7–40 (1996)
23. Kam, T., Villa, T., Brayton, R.K.: Multi-valued decision diagrams: theory and applications. Multiple-Valued Logic **4**(01), 9–62 (1998)
24. Karp, R.M., Miller, R.E.: Parallel program schemata. J. Comput. Syst. Sci. **3**(2), 147–195 (1969)
25. Kordon, F., et al.: MCC'2015 – the fifth model checking contest. In: Koutny, M., Desel, J., Kleijn, J. (eds.) Transactions on Petri Nets and Other Models of Concurrency XI. LNCS, vol. 9930, pp. 262–273. Springer, Heidelberg (2016). https://doi.org/10.1007/978-3-662-53401-4_12
26. Kosaraju, S.R.: Decidability of reachability on vector addition systems. In: ACM Symposium on Theory Computing, pp. 267–281 (1982)
27. Kristensen, L.M., Mailund, T.: A generalised sweep-line method for safety properties. In: Eriksson, L.-H., Lindsay, P.A. (eds.) FME 2002. LNCS, vol. 2391, pp. 549–567. Springer, Heidelberg (2002). https://doi.org/10.1007/3-540-45614-7_31
28. Lautenbach, K., Schmid, H.A.: Use of Petri nets for proving correctness of concurrent process systems. In: IFIP Congress, pp. 187–191 (1974)
29. Leroux, J.: The general vector addition system reachability problem by Presburger inductive invariants. In: Proceedings of the LICS, pp. 4–13. IEEE (2009)
30. Lipton, R.J.: The reachability problem requires exponential space. Technical report 62, Department of Computer Science, Yale University (1976)
31. Manna, Z., Pnueli, A.: The Temporal Logic of Reactive and Concurrent Systems - Specification. Springer, New York (1992). https://doi.org/10.1007/978-1-4612-0931-7
32. Mayr, E.W.: An algroithm for the general Petri net reachability problem. SIAM J. Comput. **13**(3), 441–460 (1984)
33. McMillan, K.L.: Using unfoldings to avoid the state explosion problem in the verification of asynchronous circuits. In: von Bochmann, G., Probst, D.K. (eds.) CAV 1992. LNCS, vol. 663, pp. 164–177. Springer, Heidelberg (1993). https://doi.org/10.1007/3-540-56496-9_14
34. Memmi, G., Roucairol, G.: Linear algebra in net theory. In: Brauer, W. (ed.) Net Theory and Applications. LNCS, vol. 84, pp. 213–223. Springer, Heidelberg (1980). https://doi.org/10.1007/3-540-10001-6_24

35. Miner, A.S., Ciardo, G.: Efficient reachability set generation and storage using decision diagrams. In: Donatelli, S., Kleijn, J. (eds.) ICATPN 1999. LNCS, vol. 1639, pp. 6–25. Springer, Heidelberg (1999). https://doi.org/10.1007/3-540-48745-X_2

36. Oanea, O., Wimmel, H., Wolf, K.: New algorithms for deciding the siphon-trap property. In: Lilius, J., Penczek, W. (eds.) PETRI NETS 2010. LNCS, vol. 6128, pp. 267–286. Springer, Heidelberg (2010). https://doi.org/10.1007/978-3-642-13675-7_16

37. Pastor, E., Cortadella, J., Peña, M.A.: Structural methods to improve the symbolic analysis of Petri nets. In: Donatelli, S., Kleijn, J. (eds.) ICATPN 1999. LNCS, vol. 1639, pp. 26–45. Springer, Heidelberg (1999). https://doi.org/10.1007/3-540-48745-X_3

38. Pastor, E., Roig, O., Cortadella, J., Badia, R.M.: Petri net analysis using boolean manipulation. In: Valette, R. (ed.) ICATPN 1994. LNCS, vol. 815, pp. 416–435. Springer, Heidelberg (1994). https://doi.org/10.1007/3-540-58152-9_23

39. Peled, D.: All from one, one for all: on model checking using representatives. In: Courcoubetis, C. (ed.) CAV 1993. LNCS, vol. 697, pp. 409–423. Springer, Heidelberg (1993). https://doi.org/10.1007/3-540-56922-7_34

40. Ridder, H.: Analyse von Petri-Netz-Modellen mit Entscheidungsdiagrammen. Ph.D. thesis, Koblenz, Landau, University (1997)

41. Roch, S., Starke, P.H.: INA: integrated net analyzer (1999)

42. Schmidt, K.: Stubborn sets for standard properties. In: Donatelli, S., Kleijn, J. (eds.) ICATPN 1999. LNCS, vol. 1639, pp. 46–65. Springer, Heidelberg (1999). https://doi.org/10.1007/3-540-48745-X_4

43. Schmidt, K.: How to calculate symmetries of Petri nets. Acta Inf. **36**(7), 545–590 (2000)

44. Schmidt, K.: Using Petri net invariants in state space construction. In: Garavel, H., Hatcliff, J. (eds.) TACAS 2003. LNCS, vol. 2619, pp. 473–488. Springer, Heidelberg (2003). https://doi.org/10.1007/3-540-36577-X_35

45. Schmidt, K.: Automated generation of a progress measure for the sweep-line method. STTT **8**(3), 195–203 (2006)

46. Starke, P.H.: Reachability analysis of Petri nets using symmetries. Syst. Anal. Model. Simul. **8**(4–5), 293–303 (1991)

47. Strehl, K., Thiele, L.: Interval diagram techniques for symbolic model checking of Petri nets. In: Proceedings of the Design, Automation and Test in Europe, pp. 756–757 (1999)

48. Tarjan, R.E.: Depth-first search and linear graph algorithms. SIAM J. Comput. **1**(2), 146–160 (1972)

49. Tarjan, R.E.: Efficiency of a good but not linear set union algorithm. J. ACM **22**(2), 215–225 (1975)

50. Triebel, M., Sürmeli, J.: Characterizing stable and deriving valid inequalities of Petri nets. Fundam. Inform. **146**(1), 1–34 (2016)

51. Valmari, A.: Stubborn sets for reduced state space generation. In: Rozenberg, G. (ed.) ICATPN 1989. LNCS, vol. 483, pp. 491–515. Springer, Heidelberg (1991). https://doi.org/10.1007/3-540-53863-1_36

52. Valmari, A.: The state explosion problem. In: Reisig, W., Rozenberg, G. (eds.) ACPN 1996. LNCS, vol. 1491, pp. 429–528. Springer, Heidelberg (1998). https://doi.org/10.1007/3-540-65306-6_21

53. Valmari, A., Hansen, H.: Stubborn set intuition explained. In: Koutny, M., Kleijn, J., Penczek, W. (eds.) Transactions on Petri Nets and Other Models of Concurrency XII. LNCS, vol. 10470, pp. 140–165. Springer, Heidelberg (2017). https://doi.org/10.1007/978-3-662-55862-1_7

54. Vardi, M.Y.: Verification of concurrent programs: the automata-theoretic framework. In: Proceedings of the LICS, pp. 167–176. IEEE (1987)

55. Vergauwen, B., Lewi, J.: A linear local model checking algorithm for CTL. In: Best, E. (ed.) CONCUR 1993. LNCS, vol. 715, pp. 447–461. Springer, Heidelberg (1993). https://doi.org/10.1007/3-540-57208-2_31

56. Wimmel, H., Wolf, K.: Applying CEGAR to the Petri net state equation. Log. Methods Comput. Sci. **8**(3) (2012)

57. Wolf, K.: Petri net model checking with LoLA 2. In: Khomenko, V., Roux, O.H. (eds.) PETRI NETS 2018. LNCS, vol. 10877, pp. 351–362. Springer, Cham (2018). https://doi.org/10.1007/978-3-319-91268-4_18

58. Wolf, K.: A simple abstract interpretation for Petri net queries. In: Proceedings of the PNSE, CEUR Workshop Proceedings, vol. 2138, pp. 163–170 (2018)

Parametric Verification: An Introduction

Étienne André[1], Michał Knapik[2], Didier Lime[3], Wojciech Penczek[2,4],
and Laure Petrucci[1(✉)]

[1] LIPN, CNRS UMR 7030, Université Paris 13, Villetaneuse, France
Laure.Petrucci@lipn.univ-paris13.fr
[2] Institute of Computer Science, PAS, Warsaw, Poland
[3] École Centrale de Nantes, LS2N, CNRS UMR 6004, Nantes, France
[4] University of Natural Sciences and Humanities, II, Siedlce, Poland

Abstract. This paper constitutes a short introduction to parametric
verification of concurrent systems. It originates from two 1-day tutorial
sessions held at the Petri nets conferences in Toruń (2016) and Zaragoza
(2017). A video of the presentation is available at https://www.youtube.
com/playlist?list=PL9SOLKoGjbeqNcdQVqFpUz7HYqD1fbFIg, consisting of 14 short sequences. The paper presents not only the basic formal
concepts tackled in the video version, but also an extensive literature to
provide the reader with further references covering the area.

We first introduce motivation behind parametric verification in general, and then focus on different models and approaches, for verifying
several kinds of systems. They include Parametric Timed Automata, for
modelling real-time systems, where the timing constraints are not necessarily known *a priori*. Similarly, Parametric Interval Markov Chains allow
for modelling systems where probabilities of events occurrences are intervals with parametric bounds. Parametric Petri Nets allow for compact
representation of systems, and cope with different types of parameters.
Finally, Action Synthesis aims at enabling or disabling actions in a concurrent system to guarantee some of its properties. Some tools implementing these approaches were used during hands-on sessions at the
tutorial. The corresponding practicals are freely available on the Web.

1 Introduction to Parametric Verification

We first introduce the motivation for performing parametric verification. Several formal models can be considered, depending on the characteristics of the
system and its properties the designer wants to address. We will also discuss the
problems of interest in such a framework. At the end of this section, we will give
pointers to some additional and complementary material.

The reader is assumed to have basic knowledge of Petri nets and/or automata
and their associated verification techniques, since they constitute the basis of the
formal models we address in a parametric setting.

This work is partially supported by the ANR national research program PACS (ANR-
14-CE28-0002).

M. Koutny et al. (Eds.): ToPNoC XIV, LNCS 11790, pp. 64–100, 2019.
https://doi.org/10.1007/978-3-662-60651-3_3

1.1   Why Parameters and of What Kind?

Parameters provide several facilities for easily modelling complex systems:

1. *Dimensioning*: systems often exhibit components that occur as *multiple copies* of the same structure or similar ones. For example, a wireless sensor network is composed of a certain number of identical sensors. At the design phase, the exact number of sensors might be unknown and would therefore be a parameter.
2. *Choice of actions*: different actions in the system might be possible in a given state. The designer having several possibilities in mind would model them and analyse the behaviour of the system with these actions enabled or disabled. In this case, the enabledness of each individual action is considered as a parameter.
3. *Design choices*: *different system characteristics* can be taken into account at the design phase so as to be evaluated. For example, when designing an electronic system, one might have the choice between different components, available of course at different prices, each providing its specific characteristics, such as a better or worse response time. In such a case, the designer may want to construct a single model, and use these timing characteristics as a parameter so as to evaluate which is the best possible choice according to his/her needs.

To handle the cases described, different kinds of parameters are used. They will influence the type of a formal model and verification techniques chosen. In the three previous cases, parameters would be:

1. *Instances numbering* allow for counting identical components in the system.
2. *Controllable actions* can be enabled or disabled, as opposed to the other ones which are always available for the system.
3. *Time or probabilities* provide means to handle different characteristics of the actual system components.

Therefore, many kinds of parameters can be considered according to the problem at hand.

1.2   Modelling Languages

Among the popular traditional languages for modelling concurrent systems are automata and Petri nets, and their numerous extensions.

Unfortunately, these modelling languages are not completely suited to handle the systems of our interest. Indeed, numbering instances of components is easily achieved with high-level Petri nets such as Coloured Petri Nets (CPNs). In CPNs, the Petri net is enriched with data carried by tokens and modified when firing transitions. These data can very well be a numbering of an instance. Nevertheless, the number of instances is fixed *a priori*, as opposed to a parameter. Therefore, when the designer wants to analyse several configurations of the system, first the model is built for a given number of instances, then analysed, the number is changed, the model analysed again, and so on.

In Petri nets or automata, there is no specific handling of controllable actions. Hence, to test several options in the design, each of them must be modelled and analysed individually.

Finally, timed versions of Petri nets and automata are widely used, but suffer from the same defaults: values must be known in advance.

Hence, with traditional modelling languages, values are to be set before the analysis is performed, and the process must be repeated for all possible values. It is thus a tedious process which boils down to testing all values one by one, taking a huge amount of time.

1.3 Problems of Interest in a Parametric Setting

The major objective for introducing parameters is to circumvent this repetition of analysis, by introducing parameters in the models. Furthermore, the analysis techniques are suited to find constraints on parameters such that desired properties are satisfied (*e.g.* the property is satisfied for all values of the parameter p between 1 and 10) or even synthesize the set of all such parameters valuations.

The first advantage of this approach is that answering these questions provides all possible values in a single analysis step. Second, the set of parameters obtained can be infinite (*e.g.* $p > 5$), a result that cannot in general be obtained with an enumerative approach.

Several sources of information are available to the reader, that provide additional details on the theoretical background, further examples, etc. Among these:

- the video of the tutorial presentation: https://www.youtube.com/watch?v= pB0AuPuw2-o&index=1&list=PL9SOLKoGjbeqNcdQVqFpUz7HYqD1fbFIg
- the slides of the tutorial: https://www.imitator.fr/tutorials/PN17/
- the exercises with IMITATOR (https://www.imitator.fr/tutorials/PN17/) and ROMÉO (http://romeo.rts-software.org/doc/tutorial.html).
- the extensive literature that is referenced.

Outline. In Sect. 2, we first consider parameters as unknown constants in timed automata, *i.e.* Parametric Timed Automata [4]. We review decidability results, and report on decidable subclasses. Then, in Sect. 3, we consider parameters as unknown probabilities in Parametric Interval Markov Chains [36]. Section 4 deals with parameters that are unknown numbers of tokens in Petri nets, yielding Parametric Time Petri Nets [32,33]. As a last formalism, we also consider in Sect. 5 the synthesis of actions (seen as Boolean parameters) [53,55]. Finally we review some verification tools in Sect. 6.

2 Parametric Timed Automata

Classical qualitative model checking, implemented in powerful tools used successfully in industry, falls short when *quantitative* aspects of systems such as time, energy, probabilities, etc., are to be verified. Timed automata [3] allow for modeling and verifying time critical concurrent systems. This seminal work [3] received the CAV conference award in 2008, and since then numerous works have extended the formalism of timed automata.

However, despite of some success, (timed) model checking can be seen as slightly disappointing. There are two main reason of that:

1. the binary response to properties satisfaction may not be informative enough, and
2. the insufficient abstraction to cater for tuning and scalability of systems.

Adding parameters offers a higher level of abstraction by allowing unknown constants in a model. Parameters can be used to model unknown *timing* constants of timed systems. This approach has the following advantages:

- it becomes possible to verify a system at an earlier design stage, when not all timing constants are known with full certainty.
- it allows designers to cope with uncertainty even at runtime: some timings constants (*e.g.* periods of a real-time system) may be known up to a given precision only (*e.g.* given with an interval of confidence), and parameters can model this imprecision.

Parametric timed automata [4] are an extension of timed automata where timing constants can become unknown, *i.e. parameters*. They represent a particularly expressive formalism: in fact, its expressiveness is Turing-complete [14] and all non-trivial problems related to parametric timed automata are undecidable. For example, the mere existence of a parameter valuation for which there exists a run reaching some location is undecidable (see *e.g.* [7] for a survey).

Parametric timed automata suffer from negative decidability results, but they still remain a quite powerful formalism. They can be used to address robustness (in the sense of possibly infinitesimal variations of timing constants [25]), to model and verify systems with uncertain constants, and to synthesize suitable (possibly unknown) valuations so that the system meets its specification.

In addition, several recent decidability results for subclasses of parametric timed automata (*e.g.* [12,23,26,28,50]) has made this formalism more promising, while new algorithmic and heuristic techniques (*e.g.* [8,15,21,50,54,56]) has made the parametric verification for some classes of problems more scalable and complete, or more often terminating.

Verification with parametric timed automata has had important outcomes in various areas, with verification of case studies such as the root contention protocol [48], Philip's bounded retransmission protocol [48], a 4-phase handshake protocol [54], the alternating bit protocol [50], an asynchronous circuit commercialised by ST-Microelectronics [30], (non-preemptive) schedulability problems [50], a distributed prospective architecture for the flight control system of the next generation of spacecrafts designed at ASTRIUM Space Transportation [43], and even analysis of music scores [40].

In this section, we recall the syntax and semantics of parametric timed automata (Sect. 2.2) and their subclasses (Sect. 2.3). We introduce theoretical problems of interest (Sect. 2.4), and review decidability results (Sect. 2.5).

2.1 Basic Notions

Let $\mathbb{N}$, $\mathbb{Z}$, $\mathbb{Q}_+$, and $\mathbb{R}_+$ denote the sets of non-negative integers, integers, non-negative rational numbers, and non-negative real numbers, respectively.

We first define the notions necessary to deal with clocks. We begin with clock valuations.

Definition 1 (clock valuation). *Let* $X = \{x_1, \ldots, x_H\}$ *be a finite set of clocks, i.e. real-valued variables that evolve at the same rate.*

A clock valuation is a function $w : X \to \mathbb{R}_+$.

We identify a clock valuation w *with the* point $(w(x_1), \ldots, w(x_H))$.

The following notations allow for specifying null clocks, and adding simultaneously the same delay to all clocks.

Notation 1 (clock operations)

- *We write* $X = \mathbf{0}$ *for* $\bigwedge_{1 \le i \le H} x_i = 0$.
- *We also use a special zero-clock* x_0, *always equal to 0 (as in e.g. [48]).*
- *Given* $d \in \mathbb{R}_+$, $w + d$ *denotes the valuation such that* $(w + d)(x) = w(x) + d$, *for all* $x \in X$.
- *Given* $R \subseteq X$, *we define the* reset *of a valuation* w, *denoted by* $[w]_R$, *as follows:* $[w]_R(x) = 0$ *if* $x \in R$, *and* $[w]_R(x) = w(x)$ *otherwise.*

The systems considered comprise *a priori* unknown timing constants that are thus parameters to be synthesized according to the targeted property.

Definition 2 (timing parameter valuation). *Let* $P = \{p_1, \ldots, p_M\}$ *be a set of* timing parameters, *i.e. unknown timing constants.*

A timing parameter valuation v *is a function* $v : P \to \mathbb{Q}_+$.

We identify a valuation v *with the* point $(v(p_1), \ldots, v(p_M))$.

Clocks and parameters are used together and thus can be combined.

Notation 2 (clocks and parameters valuations combined). *Given a timing parameter valuation* v *and a clock valuation* w, *we denote by* $w|v$ *the valuation over* $X \cup P$ *such that for all clocks* x, $w|v(x) = w(x)$ *and for all timing parameters* p, $w|v(p) = v(p)$.

The expressions on clocks can concern clock themselves, but also involve parameters and constant delays.

Notation 3 (linear terms)

- *In the following, let* lt *denote a linear term over* $X \cup P$ *of the form* $\sum_{1 \le i \le H} \gamma_i x_i + \sum_{1 \le j \le M} \beta_j p_j + d$, *with* $x_i \in X$, $p_j \in P$, *and* $\gamma_i, \beta_j, d \in \mathbb{Z}$.

– *Let plt denote a parametric linear term over P, that is a linear term without clocks (i.e., $\gamma_i = 0$ for all i).*

The synthesis of parameters leads to expressing *constraints* on their values in order to guarantee that the model satisfies the expected properties.

Definition 3 (constraints on clocks and timing parameters). *A constraint C over $X \cup P$ is defined by the following grammar:*

$$\phi := \phi \wedge \phi \mid \neg \phi \mid lt \bowtie 0,$$

where $\bowtie \in \{<, \leq, \geq, >\}$, lt is a linear term.

Definition 4 (constraint satisfaction). *A valuation $w|v$ satisfies a constraint C, denoted $w|v \models C$, if the expression obtained by replacing in C each timing parameter by its valuation as in v evaluates to true.*

Zones allow for defining convex sets of clocks and timing parameters values.

Definition 5 (zones and parametric guards). *A zone C is a constraint such that each of its linear conjuncts can be written in the form $x_i - x_j \bowtie plt$, where $x_i, x_j \in X \cup \{x_0\}$. A parametric guard g is a zone such that each of its linear conjuncts can be written in the form $x_i \bowtie plt$.*

Definition 6 (satisfiability). *Given a zone C, $w|v \models C$ indicates that valuating each clock variable x with $w(x)$ and each timing parameter p with $v(p)$ within C, evaluates to true. Zone C is satisfiable if $\exists w, v$ s.t. $w|v \models C$.*

Time elapsing can be obtained by adding a new variable to all clocks, ensuring that this variable is non-negative, and eliminating it (see, *e.g.* [9]).

Definition 7 (time elapsing of a zone). *The time elapsing of a zone C, denoted by $C^{\nearrow}$, is the constraint over $X \cup P$ obtained from C by delaying all clocks by any arbitrary amount of time.*
    That is,

$$w'|v \models C^{\nearrow} \text{ iff } \exists w : X \to \mathbb{R}_+, \exists d \in \mathbb{R}_+ \text{ s.t. } w'|v \models C \wedge w' = w + d.$$

2.2   Syntax and Semantics

Syntax

Definition 8 (parametric timed automaton [4]). *A parametric timed automaton (PTA) is a tuple $\mathcal{A} = (\Sigma, L, l_0, F, X, P, I, E)$, where:*

1. *Σ is a finite set of actions,*
2. *L is a finite set of locations,*
3. *$l_0 \in L$ is the initial location,*
4. *$F \subseteq L$ is a set of final or accepting locations,*
5. *X is a finite set of clocks,*

6. P is a finite set of parameters,
7. I is the invariant, assigning to every $l \in L$ a parametric guard $I(l)$,
8. E is a finite set of edges $e = (l, g, \sigma, R, l')$ where $l, l' \in L$ are the source and target locations, $\sigma \in \Sigma$, $R \subseteq X$ is a set of clocks to be reset, and g is a parametric guard called the transition guard.

Given a parameter valuation v, we denote by $v(\mathcal{A})$ the non-parametric timed automaton where all occurrences of each parameter p_i have been replaced by $v(p_i)$. If $v(\mathcal{A})$ is such that all constants in guards and resets are integers, then $v(\mathcal{A})$ is a *timed automaton* [3]. In the following, we refer to any structure $v(\mathcal{A})$ as a timed automaton, by assuming a rescaling of the constants: by multiplying all constants in $v(\mathcal{A})$ by the least common multiple of their denominators, we obtain an equivalent (integer-valued) timed automaton.

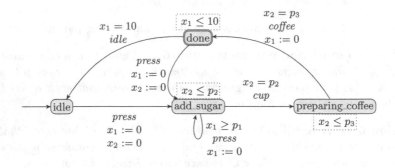

Fig. 1. A coffee machine modeled using a PTA

Example 1. Consider the coffee machine in Fig. 1, modelled using a PTA with 4 locations, 2 clocks (x_1 and x_2) and 3 parameters (p_1, p_2, p_3). Invariants are boxed. The only accepting location (with a double border) is done. Observe that all guards and invariants are simple constraints.

The machine can initially be idle for an arbitrarily long time. Then, whenever the user presses the (unique) button (action *press*), the PTA enters location add_sugar, resetting both clocks. The machine can remain in this location as long as the invariant ($x_2 \leq p_2$) is satisfied; there, the user can add a dose of sugar by pressing the button (action *press*), provided the guard ($x_1 \geq p_1$) is satisfied, which resets x_1. That is, the user cannot press twice the button (and hence add two doses of sugar) within a time less than p_1. Then, p_2 time units after the machine left the idle mode, a cup is delivered (action *cup*), and the coffee is being prepared; eventually, p_2 time units after the machine left the idle mode, the coffee (action *coffee*) is delivered. Then, after 10 time units, the machine returns to the idle mode—unless a user again requests a coffee by pressing the button.

Concrete Semantics

Definition 9 (Concrete semantics of a TA). *Given a PTA $\mathcal{A} = (\Sigma, L, l_0, F, X, P, I, E)$, and a parameter valuation v, the concrete semantics of $v(\mathcal{A})$ is given by the timed transition system $(S, s_0, \rightarrow)$, with*

- *$S = \{(l, w) \in L \times \mathbb{R}_+^H \mid w|v \models I(l)\}$,*
- *$s_0 = (l_0, \mathbf{0})$, and*
- *$\rightarrow$ consists of the discrete and (continuous) delay transition relations:*
  - *discrete transitions: $(l, w) \xrightarrow{e} (l', w')$, if $(l, w), (l', w') \in S$, there exists $e = (l, g, \sigma, R, l') \in E$, $w' = [w]_R$, and $w|v \models g$.*
  - *delay transitions: $(l, w) \xrightarrow{d} (l, w + d)$, with $d \in \mathbb{R}_+$, if $\forall d' \in [0, d], (l, w + d') \in S$.*

Moreover we write $(l, w) \xmapsto{e} (l', w')$ for a combination of a delay and discrete transition where $((l, w), e, (l', w')) \in \mapsto$ if $\exists d, w'' : (l, w) \xrightarrow{d} (l, w'') \xrightarrow{e} (l', w')$.

Given a TA $v(\mathcal{A})$ with concrete semantics $(S, s_0, \rightarrow)$, we refer to the states of S as the *concrete states* of $v(\mathcal{A})$. A (concrete) *run* of $v(\mathcal{A})$ is a possibly infinite alternating sequence of concrete states of $v(\mathcal{A})$ and edges starting from the initial concrete state s_0 of the form $s_0 \xmapsto{e_0} s_1 \xmapsto{e_1} \cdots \xmapsto{e_{m-1}} s_m \xmapsto{e_m} \cdots$, such that for all $i = 0, 1, \ldots : e_i \in E$, and $(s_i, e_i, s_{i+1}) \in \mapsto$. Given a state $s = (l, w)$, we say that s is reachable (or that $v(\mathcal{A})$ reaches s) if s belongs to a run of $v(\mathcal{A})$. By extension, we say that l is reachable in $v(\mathcal{A})$, if there exists a state (l, w) that is reachable. By extension, given a set of locations $T \subseteq L$ (T stands for "target"), we say that T is reachable in $v(\mathcal{A})$, if there exists a location $l \in T$ that is reachable in $v(\mathcal{A})$. Given a set of locations $T \subseteq L$, we say that a run *stays* in T if all of its states (l, w) are such that $l \in T$.

A *maximal run* is a run that is either infinite (*i.e.* contains an infinite number of discrete transitions), or that cannot be extended by a discrete transition. A maximal run is *deadlocked* if it is finite, *i.e.* contains a finite number of discrete transitions. By extension, we say that a TA is deadlocked if it contains at least one deadlocked run.

Example 2. Consider again the PTA modeling a coffee machine in Fig. 1. Let v be the parameter valuation such that $v(p_1) = 1$, $v(p_2) = 5$ and $v(p_3) = 8$.

Given a clock valuation w, we denote it by $(w(x_1), w(x_2))$. For example, $(0, 4.2)$ denotes that $w(x_1) = 0$ and $w(x_2) = 4.2$.

The following sequence is a concrete run of $v(\mathcal{A})$.

$(\text{idle}, (0, 0)) \xmapsto{press} (\text{add\_sugar}, (0, 0)) \xmapsto{press} (\text{add\_sugar}, (0, 1.78)) \xmapsto{press}$
$(\text{add\_sugar}, (0, 4.2)) \xmapsto{cup} (\text{preparing\_coffee}, (0.8, 5)) \xmapsto{coffee} (\text{done}, (0, 8)) \xmapsto{press}$
$(\text{add\_sugar}, (0, 0))$

As an abuse of notation, we write above each arrow the action name (instead of the edge), as edges are unnamed in Fig. 1.

This concrete run is not maximal (it could be extended).

Language of Timed Automata. Let $(l_0, w_0) \overset{e_0}{\mapsto} (l_1, w_1) \overset{e_1}{\mapsto} \cdots \overset{e_{m-1}}{\mapsto}$
$(l_m, w_m) \overset{e_m}{\mapsto} \cdots$ be a (finite or infinite) run of a TA $v(\mathcal{A})$. The associated *untimed word* is $\sigma_0 \sigma_1 \cdots \sigma_m \cdots$, where σ_i is the action of edge e_i, for all $i \geq 0$; the associated *trace*[1] is $l_0 \sigma_0 l_1 \sigma_1 l_2 \cdots \sigma_m l_{m+1} \cdots$

Given a run $(l_0, w_0) \overset{e_0}{\mapsto} (l_1, w_1) \overset{e_1}{\mapsto} \cdots \overset{e_{m-1}}{\mapsto} (l_m, w_m)$, we say that this run is *accepting* if $l_m \in F$.

We define the untimed language as the set of all untimed words associated with accepting runs of a TA.

Definition 10 (untimed language of a TA). *Given a PTA $\mathcal{A} = (\Sigma, L, l_0, F, X, P, I, E)$, and a parameter valuation v, the* untimed language *of $v(\mathcal{A})$, denoted by $\mathsf{UL}(v(\mathcal{A}))$, is the set of untimed words associated with all accepting runs of $v(\mathcal{A})$.*

We define the trace set as the set of traces associated with the accepting runs.

Definition 11 (trace set of a TA). *Given a PTA $\mathcal{A} = (\Sigma, L, l_0, F, X, P, I, E)$, and a parameter valuation v, the* trace set *of $v(\mathcal{A})$ is the set of traces associated with all accepting runs of $v(\mathcal{A})$.*

Example 3. Consider again the PTA $\mathcal{A}$ modeling a coffee machine in Fig. 1. Let v be the parameter valuation such that $v(p_1) = 1$, $v(p_2) = 5$ and $v(p_3) = 8$.
The untimed language of $v(\mathcal{A})$ can be described as follows:

$$press^{[1..6]} \; cup \; coffee \left(idle^? \; press^{[1..6]} \; cup \; coffee \right)^*$$

where $\sigma^{[a,b]}$, $\sigma^?$, σ^* denote between a and b occurrences, zero or one occurrence, and zero or more occurrence(s) of σ, respectively.
The trace set of $v(\mathcal{A})$ can be described as follows:

$$idle \; (press \; add\_sugar)^{[1..6]} \; cup \; preparing\_coffee \; coffee \; done$$

$$\left((idle \; idle)^? \; (press \; add\_sugar)^{[1..6]} \; cup \; preparing\_coffee \; coffee \; done \right)^*$$

Symbolic Semantics. Let us now recall the symbolic semantics of PTAs (see *e.g.* [9, 48, 50]).

Definition 12 (Symbolic state). *A symbolic state is a pair (l, C) where $l \in L$ is a location, and C its associated parametric zone.*

Definition 13 (Symbolic semantics). *Given a PTA $\mathcal{A} = (\Sigma, L, l_0, F, X, P, I, E)$, the symbolic semantics of $\mathcal{A}$ is the labelled transition system called parametric zone graph $\mathcal{PZG} = (E, \mathbf{S}, \mathbf{s_0}, \Rightarrow)$, with*

- $\mathbf{S} = \{(l, C) \mid C \subseteq I(l)\}$,

---

[1] This is a non-standard definition of traces (compared to *e.g.* [45]), but we keep this term as it is used in *e.g.* [9,17].

- $s_0 = \left(l_0, (\bigwedge_{1 \leq i \leq H} x_i = 0)^{\nearrow} \wedge I(l_0)\right)$, and
- $\left((l, C), e, (l', C')\right) \in \Rightarrow if\ e = (l, g, \sigma, R, l')$ and

$$C' = \left([[(C \wedge g)]_R \wedge I(l')\right)^{\nearrow} \wedge I(l')$$

with C' satisfiable.

That is, in the parametric zone graph, nodes are symbolic states, and arcs are labelled by *edges* of the original PTA.

If $\left((l, C), e, (l', C')\right) \in \Rightarrow$, we write $\mathsf{Succ}(s, e) = (l', C')$.

A graphical illustration of the computation of Succ is given in Fig. 2.[2] Starting from the parametric zone C, it is intersected with guard g, leading to the parametric values that allow for taking the transition. Then the necessary clocks are reset. However, for the transition to be taken, the new values thus obtained must satisfy the invariant of the target location, $I(l')$. After this, when in l', time can elapse as long as the invariant still holds, leading to the new zone C'.

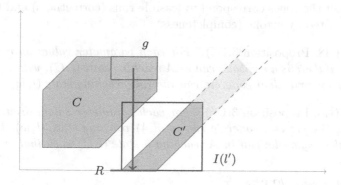

Fig. 2. Computing the successor of a symbolic state

A symbolic run of a PTA is an alternating sequence of symbolic states and edges starting from the initial symbolic state, of the form $s_0 \overset{e_0}{\Rightarrow} s_1 \overset{e_1}{\Rightarrow} \cdots \overset{e_{m-1}}{\Rightarrow} s_m$, such that for all $i = 0, \ldots, m-1$, $e_i \in E$, $s_i, s_{i+1} \in S$ and $(s_i, e, s_{i+1}) \in \Rightarrow$. Given a symbolic state s, we say that s is reachable if s belongs to a symbolic run of $\mathcal{A}$. In the following, we simply refer to symbolic states belonging to a run of $\mathcal{A}$ as symbolic states of $\mathcal{A}$.

Example 4. Consider again the coffee machine example in Fig. 1. A (non-maximal) symbolic run is as follows:

$(\text{idle}, x_1 = x_2 \wedge x_1 \geq 0) \overset{press}{\Rightarrow} (\text{add\_sugar}, x_1 = x_2 \wedge 0 \leq x_2 \leq p_2)$

$\overset{press}{\Rightarrow} (\text{add\_sugar}, p_1 \leq x_2 - x_1 \leq p_2 \wedge 0 \leq x_2 \leq p_2)$

---

[2] This figure comes from [19], itself coming from an adaptation of a figure by Ulrich Kühne.

$\overset{press}{\Rightarrow} \left(\text{add\_sugar}, 2 \times p_1 \leq x_2 - x_1 \leq p_2 \wedge 0 \leq x_2 \leq p_2\right)$

$\overset{cup}{\Rightarrow} \left(\text{preparing\_coffee}, 2 \times p_1 \leq x_2 - x_1 \leq p_2 \wedge p_2 \leq x_2 \leq p_3\right)$

$\overset{coffee}{\Rightarrow} \left(\text{done}, 0 \leq x_1 \leq 10 \wedge x_2 - x_1 = p_3 \wedge 2 \times p_1 \leq p_2 \leq p_3\right)$

$\overset{press}{\Rightarrow} \left(\text{add\_sugar}, x_1 = x_2 \wedge 0 \leq x_2 \leq p_2 \wedge 2 \times p_1 \leq p_2 \leq p_3\right)$

(For sake of readability, we use action names instead of edges along the transitions.)

The parametric zone graph of this example is infinite.

Given a concrete (respectively symbolic) run $(l_0, \mathbf{0}) \overset{e_0}{\mapsto} (l_1, w_1) \overset{e_1}{\mapsto} \cdots \overset{e_{m-1}}{\mapsto}$ (l_m, w_m) (respectively $(l_0, C_0) \overset{e_0}{\Rightarrow} (l_1, C_1) \overset{e_1}{\Rightarrow} \cdots \overset{e_{m-1}}{\Rightarrow} (l_m, C_m)$), we define the corresponding discrete sequence as $l_0 \overset{e_0}{\Rightarrow} l_1 \overset{e_1}{\Rightarrow} \cdots \overset{e_{m-1}}{\Rightarrow} l_m$. Two runs (concrete or symbolic) are said to be *equivalent* if their associated discrete sequences are equal.

Two important results (see *e.g.* [48]) relate the concrete and the symbolic semantics, and are recalled below using our syntax. They provide a sort of equivalence between symbolic parametric zones and concrete runs. That is, they guarantee that the zones correspond to feasible runs (correctness) and that each run is represented by a zone (completeness).

Lemma 1 ([48, Proposition 3.17]). *For each parameter valuation v and clock valuation w, if there is a symbolic run in $\mathcal{A}$ reaching state (l, C), with $w \models v(C)$, then there is an equivalent concrete run in $v(\mathcal{A})$ reaching state (l, w).*

Lemma 2 ([48, Proposition 3.18]). *For each parameter valuation v and clock valuation w, if there is a concrete run in $v(\mathcal{A})$ reaching state (l, w), then there is an equivalent symbolic run in $\mathcal{A}$ reaching a state (l, C) such that $w \models v(C)$.*

2.3   Subclasses of PTAs

Lower-bound/upper-bound parametric timed automata (L/U-PTAs), proposed in [48], restrict the use of parameters in the model.

Definition 14 (L/U-PTA). *An L/U-PTA is a PTA where the set of parameters is partitioned into lower-bound parameters and upper-bound parameters, where an upper-bound (resp. lower-bound) parameter p_i is such that, for every guard or invariant constraint $x \bowtie \sum_{1 \leq j \leq M} \beta_j p_j + d$, we have: $\beta_j > 0$ implies $\bowtie \in \{\leq, <\}$ (resp. $\bowtie \in \{\geq, >\}$), and $\beta_j < 0$ implies $\bowtie \in \{\geq, >\}$ (resp. $\bowtie \in \{\leq, <\}$).*

In [26], two additional subclasses are introduced: L-PTAs (resp. U-PTAs) are PTAs with only lower-bound (resp. upper-bound) parameters.

L/U-PTAs enjoy a well-known monotonicity property [48]: increasing upper-bound parameters or decreasing lower-bound parameters can only add behaviours.

Example 5. Consider again the coffee machine in Fig. 1, modelled using a PTA $\mathcal{A}$. This PTA is not an L/U-PTA; indeed, in the guard $x_2 = p_2$ (resp. $x_2 = p_3$), p_2 (resp. p_3) is compared with clocks both as a lower-bound and as an upper-bound. (Recall that = stands for $\leq$ and $\geq$.)

However, if one replaces $x_2 = p_2$ with $x_2 \leq p_2$ and $x_2 = p_3$ with $x_2 \leq p_3$, then $\mathcal{A}$ becomes an L/U-PTA with lower-bound parameter p_1 and upper-bound parameters $\{p_2, p_3\}$. Note that equalities are not forbidden in L/U-PTAs (*e.g.* $x_1 = 10$), but only equalities involving parameters are.

Several case studies fit into the class of L/U-PTAs: the root contention protocol, the bounded retransmission protocol and the Fischer mutual exclusion protocol are all modelled with L/U-PTAs in [48]; in [48,54], both the Fischer mutual exclusion protocol and a producer-consumer are verified using L/U-PTAs. Interestingly, the two case studies of the seminal paper on PTAs [4] (viz. a toy train gate controller model and a model of Fischer mutual exclusion protocol) are also L/U-PTAs, although the concept of L/U-PTAs had not yet been proposed at that time. In addition, most models of asynchronous circuits with bi-bounded delays (*i.e.* where each delay between the change of an input signal and the change of the corresponding output is a parametric interval) can be modelled using L/U-PTAs.

We will also consider *bounded* PTAs, *i.e.* PTAs with a bounded parameter domain that assigns to each parameter an infimum and a supremum, both integers.

Definition 15 (bounded PTA). *A bounded PTA is $\mathcal{A}_{|bounds}$, where $\mathcal{A}$ is a PTA, and bounds assigns to each parameter p an interval* [inf, sup], (inf, sup], [inf, sup), *or* (inf, sup), *with* inf, sup $\in \mathbb{N}$. *We use* inf$(p, bounds)$ *and* sup$(p, bounds)$ *to denote the infimum and the supremum of p, respectively. (Note that we rule out ∞ as a supremum.)*

We say that a bounded PTA is a closed bounded PTA *if, for each parameter p, its ranging interval* bounds(p) *is of the form* [inf, sup]; *otherwise it is an* open bounded PTA.

We define similarly bounded L/U-PTAs.

2.4 Decision and Computation Problems

TCTL. TCTL [2] is the quantitative extension of CTL where temporal modalities are augmented with constraints on duration. Formulae are interpreted over timed transition systems.

Given $ap \in AP$ and $c \in \mathbb{N}$, the language of TCTL is given by the following grammar:

$$\varphi ::= \top \mid ap \mid \neg\varphi \mid \varphi \wedge \varphi \mid \mathsf{E}\varphi\mathsf{U}_{\bowtie c}\varphi \mid \mathsf{A}\varphi\mathsf{U}_{\bowtie c}\varphi$$

A reads "always", E reads "exists", and U reads "until".

Standard abbreviations include Boolean operators as well as $\mathsf{EF}_{\bowtie c}\varphi$ for $\mathsf{E}\top\mathsf{U}_{\bowtie c}\varphi$, $\mathsf{AF}_{\bowtie c}\varphi$ for $\mathsf{A}\top\mathsf{U}_{\bowtie c}\varphi$ and $\mathsf{EG}_{\bowtie c}\varphi$ for $\neg\mathsf{AF}_{\bowtie c}\neg\varphi$. F reads "eventually" while G reads "globally".

Definition 16 (Semantics of TCTL). *Given a TA $v(\mathcal{A})$, the following clauses define when a state s_i of its timed transition system $(S, s_0, \rightarrow)$ satisfies a TCTL formula φ, denoted by $s_i \models \varphi$, by induction over the structure of φ (semantics of Boolean operators is omitted:*

1. $s_i \models E\varphi U_{\bowtie c}\psi$ *if there is a maximal run ρ in $v(\mathcal{A})$ with $\sigma = s_i \overset{e_i}{\mapsto} \cdots \overset{e_{j-1}}{\mapsto} s_j$ ($i < j$) a prefix of ρ s.t. $s_j \models \psi$, $\mathsf{time}(\sigma) \bowtie c$, and $\forall k, i \leq k < j : s_k \models \varphi$*
2. $s_i \models A\varphi U_{\bowtie c}\psi$ *if for each maximal run ρ in $v(\mathcal{A})$ there exists $\sigma = s_i \overset{e_i}{\mapsto} \cdots \overset{e_{j-1}}{\mapsto} s_j$ ($i < j$) a prefix of ρ s.t. $s_j \models \psi$, $\mathsf{time}(\sigma) \bowtie c$, and $\forall k, i \leq k < j : s_k \models \varphi$.*

where, given a concrete run ρ, $\mathsf{time}(\rho)$ gives the total sum of the delays d along ρ.

In $E\varphi U_{\bowtie c}\psi$ the classical until is extended by requiring that φ be satisfied within a duration (from the current state) verifying the constraint "$\bowtie c$". Given v, a PTA $\mathcal{A}$ and a TCTL formula φ, we write $v(\mathcal{A}) \models \varphi$ when $s_0 \models \varphi$.

Decision problems

Emptiness and universality of the valuations set. Let $\mathcal{P}$ be a given a class of decision problems (reachability, unavoidability, etc.).

$\mathcal{P}$-**emptiness problem:**
INPUT: A PTA $\mathcal{A}$ and an instance ϕ of $\mathcal{P}$
PROBLEM: Is the set of parameter valuations v such that $v(\mathcal{A})$ satisfies ϕ empty?

$\mathcal{P}$-**universality problem:**
INPUT: A PTA $\mathcal{A}$ and an instance ϕ of $\mathcal{P}$
PROBLEM: For all parameter valuations v, does $v(\mathcal{A})$ satisfy ϕ?

In this section, we mainly focus on the following decision problems:

- reachability (EF[3]): given a TA $v(\mathcal{A})$, is there at least one run of $v(\mathcal{A})$ that reaches a given location? That is, EF-emptiness asks: "is the set of parameter valuations v such that the TA $v(\mathcal{A})$ reaches a given location empty?" And EF-universality asks: "are all parameter valuations such that the corresponding TA reaches a given location?"
- unavoidability (AF): given a TA $v(\mathcal{A})$, do all runs of $v(\mathcal{A})$ eventually reach a given location?
- EG: given a TA $v(\mathcal{A})$ and a subset T of its locations, is there at least one maximal run of $v(\mathcal{A})$ that always stays in T?
- AG: given a TA $v(\mathcal{A})$ and a subset T of its locations, do all runs of $v(\mathcal{A})$ stay in T?
- deadlock-existence (ED): given a TA $v(\mathcal{A})$, is there at least one maximal run of $v(\mathcal{A})$ that is deadlocked, *i.e.* has no discrete successor (possibly after some delay)?

---

[3] The names "EF", "AF", "EG" come from the TCTL syntax, and are consistent with the notations introduced in [50] and subsequently used in further papers (such as [12,14]).

– cycle-existence (EC): given a TA $v(\mathcal{A})$, is there at least one run of $v(\mathcal{A})$ with an infinite number of discrete transitions?

Note that AF-emptiness is equivalent to EG-universality, while AG-emptiness is equivalent to EF-universality.

We will finally consider the following two additional emptiness problems:

Language-preservation-emptiness problem:
INPUT: A PTA $\mathcal{A}$ and a parameter valuation v'
PROBLEM: Is the set of parameter valuations v such that $v \neq v'$ and for which $v(\mathcal{A})$ has the same untimed language as $v'(\mathcal{A})$ empty?

Trace-preservation-emptiness problem:
INPUT: A PTA $\mathcal{A}$ and a parameter valuation v'
PROBLEM: Is the set of parameter valuations v such that $v \neq v'$ and for which $v(\mathcal{A})$ has the same set of traces as $v'(\mathcal{A})$ empty?

Computation Problem. Additionally, we define the following computation problem:

$\mathcal{P}$-synthesis problem:
INPUT: A PTA $\mathcal{A}$ and an instance ϕ of $\mathcal{P}$
PROBLEM: Compute the parameter valuations such that $v(\mathcal{A})$ satisfies ϕ.

Example 6. Let us exemplify some decision and computation problems for the coffee machine PTA in Fig. 1. Assume the unique target location is done, *i.e.* $T = \{done\}$.

EF-emptiness asks whether the set of parameter valuations that can reach location done for some run is empty, *i.e.* there is an execution in which the coffee is not delivered. This is false (*e.g.* $p_1 = 1$, $p_2 = 2$, $p_3 = 3$ can reach done).

EF-universality asks whether all parameter valuations can reach location done for some run, *i.e.* all executions allow for delivering the coffee, regardless of the parameters valuation. This is false (no parameter valuation such that $p_2 > p_3$ can reach done).

AF-emptiness asks whether the set of parameter valuations that can reach location done for all runs is empty, *i.e.* if there is some parameters valuation for which coffee delivery is not guaranteed. This is false (*e.g.* $p_1 = 1$, $p_2 = 2$, $p_3 = 3$ cannot avoid done).

EF-synthesis consists in synthesizing all valuations for which a run reaches location done, *i.e.* identifies all parameters valuations for which a coffee will eventually be delivered. The resulting set of valuations is $0 \leq p_2 \leq p_3 \leq 10 \wedge p_1 \geq 0$.

2.5 Decidability

The General Class of PTAs. With the rule of thumb that all problems are undecidable for PTAs, we review the decidability of the aforementioned problems.

- EF-emptiness was shown to be undecidable [4], with different flavours and settings: for a single bounded parameter [59], for a single rational-valued or integer-valued parameter [23], with only one clock compared to parameters [59], or with strict constraints only [39].
- AF-emptiness was shown undecidable in [50].
- AG-emptiness was shown undecidable in [14].
- EG-emptiness (as well as EC and ED) were shown undecidable in [12].
- The language and trace-preservation problems were shown undecidable in [17].

A complete survey is available in [7].

Following the very negative results for PTAs, subclasses have been proposed. We review some in the following.

The class of L/U-PTAs

A Main Decidability Result. The first (and main) positive result for L/U-PTAs is the decidability of the EF-emptiness problem [48]. L/U-PTAs benefit from the following interesting monotonicity property: increasing the value of an upper-bound parameter or decreasing the value of a lower-bound parameter necessarily relaxes the guards and invariants, and hence can only add behaviours. Therefore, checking the EF-emptiness of an L/U-PTA can be achieved by replacing all lower-bound parameters with 0, and all upper-bound parameters with a sufficiently large constant; this yields a non-parametric TA, for which emptiness is PSPACE-complete [3]. This procedure is not only sound but also complete.

Undecidability Results. The first undecidability results for L/U-PTAs are shown in [26]: the *constrained* EF-emptiness problem and constrained EF-universality problem (for infinite runs acceptance properties) are undecidable for L/U-PTAs. By constrained it is meant that some parameters of the L/U-PTA can be constrained by an initial linear constraint, *e.g.* $p_1 \leq 2 \times p_2 + p_3$. Indeed, using linear constraints, one can constrain an upper-bound parameter to be equal to a lower-bound parameter, and hence build a 2-counter machine using an L/U-PTA. However, when no upper-bound parameter is compared to a lower-bound parameter (*i.e.* when no initial linear inequality contains both an upper-bound and a lower-bound parameter), these two problems become decidable [26]. The exact decidability frontier may have not been found yet: the case where a lower-bound parameter is constrained to be less than or equal to an upper-bound parameter fits in none of the considered cases.

A second negative result is shown in [50]: the AF-emptiness problem is undecidable for L/U-PTAs. This restricts again the use of L/U-PTAs, as AF is essential to show that all possible runs of a system eventually reach a (good) state.

Third, in [17], the language- and trace-preservation problems were shown to be undecidable for L/U-PTAs.

Model-checking L/U-PTAs. In [26], a parametric extension of the dense-time linear temporal logic $MITL_{0,\infty}$ (denoted "$PMITL_{0,\infty}$") is proposed; when parameters are used only as lower or upper bound in the formula (to which we refer as $L/U\text{-}PMITL_{0,\infty}$), satisfiability and model checking are PSPACE-complete; this is obtained by translating the formula into an L/U-automaton and checking an infinite acceptance property.

Then, in [38], an extension of MITL allowing parametric linear expressions in bounds is proposed (yielding PMITL). Two sets of (integer-valued) parameter valuations are considered:

1. the set of valuations for which a PMITL formula is satisfiable, *i.e.* for which there exists a timed sequence (possibly belonging to a given L/U-PTA) satisfying it, and
2. the set of valuations for which a PMITL formula is valid, *i.e.* for which all timed sequences (possibly belonging to a given L/U-PTA) satisfy it.

Under some assumptions, the emptiness and universality of the valuation set for which a PMITL property is satisfiable or valid (possibly *w.r.t.* a given L/U-PTA) are decidable, and EXPSPACE-complete. Essential assumptions for decidability include the fact that parameters should be used with the same polarity (positive or negative coefficient, as lower or upper bound in the intervals) within the entire PMITL formula, and each interval can only use parameters in one of the endpoints. Additional assumptions include that no interval of the PMITL formula should be punctual (nor empty), and linear parametric expressions are only used in right endpoints of the intervals (single parameters can still be used as left endpoints). In addition, two fragments of PMITL are showed to be in PSPACE, including one that allows for expressing parameterized response ("if an event occurs, then another event shall occur within some possibly parametric time interval").

Finally, we showed that the emptiness-problem using *nested* quantifiers (*i.e.* beyond EF, EG, AF, AG) automatically leads to the undecidability, even for the very restricted class of U-PTAs with a single parameter (that can even be integer-valued) [13]. In other words, the nested TCTL emptiness problem is undecidable for U-PTAs. We may wonder if the *timed* aspect of TCTL (and notably the urgency required by the TCTL formula $EGAG_{=0}$) is responsible for the undecidability. In fact, it is not, and we could modify the proof to show that CTL itself leads to undecidability, *i.e.* that EGAX-emptiness is undecidable.

Intractability of the Synthesis. A very disappointing result concerning L/U-PTAs is shown in [50]: despite decidability of the underlying decision problems (EF-emptiness and EF-universality), the solution to the EF-synthesis problem for L/U-PTAs, if it can be computed, cannot be represented using a formalism for which the emptiness of the intersection with equality constraints is decidable. A very annoying consequence is that such a solution cannot be represented as a finite union of polyhedra (since the emptiness of the intersection with equality constraints is decidable).

Liveness. The EG-emptiness problem stands at the frontier between decidability and undecidability for the class of L/U-PTAs: while this problem is decidable for L/U-PTAs with a bounded parameter domain with closed bounds, it becomes undecidable if either the assumption of boundedness or of closed bounds is lifted [12].

The deadlock-existence emptiness problem is undecidable, even for the restricted class of closed bounded L/U-PTAs [12].

In contrast to deadlock-freeness that is consistently undecidable, and to EG-emptiness for which the frontier between decidability and undecidability is thin, the existence of a parameter valuation for which there exists at least one infinite run (EC-emptiness) is consistently decidable for L/U-PTAs [12].

The Power of Integer Points. Following works related integer clock and parameter valuations with decidability in [50], we introduced in [14] *integer-points parametric timed automata* (IP-PTAs for short), *i.e.* a subclass of PTAs in which any symbolic state contains at least one integer point.

Definition 17. *A PTA $\mathcal{A}$ is an* integer points PTA *(in short IP-PTA) if, in any reachable symbolic state (l, C) of $\mathcal{A}$, C contains at least one integer point, i.e. $\exists v : P \to \mathbb{N}, \exists w : X \to \mathbb{N}$ s.t. $w|v \models C$.*

Example 7. Consider the PTA in Fig. 3a, containing two clocks x_1 and x_2, and one parameter p. This PTA is not an IP-PTA. Indeed, as can be seen on its parametric zone graph, the (unique) symbolic state with location l_3 contains only $p = \frac{1}{2}$, and this symbolic state therefore contains no integer point.

In contrast, the PTA in Fig. 3b is an IP-PTA: each zone in its parametric zone graph contains an integer valuation of all parameters. The coffee machine in Fig. 1 (which has an infinite parametric zone graph) is also an IP-PTA.

In [14], we studied the expressiveness of IP-PTAs: while the class of IP-PTAs is incomparable with the class of L/U-PTAs, any *non-strict* L/U-PTA, *i.e.* with only non-strict inequalities, is an IP-PTA.

Concerning decidability, the only non-trivial general class with a decidability result for EF-emptiness is L/U-PTAs [48]. We extended this class, by proving that EF-emptiness is decidable for bounded IP-PTAs [14]. However, other studied problems turned out to be undecidable.

Summary. Table 1 summarises the decidability results. It gives from left to right the (un)decidability for U-PTAs, bounded L/U-PTAs (with either closed or open bounds), L/U-PTAs, bounded IP-PTAs, IP-PTAs, bounded PTAs, and PTAs. We review the emptiness of TCTL subformulas (EF, AF, EG, AG), full TCTL, cycle-existence, deadlock-existence and language- and trace-preservation. Decidability is given in green, whereas undecidability is given in italic red. Our

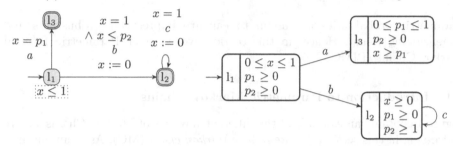

(a) A PTA which is not an IP-PTA, and its Parametric Zone Graph

(b) A PTA which is an IP-PTA, and its Parametric Zone Graph

Fig. 3. Examples of PTA

Table 1. Decidability of the emptiness problems for PTAs and subclasses

Class	U-PTAs	bL/U-PTAs closed	open	L/U-PTAs	bIP-PTAs	IP-PTAs	bPTAs	PTAs
EF	[48]	[14]	open	[48]	[14]	*[14]*	*[59]*	*[4]*
AF	open	*[14]*		*[50]*	[14]	[14]	[14]	*[50]*
EG	open	[12]	*[12]*	*[12]*	open			*[12]*
AG	[12]	[14]	open	[12]		*[14]*		
TCTL	*[13]*	*[14]*		*[50]*	[14]		*[59]*	*[4]*
EC	[12]	[12]	open	[12]	open			*[12]*
ED	open	*[12]*			open		*[12]*	*[5]*
LgP	open	*[17]*			open		*[17]*	
TrP	open	*[17]*			open		*[17]*	

contributions are emphasized in bold using a plain background, whereas existing results are depicted using a light background. When several papers in the literature proved the same result, we only give the earliest result, and not necessarily the best (in terms of number of clocks and parameters, or complexity).

Perspective: Open Subclasses. L-PTAs and U-PTAs [26] are very open classes, in the sense that the only known decidability results come from the larger class of PTAs, and no undecidability result was known—with the exception of our recent result concerning TCTL-emptiness [13]. To summarize, the EG-emptiness, AG-emptiness and AF-emptiness problems, as well as the language- and trace-preservation problems, are all undecidable for (general) L/U-PTAs, but remain open for L-PTAs and U-PTAs. Similarly, the EF-synthesis problem (shown intractable for L/U-PTAs in [50] despite the decidability of the EF-emptiness problem) remains open for rational-valued L- and U-PTAs, and would significantly increase the interest of these subclasses if it was shown to be computable.

3 Parametric Interval Markov Chains

Parametric probabilities are useful to capture imprecisions, robustness and dimensioning issues. Hence, in this section we consider Parametric Interval Markov Chains (PIMC).

3.1 Introduction to Parametric Markov Chains

Figure 4 contains an example of the different flavours of Markov Chains we are addressing in this section. Figure 4a is a *Markov chain* (MC). As in an automaton, there are states and transitions between them, but these are labelled by probabilities for the transition to occur. For example, from state s_0, there is a probability of 0.7 to go to state s_1 and of 0.3 to go to state s_2. Therefore, the sum of all probabilities labelling transitions exiting a state must be 1.

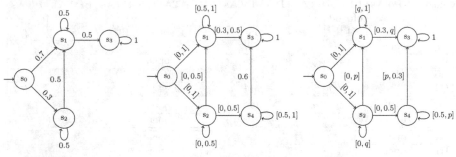

(a) Markov Chain (MC)   (b) Interval MC (IMC)   (c) Parametric IMC (PIMC)

Fig. 4. Markov chains and their extensions

When probabilities are not known in advance, it might still be possible to know an interval to which they belong. Hence, we introduce *Interval Markov Chains* (IMC), as pictured in Fig. 4b. The transitions are then no more labelled by a fixed probability but an interval meaning that the probability should be between the lower and the upper bound of the interval. For example, the probability to move from state s_1 to state s_3 is between 0.3 and 0.5. The MC in Fig. 4a can be seen as an implementation of the IMC in Fig. 4b which stands as a specification. Notice that state s_4 does not appear in Fig. 4a, which is equivalent to having a transition from state s_2 to state s_4 with a probability 0. Once a probability is chosen in an interval, it imposes constraints on the other probabilities outgoing the same state since they must add up to 1. An IMC is said to be *consistent* if it admits at least one implementation.

When the upper or lower bounds are unknown, it is convenient to use parameters. Figure 4c shows a *Parametric Interval Markov Chain* (PIMC). As compared with the IMC of Fig. 4b, some of the bounds are replaced with parameters p and q. Notice that the same parameter occurs at several places in the PIMC, therefore imposing constraints on the interval. For example, from state s_1, it is possible to stay in this state with a probability between q and 1, or to move to state s_3 with a probability between 0.3 and the same q. When assigning a valuation to all parameters in a PIMC, we obtain an IMC.

3.2 Markov Chains Definitions

We now formally define the different Markov Chain models. These definitions are detailed in [36].

Definition 18 (Markov Chain). *A Markov Chain is a tuple* $\mathcal{M} = (S, s_0, M, A, V)$, *where S is a finite set of states containing the initial state s_0, A is a set of atomic propositions, $V : S \to 2^A$ is a labelling function, and $M : S \times S \to [0, 1]$ is a probabilistic transition function such that* $\forall s \in S, \sum_{t \in S} M(s, t) = 1$.

We now introduce the notation of parameters, and interval ranges that will be used throughout this section. A parameter $p \in P$ is a variable ranging through the interval $[0, 1]$. A valuation for P is a function $\psi : P \to [0, 1]$ that associates values with each parameter in P. We write $\text{Int}_{[0,1]}(P)$ for the set of all closed parametric intervals of the form $[x, y]$ with $x, y \in [0, 1] \cup P$. When $P = \emptyset$, we write $\text{Int}_{[0,1]} = \text{Int}_{[0,1]}(\emptyset)$ to denote closed intervals with real-valued endpoints. Given an interval I of the form $I = [a, b]$, $\text{Low}(I)$ and $\text{Up}(I)$ respectively denote the lower and upper endpoints of I, i.e. a and b. Given an interval $I = [a, b] \in \text{Int}_{[0,1]}$, we say that I is well-formed whenever $a \leq b$.

The definition of Interval Markov Chains is adapted from [35].

Definition 19 (Interval Markov Chain [35]). *An Interval Markov Chain is a tuple* $\mathcal{I} = (S, s_0, \phi, A, V)$, *where S, s_0, A and V are as for MCs, and $\phi : S \times S \to \text{Int}_{[0,1]}$ is a transition constraint that associates with each potential transition an interval of probabilities.*

The following definition recalls the notion of satisfaction introduced in [35]. Satisfaction (also called implementation in some cases) allows to characterise the set of MCs represented by a given IMC specification. Satisfaction abstracts from the syntactic structure of transitions in IMCs: a single transition in the implementation MC can contribute to satisfaction of more than one transition in the specification IMC, by distributing its probability mass against several transitions. Similarly many MC transitions can contribute to the satisfaction of just one specification transition. This crucial notion is embedded in the so-called *correspondence function* δ introduced below. Informally, such a function is given for all pairs of states (t, s) in the satisfaction relation, and associates with each successor state t' of t – in the implementation MC – a distribution over potential successor states s' of s – in the specification IMC – specifying how the transition $t \to t'$ contributes to the transition $s \to s'$.

Definition 20 (Satisfaction Relation [35]). *Let $\mathcal{I} = (S, s_0, \phi, A, V^I)$ be an IMC and $\mathcal{M} = (T, t_0, M, A, V^M)$ be a MC. A relation $\mathcal{R} \subseteq T \times S$ is a* satisfaction relation *if whenever $t\mathcal{R}s$,*

1. *the labels of s and t agree: $V^M(t) = V^I(s)$,*
2. *there exists a* correspondence *function $\delta : T \to (S \to [0, 1])$ such that*
   (a) *for all $t' \in T$ such that $M(t, t') > 0$, $\delta(t')$ is a distribution on S,*
   (b) *for all $s' \in S$, we have $(\sum_{t' \in T} M(t, t') \cdot \delta(t')(s')) \in \phi(s, s')$, and*
   (c) *for all $t' \in T$ and $s' \in S$, if $\delta(t')(s') > 0$, then $(t', s') \in \mathcal{R}$.*
   We say that state $t \in T$ satisfies state $s \in S$ (written $t \models s$) iff there exists a (minimal) satisfaction relation containing (t, s) and that $\mathcal{M}$ satisfies $\mathcal{I}$ (written $\mathcal{M} \models \mathcal{I}$) iff $t_0 \models s_0$.

The set of MCs satisfying a given IMC $\mathcal{I}$ is written $[\![\mathcal{I}]\!]$. Formally, $[\![\mathcal{I}]\!] = \{\mathcal{M} \mid \mathcal{M} \models \mathcal{I}\}$.

Definition 21. *An IMC $\mathcal{I}$ is* consistent *iff $[\![\mathcal{I}]\!] \neq \emptyset$.*

Although the satisfaction relation abstracts from the syntactic structure of transitions, we recall the following result from [34], that states that whenever a given IMC is consistent, it admits at least one implementation that strictly respects its structure.

Theorem 1 ([34]). *An IMC $\mathcal{I} = (S, s_0, \phi, A, V)$ is consistent iff it admits an implementation of the form $\mathcal{M} = (S, s_0, M, A, V)$ where, for all reachable states s in $\mathcal{M}$, it holds that $M(s, s') \in \phi(s, s')$ for all s'.*

In the following, we say that state s is consistent in the IMC $\mathcal{I} = (S, s_0, \phi, A, V)$ if there exists an implementation $\mathcal{M} = (S, s_0, M, A, V)$ of $\mathcal{I}$ in which state s is reachable with a non-zero probability.

We now recall to the notion of Parametric Interval Markov Chain, previously introduced in [34].

Definition 22 (Parametric Interval Markov Chain). *A Parametric Interval Markov Chain is a tuple $\mathcal{I}^P = (S, s_0, \phi_P, A, V, P)$, where S, s_0, A and V are as for IMCs, P is a set of variables (parameters) ranging over $[0, 1]$ and $\phi_P : S \times S \to \textit{Int}_{[0,1]}(P)$ associates with each potential transition a (parametric) interval.*

Given a pIMC $\mathcal{I}^P = (S, s_0, \phi_P, A, V, P)$ and a parameter valuation $\psi : P \to [0, 1]$, we write $\psi(\mathcal{I}^P)$ for the IMC obtained by replacing ϕ_P by the function $\phi : S \times S \to \texttt{Int}_{[0,1]}$ defined by $\forall s, s' \in S, \phi(s, s') = \psi(\phi_P(s, s'))$. The IMC $\psi(\mathcal{I}^P)$ is called an *instance* of pIMC $\mathcal{I}^P$.

Finally, we say that a MC $\mathcal{M} = (T, t_0, M, A, V^M)$ *implements* pIMC $\mathcal{I}^P$, written $\mathcal{M} \models \mathcal{I}^P$, iff there exists an instance $\mathcal{I}$ of $\mathcal{I}^P$ such that $\mathcal{M} \models \mathcal{I}$. We write $[\![\mathcal{I}^P]\!]$ for the set of MCs implementing $\mathcal{I}^P$ and say that a pIMC is *consistent* iff its set of implementations is not empty.

3.3 Consistency of PIMCs

When considering IMCs, one question of interest is to decide whether it is consistent without computing its set of implementations. This problem has been addressed in [34,35], yielding polynomial decision algorithms and procedures that produce one implementation when the IMC is consistent. The same question holds for pIMCs, although in a slightly different setting. [34] proposed a polynomial algorithm for deciding whether a given pIMC is consistent, in the sense that it admits at least one parameter valuation for which the resulting IMC is consistent.

In order to decide whether a given IMC is consistent, we need to address the set of potential successors of a given state s. Let $\texttt{Succ}(s)$ be the set of states that can be reached from s with a probability interval not reduced to $[0, 0]$: $\texttt{Succ}(s) = \{s' \in S \mid \phi_P(s, s') \neq [0, 0]\}$.

We now introduce the notion of n-consistency in the IMC setting and then adapt this notion to pIMCs. In practice, n-consistency is defined by induction over the structure of $\mathcal{I}$.

Definition 23 (n-consistency). *Let $\mathcal{I} = (S, s_0, \phi, A, V)$ be an IMC and let $D : S \to \texttt{Dist}(S)$ be a function that assigns a distribution on S to each state of $\mathcal{I}$. State $s \in S$ is (n, D)-consistent iff for all $s' \in S$, $D(s)(s') \in \phi(s, s')$, and, for $n > 0$, $D(s)(s') > 0$ implies s' is $(n - 1, D)$-consistent.*

We say that s is n-consistent if there exists $D : S \to \texttt{Dist}(S)$ such that s is (n, D)-consistent.

Definition 23 is thus equivalent to the following intuitive inductive definition: a state s is n-consistent iff there exists a distribution ρ satisfying all of its outgoing probability intervals and such that for all $s' \in S$, $\rho(s') > 0$ implies that s' is $(n - 1)$-consistent.

Theorem 2. *Given an IMC $\mathcal{I} = (S, s_0, \phi, A, V)$, $\mathcal{I}$ is consistent iff s_0 is $|S|$-consistent.*

Example 8. Let us consider the example in Fig. 4b. All states but s_4 are 0-consistent. Indeed it is possible to find probabilities within the exiting intervals that add up to 1. It is not the case for state s_4, which has a probability of 0.6 to which we should add one that is at least 0.5, so the sum is at least 1.1. Then, for state s_2 to be 1-consistent, it must not have s_4 has a successor. This is possible by choosing 0 as the probability to go from s_2 to s_4. In the remaining two intervals, choosing probability 0.5 leads to a sum of 1. Therefore, state s_2 is 1-consistent. One can check that all states but s_4 are n-consistent, for all n.

For the problem of consistency of pIMCs, the aim is not only to decide whether a given pIMC is consistent, but also to synthesise all parameter valuations that ensure consistency of the resulting IMC. For this purpose, we adapt the notion of n-consistency defined above to pIMCs.

We first define the *local consistency* of a state *w.r.t.* some subset S' of its successors: the sum of upper bounds should be greater than 1, the sum of lower bounds smaller than 1, and all successors in S' have a valid interval.

$$LC(s, S') = \left[\sum_{s' \in S'} \mathrm{Up}(\phi_P(s, s')) \geq 1\right] \cap \left[\sum_{s' \in S'} \mathrm{Low}(\phi_P(s, s')) \leq 1\right]$$
$$\cap \left[\bigcap_{s' \in S'} \mathrm{Low}(\phi_P(s, s')) \leq \mathrm{Up}(\phi_P(s, s'))\right]$$

Let us start by fixing a set of states X that we want to avoid and then compute the set of valuations $\mathrm{Cons}_n^X(s)$ that ensure n-consistency of s through a distribution ρ that avoids states from X. Formally, $\mathrm{Cons}_n^X(s)$ is defined as: let $\mathrm{Cons}_0^X(s) = LC(s, \mathrm{Succ}(s) \setminus X) \cap \left[\bigcap_{s' \in X} \mathrm{Low}(\phi_P(s, s')) = 0\right]$ and for $n \geq 1$,

$$\mathrm{Cons}_n^X(s) = \left[\bigcap_{s' \in \mathrm{Succ}(s) \setminus X} \mathrm{Cons}_{n-1}(s')\right] \cap [LC(s, \mathrm{Succ}(s) \setminus X)]$$
$$\cap \left[\bigcap_{s' \in X} \mathrm{Low}(\phi_P(s, s')) = 0\right]$$

The set of valuations ensuring n-consistency is then the union, for all potential choices of X, of $\mathrm{Cons}_n^X(s)$. We need to choose X as a subset of the set $Z(s)$ of states which can be avoided, by transitions that have 0 or a parameter as lower bound. Therefore, we define $\mathrm{Cons}_n(s) = \bigcup_{X \subseteq Z(s)} \mathrm{Cons}_n^X(s)$.

Theorem 3. *Given a pIMC $\mathcal{I}^P = (S, s_0, \phi_P, A, V, P)$ and a parameter valuation $\psi : P \to [0, 1]$, we have $\psi \in \mathrm{Cons}_{|S|}(s_0)$ iff the IMC $\psi(\mathcal{I}^P)$ is consistent.*

Example 9. It is easily shown that the consistency of the PIMC in Fig. 4c is $[(q \leq 0.7) \cap (q \geq 0.3)] \cup (q = 1)$.

This section has shown the crucial property of consistency in both parametric and non-parametric interval Markov chains. It thus sets the necessary elements before model checking such probabilistic models.

3.4   Further Reading

The definitions and properties stated in this section are detailed in [36]. They were revised in [63], where both inductive and co-inductive definitions of consistency are given, implemented with forward and backward algorithms. [36]

also addresses some properties: consistent avoidability, existential and universal consistent reachability.

4   Parametric Petri Nets

We now consider a parametric extension of Petri nets in which the weights of the arcs can be parameters. This extension was mainly studied in the Ph.D. thesis of Nicolas David [31] and many of those results can also be found in [32, 33].

Example 10. In order to illustrate the usefulness of this parameterised formalism, we consider the example, taken from [31], of a financial loan. It is modelled in Fig. 5.

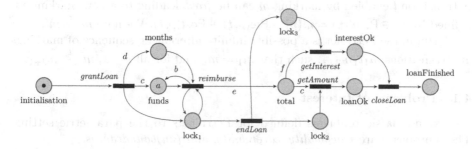

Fig. 5. Modelling a financial loan with parametric Petri nets

In the general case, a client has a certain amount of money, say a, and is ensured to get income b every month. To finance his project, the client needs to define with the bank the amount loaned c, the duration of reimbursement d (in months) and the amount of each reimbursement e. At the end of the process, the bank expects to get back the initial amount loaned c plus the interest f. The amount of money possessed by the client is depicted in place funds. The signature of the contract is symbolised through the firing of transition *grantLoan* which imposes to the bank to loan an amount c for a term of d months. Each month, we can fire *reimburse*: the client receives its own income b that is added to its capital in funds and in parallel an amount e is reimbursed to the bank. When we consider that the reimbursement is finished, we fire *endLoan* that removes the token from $lock_1$ and allows us to enable the transitions of the second part of this example. We can then check if the bank can get its money back by testing if loanOk can be marked and if the bank can get the interest by checking interestOk can be marked or both by checking if loanFinished can be marked.

With this example in mind, we now proceed to the corresponding formal definitions.

Definition 24 (Parametric Petri Net [32]**).** *A (marked) parametric Petri Net (PPN) is a tuple* $\mathcal{N} = (\mathsf{P}, \mathsf{T}, \mathbb{P}, \mathsf{Pre}, \mathsf{Post}, m_0)$ *where* P *is a finite set of places,* T *is a finite set of* transitions *such that* $\mathsf{P} \cap \mathsf{T} = \emptyset$, $\mathbb{P}$ *is a finite set of* parameters, $\mathsf{Pre} : \mathsf{P} \times \mathsf{T} \to \mathbb{N} \cup \mathbb{P}$ *is the* backward incidence function, $\mathsf{Post} : \mathsf{P} \times \mathsf{T} \to \mathbb{N} \cup \mathbb{P}$ *is the* forward incidence function, $m_0 \in \mathbb{N}^P$ *is the* initial marking.

Let $\mathcal{N}$ be a parametric Petri net. A valuation v of the parameters of $\mathcal{N}$ is a mapping from $\mathbb{P}$ to $\mathbb{N}$.

We denote by $v(\mathcal{N})$ the *Petri net* obtained by replacing all parameters by the value they are given by v.

A *marking* of a (non-parametric) Petri net is a mapping from P to $\mathbb{N}$. Markings can be compared component by component: $m \geq m'$ if $\forall p \in \mathsf{P}, m(p) \geq m'(p)$.

A transition $t \in \mathsf{T}$ is *enabled* by marking m if for all $p \in \mathsf{P}$, $m(p) \geq \mathsf{Pre}(p, t)$. A transition t enabled by marking m can be *fired*, leading to a new marking m' defined by $\forall p \in \mathsf{P}, m'(p) = m(p) - \mathsf{Pre}(p, t) + \mathsf{Post}(p, t)$. We note $m \xrightarrow{t} m'$.

A run in the Petri net is a possibly infinite alternating sequence of markings and transitions $m_1 t_1 m_2 \ldots$ such that $m_1 = m_0$ and for all $i \geq 1$, $m_i \xrightarrow{t_i} m_{i+1}$.

4.1 Problems of Interest

We extend classic problems defined for Petri nets to the parametric setting. These problems are *reachability*, *coverability*, and *(un)boundedness*.

A marking m is *reachable* if there exists a finite run $m_1 t_1 m_2 \ldots t_n m_{n+1}$ such that $m_{n+1} = m$.

A marking m is *coverable* if there exists a reachable marking m' such that $m' \geq m$.

A (non-parametric) Petri net is k-*bounded* if for all reachable markings m, we have $\forall p \in \mathsf{P}, m(p) \leq k$. It is *bounded* if there exists some k such that it is k-bounded. If a net is not bounded, we say it is *unbounded* and then for all $B \geq 0$ there exists a place p and a reachable marking m such that $m(p) > B$.

Similarly a Petri net is *simultaneously* X *unbounded* [37], for some subset X of P, if for all $B \geq 0$, there exists a reachable marking m such that for all places $p \in X$, we have $m(p) > B$.

We consider two associated *parametric* decision problems: the *existential* and the *universal* problems. In the former we want to decide the existence of a parameter valuation for which some property holds, and in the latter we want to decide if it holds for all the possible parameter valuations. The property in question can be any of those defined above.

For instance, the existential parametric reachability problem asks, given a target marking m, if there exists a parameter valuation v such that m is reachable in $v(\mathcal{N})$. Similarly, the universal coverability problem asks, given a target marking m, if for all valuations v of the parameters, m is coverable in $v(\mathcal{N})$.

We finally define *synthesis* problems, in which we want to effectively compute the set of all parameter valuations for which some property holds. Note that if we can effectively compute this set, and check its emptiness or universality, then we can also solve the two decision problems above.

4.2   Undecidability Results for Parametric Petri Nets

We start with a few negative results.

Theorem 4 ([32]). *The existential and universal parametric coverability, reachability, and (simultaneous) unboundedness problems are undecidable.*

This theorem is proved by reducing the halting, and counter-boundedness problems for 2-counter machines [60] to those parametric problems for parametric Petri nets. We encode the value of each counter as the number of tokens in a place and we use parametric arcs to test for emptiness of that place (*i.e.* counter value 0).

As a consequence, we need to consider meaningful subclasses for which we might obtain some decidability results.

4.3   Subclasses of Parametric Petri Nets

The basic observation guiding us in defining interesting subclasses of PPNs is that, in the 2-counter machine reduction briefly outlined above, we need both a post arc with a parametric weight a and a pre arc with *the same* parametric weight a.

We thus define preT-PPNs, postT-PPNs, and distinctT-PPNs according to whether the parameters are allowed only in pre arcs, in post arcs, or in both but not with the same parameters.

Definition 25 (preT- and postT-PPNs). *A preT-PPN (resp. postT-PPN) is a PPN in which the* Post *(resp.* Pre*) function has the form* $P \times T \to \mathbb{N}$.

Definition 26 (distincT-PPNs). *A distinctT-PPN is a PPN in which the set of parameters used in the* Pre *function and the set of those used in the* Post *function are disjoint.*

We could also consider classic Petri nets in which the initial marking is parameterised. The corresponding formalism is called P-PPN. Such a parametric initial marking is easily simulated with an initial transition that has parametric weights in its post arcs and sets the initial marking. Interestingly, it can also be proved that postT-PPNs can be (weakly) simulated by P-PPNs [32].

Figure 6 [32] summarises this hierarchy of subclasses.

4.4   Global Results

We now summarise the current state-of-the art for the study of PPNs. Table 2 gives the decidability results and whenever relevant the complexities for the universal problems, while Table 3 gives them for the existential problems.

In order to establish the decidability results, we use a variety of techniques. The most basic is that preT- and postT-PPNs have a strong monotonicity property [32]: increasing (resp. decreasing) the value of parameters in a postT-PPN (resp. preT-PPN) can only add behaviours. Second we can reduce universal

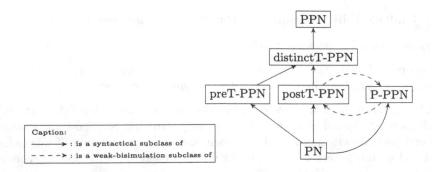

Fig. 6. Subclasses of PPNs

Table 2. Un(decidability) and complexity results for the universal parametric problems

Class	Reachability	S. Unboundness	Coverability
PPN	Undecidable [32]	Undecidable [32]	Undecidable [32]
preT-PPN	Undecidable [31]	ExpSpace-c [33]	ExpSpace-c [33]
postT-PPN	Undecidable [31]	ExpSpace-c[31]	ExpSpace-c [32]
P-PPN	open	ExpSpace-c[31]	ExpSpace-c [32]
distinctT-PPN	Undecidable [31]	ExpSpace-c[31]	ExpSpace-c [32]

Table 3. Un(decidability) and complexity results for the existential parametric problems

Class	Reachability	S. Unboundedness	Coverability
PPN	Undecidable [32]	Undecidable [32]	Undecidable [32]
preT-PPN	Undecidable [31]	ExpSpace-c[31]	ExpSpace-c [32]
postT-PPN	Undecidable [31]	ExpSpace-h[31]	ExpSpace-c [32]
P-PPN	Decidable [32]	ExpSpace-h[31]	ExpSpace-c [32]
distinctT-PPN	Undecidable [31]	ExpSpace-h[31]	ExpSpace-c [33]

coverability in preT-PPNs to simultaneous unboundedness, and existential coverability in postT-PPNs to coverability in the ω Petri nets of [44]. Both of these reductions can be found in [33]. Finally, we can adapt the Karp & Miller algorithm [51] to preT- and postT-PPNs [31].

Table 4 presents results for the synthesis problem [33]. In the cases where we can compute the set of adequate valuations ($\checkmark$), we mostly rely on the use of an algorithm to compute upward-closed sets by Valk and Jantzen [71]. The negative results ($\times$) come from the fact that the emptiness or the universality of the set cannot be decided as a direct consequence of the undecidability results above, so there is little hope to find a useful representation of that set. The case of distinctT-PPNs is similar in so far as if we can compute the solution and test its intersection with equality constraints we can solve the synthesis problem for any PPN, by replacing parameters used both in pre and post arcs by different parameters (which gives a distinctT-PPN) and then constraining the solution set with equality constraints on these different parameters.

Table 4. Results for the synthesis problem

Class	Reachability	S. Unboundedness	Coverability
PPN	✗	✗	✗
preT-PPN	✗	✓	✓
postT-PPN	✗	✓	✓
P-PPN	open	✓	✓
distinctT-PPN	✗	✗	✗

4.5   Conclusion

Parametric Petri nets are a powerful formalism to model flexible systems. In the general case, the interesting problems are undecidable but still useful sub-classes can be obtained by restricting the use of parameters. For most of these subclasses, it is possible to actually synthesise the values of the parameters such that the net is unbounded, or such that some marking is coverable. It would nevertheless be interesting to design semi-algorithms or incomplete algorithms for the most expressive cases that do not fit in this restricted setting. A problem also remains open, *i.e.* the decidability of universal reachability for Petri nets with a parameterised initial marking.

5   Action Synthesis

One of the classical approaches to verification and specification of concurrent systems employs Kripke structures as models and branching-time logics such as CTL as property description languages. In contrast to the models presented earlier, Kripke structures allow only for specifying sequential behaviours. Advanced data structures such as Binary Decision Diagrams [27] (BDDs) together with algorithms based on fixed-point specification of CTL enable efficient verification of models whose state spaces exceed 10^{20} [29]. Here, we extend Action-Restricted Computation Tree Logic ARCTL [62] and its models with parameters, to obtain a framework that benefits from BDD-based fixed-point algorithms.

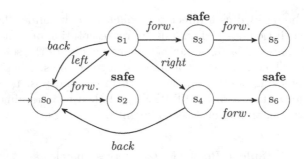

Fig. 7. A simple mixed transition system

5.1 Mixed Transition Systems

Mixed Transition Systems [62] (MTS) are essentially Kripke structures with transitions labelled by actions.

Definition 27 (MTS). *Let AP be a set of propositional variables. A Mixed Transition System is a 5-tuple $\mathcal{M} = (\mathcal{S}, s^0, \mathcal{A}, \mathcal{T}, \mathcal{L})$, where:*

- *$\mathcal{S}$ is a finite set of states, and $s^0 \in \mathcal{S}$ is the initial state,*
- *$\mathcal{A}$ is a non-empty finite set of actions,*
- *$\mathcal{T} \subseteq \mathcal{S} \times \mathcal{A} \times \mathcal{S}$ is a transition relation,*
- *$\mathcal{L} : \mathcal{S} \to 2^{AP}$ is a (state) valuation function.*

We write $s \xrightarrow{a} s'$ if $(s, a, s') \in \mathcal{T}$. Let $B \subseteq \mathcal{A}$ and $\pi = (s_0, a_0, s_1, a_1, \ldots)$ be a finite or infinite sequence of interleaved states and actions. By $|\pi|$ we denote the number of states in π if π is finite, and ω if π is infinite. A sequence π is a *path* over B iff $s_i \xrightarrow{a_i} s_{i+1}$ and $a_i \in B$ for each $i < |\pi|$ and either π is infinite or its final state does not have a B-successor state in $\mathcal{S}$, i.e. $\pi = (s_0, a_0, s_1, a_1, \ldots, s_m)$ for some $m \in \mathbb{N}$ and there is no $s' \in \mathcal{S}$ and $a \in B$ such that $s_m \xrightarrow{a} s'$. By π_i we denote the i-th state of π for all $i \in \mathbb{N}$.

The set of all paths over B in $\mathcal{M}$ is denoted by $\Pi(\mathcal{M}, B)$, and the set of all paths $\pi \in \Pi(\mathcal{M}, B)$ starting from a given state $s \in \mathcal{S}$ is denoted by $\Pi(\mathcal{M}, B, s)$. We typically omit the symbol $\mathcal{M}$, writing $\Pi(B)$ and $\Pi(B, s)$. By $\Pi^\omega(B)$ and $\Pi^\omega(B, s)$ we mean the corresponding sets restricted to the infinite paths only.

Example 11 (MTS). A MTS with $AP = \{\mathbf{p}, \mathbf{safe}\}$, $\mathcal{A} = \{left, right, forw., back\}$, and initial state s_0 is shown in Fig. 7. The path $\pi = (s_0, left, s_1, right, s_4)$ belongs to $\Pi(\{left, right\})$, but not to the set $\Pi(\{left, right, back\})$. The reason is that while π is a maximal path over $\{left, right\}$, it is not maximal over $\{left, right, back\}$ as it can be extended e.g. into the infinite path $\pi' = (s_0, left, s_1, right, s_4, back, s_0, left, s_1, right, s_4, back, s_0, \ldots) \in \Pi(\{left, right, back\})$.

5.2   Parametric Action-Restricted CTL

The main difference between ARCTL and CTL is that in ARCTL each path quantifier is subscripted with a set of actions, *e.g.* $E_{\{left, right\}} G(E_{\{forw.\}} F\mathbf{safe})$ may be read as *"there exists a path over left and right, on which it holds globally that a state satisfying* **safe** *is reachable along some path over forw."*.

Parametric ARCTL (pmARCTL) extends ARCTL by allowing free variables in place of sets of actions, *e.g.* $E_Y G(E_Z F\mathbf{safe})$ is a formula of pmARCTL, where Y and Z are free variables.

Definition 28 (pmARCTL syntax). *Let A be a finite set of actions, X a finite set of variables, AP a set of propositional variables, and $ActSets = 2^A \setminus \{\emptyset\}$. The set of formulae of pmARCTL is defined by the following grammar:*

$$\phi := p \mid \neg\phi \mid \phi \vee \phi \mid E_\alpha X\phi \mid E_\alpha G\phi \mid E_\alpha^\omega G\phi \mid E_\alpha(\phi U\phi),$$

where $p \in AP$ and $\alpha \in ActSets \cup X$.

The basic path quantifiers and modalities of pmARCTL have the same meaning as in CTL. The superscript $^\omega$ restricts the quantification to the infinite paths, whereas the subscript $_\alpha$ restricts the quantification to the paths over α.

The semantics of pmARCTL is defined *w.r.t.* parameter valuations, *i.e.* functions $\upsilon \colon X \to ActSets$. *ParVals* denotes the set of all parameter valuations. For conciseness, for $\upsilon \in ParVals$ we write $\upsilon(\alpha) = B$ if $\alpha = B \subseteq A$ and $\upsilon(\alpha) = \upsilon(Y)$ if $\alpha = Y \in X$. Moreover, we assume that $O^\epsilon = O$, for $O \in \{E, A, \Pi\}$.

Definition 29 (pmARCTL semantics). *Let $\mathcal{M} = (\mathcal{S}, s^0, \mathcal{A}, \mathcal{T}, \mathcal{L})$ be an MTS and $\upsilon \in ParVals$ a parameter valuation. The relation $\models_\upsilon$ is defined as follows:*

- $s \models_\upsilon p$ *iff* $p \in \mathcal{L}(s)$,
- $s \models_\upsilon \neg\phi$ *iff* $s \not\models_\upsilon \phi$,
- $s \models_\upsilon \phi \vee \psi$ *iff* $s \models_\upsilon \phi$ *or* $s \models_\upsilon \psi$,
- $s \models_\upsilon E_\alpha X\phi$ *iff* $|\pi| > 1$ *and* $\pi_1 \models_\upsilon \phi$, *for some* $\pi \in \Pi(\upsilon(\alpha), s)$,
- $s \models_\upsilon E_\alpha^r G\phi$ *iff* $\pi_i \models_\upsilon \phi$ *for all* $i < |\pi|$, *for some* $\pi \in \Pi^r(\upsilon(\alpha), s)$,
- $s \models_\upsilon E_\alpha(\phi U\psi)$ *iff* $\pi_i \models_\upsilon \psi$ *for some* $i < |\pi|$ *and* $\pi_j \models_\upsilon \phi$ *for all* $0 \leq j < i$, *for some* $\pi \in \Pi(\upsilon(\alpha), s)$,

where $p \in AP$, $\phi, \psi \in pmARCTL$, $r \in \{\omega, \epsilon\}$, and $\alpha \in ActSets \cup X$.

Example 12. For the MTS in Fig. 7 we have $s_0 \models_\upsilon E_Y G(E_Z F\mathbf{safe})$ iff *forw.* $\in \upsilon(Z)$.

5.3   Parameter Synthesis for pmARCTL

In this subsection we show how to recursively characterise pmARCTL using the basic operator of parametric pre-image. These equivalences give rise to fixed-point algorithms that can be implemented using BDDs.

Let $\phi \in pmARCTL$. Our goal is to construct the function $f_\phi \colon \mathcal{S} \to 2^{ParVals}$ s.t. for all $s \in \mathcal{S}$ we have $s \models_\upsilon \phi$ iff $\upsilon \in f_\phi(s)$. In other words, the set $f_\phi(s)$ consists of all the parameter valuations that make ϕ true in state s. Let us show how to build this function recursively, case by case. We omit the treatment of non-parametric modalities as the classical non-parametric methods of symbolic verification carry here with minimal alterations [49,64].

Boolean Connectives and Non-parametric Modalities. For each $p \in AP$, $\phi, \psi \in$ pmARCTL we have $f_p(s) = ParVals$ if $p \in \mathcal{L}(s)$ and $f_p(s) = \emptyset$ otherwise; $f_{\neg\phi}(s) = ParVals \setminus f_\phi(s)$; and $f_{\phi\vee\psi}(s) = f_\phi(s) \cup f_\psi(s)$.

Parametric Pre-image and NeXt. Let $f : \mathcal{S} \to 2^{ParVals}$ be a function. The *parametric pre-image* of f w.r.t. $Y \in \mathcal{X}$ is defined as the function $\mathrm{parPre}_Y^\exists(f) : \mathcal{S} \to 2^{ParVals}$ s.t. $\mathrm{parPre}_Y^\exists(f)(s) = \left\{ v \mid \exists_{s' \in \mathcal{S}} \exists_{a \in v(Y)} \ s \xrightarrow{a} s' \wedge v \in f(s') \right\}$ for each $s \in \mathcal{S}$. Intuitively, for each $\phi \in$ pmARCTL in $\mathrm{parPre}_Y^\exists(f_\phi)(s)$ we collect all the parameter valuations v s.t. some state s' satisfying $s' \models_v \phi$ can be reached by firing in $s \in \mathcal{S}$ an action from $v(Y)$. We therefore have $f_{E_Y X\phi} = \mathrm{parPre}_Y^\exists(f_\phi)(s)$

Parametric Temporal Modalities. We employ the following equations to deal with two versions of the Globally modality:

$$f_{E_Y^\omega G\phi}(s) = f_\phi(s) \cap \mathrm{parPre}_Y^\exists(f_{E_Y^\omega G\phi})(s),$$

$$f_{E_Y G\phi}(s) = f_\phi(s) \cap (\mathrm{parPre}_Y^\exists(f_{E_Y^\omega G\phi})(s) \cup f_{\neg E_Y X\mathrm{true}}(s)).$$

The following equation characterises the Until modality:

$$f_{E_Y(\phi U\psi)}(s) = f_\psi(s) \cup (f_\phi(s) \cap \mathrm{parPre}_Y^\exists(f_{E_Y(\phi U\psi)})(s)).$$

We refer to [53,55] on how to turn the above equations into fixed-point algorithms and implement them using BDDs. In all the cases, the returned result of running the overall synthesis algorithm for $\phi \in$ pmARCTL is the BDD that represents f_ϕ. This structure can be then queried for individual parameter valuations; for a certain class of formulae it is possible to synthesise minimal parameter valuations using prime implicants.

As a closing note let us mention that the emptiness problem for pmARCTL, *i.e.* the question whether $f_\phi(s^0) \neq \emptyset$ for $\phi \in$ pmARCTL is known to be NP-complete [53].

6   Tools

In this section, we finally briefly review tools related to the aforementioned formalisms.

6.1   IMITATOR

IMITATOR [10] is a software tool for parametric verification and robustness analysis of PTAs augmented with integer variables and stopwatches. Parameters can be used both in the model and in the properties. Verification capabilities include reachability-synthesis, deadlock-freeness-synthesis [5], non-Zeno model checking [18], minimal-time synthesis [8], and trace-preservation-synthesis. IMITATOR is fully written in OCaml, and makes use of the Parma Polyhedra Library [22]. It also features distributed capabilities to run over a cluster.

IMITATOR comes with a benchmarks library available under an open source license [6].

IMITATOR was successfully used in several application domains such as parametric schedulability analysis of a prospective architecture for the flight control system of the next generation of spacecrafts designed at ASTRIUM Space Transportation [43], formal timing analysis of music scores [40], verification of software product lines [58], monitoring logs from the automotive domain against parametric properties [11], and was used to propose a solution to a challenge related to a distributed video processing system by Thales [69].

Related Tools. The first tool to support modelling and verification using parametric timed automata was HYTECH [46]. In fact, HYTECH supports linear hybrid automata (including clocks, parameters, stopwatches and general continuous variables); it can compute the state space, and perform operations (such as intersection, convex hull, difference) between sets of symbolic states. Therefore, it can be used to perform parametric model checking using reachability checking [1]. HYTECH is not maintained anymore, but can still be found online in the form of a standalone binary for Linux.[4]

In [48], an extension of UPPAAL implementing parametric difference bound matrices (PDBMs) and hence allowing for verification using PTAs is mentioned. However, this tool does not seem to be available anywhere online.

PHAVer [41] is a tool for verifying safety properties of hybrid systems. It notably relies on exact arithmetic (with unlimited precision) using the Parma Polyhedra Library [22]; on the other hand, it also supports approximations.

SpaceEx [42] can be seen as a successor of PHAVer, and also tackles verification of reachability properties for hybrid systems. Parameters are not natively supported, but can be encoded using variables that are arbitrarily set up upon system start, and then remain subsequently constant (with a 0-slope). SpaceEx seems to have a lot of interesting recent developments.

PSyHCoS [16] allows the synthesis of parameters for a parametric extension of the process algebra Stateful timed CSP [68], itself a timed extension of Hoare's communicating sequential processes [47]. When compared to other formalisms such as (parametric) timed automata, (parametric) stateful timed CSP has the advantage of giving the designer the ability to specify hierarchical systems.

Finally, Symrob is not strictly a tool for synthesis, but allows robustness measurement for timed automata [65].

6.2 Roméo

ROMÉO [57] is a model-checking tool for a selection of hybrid extensions of Petri nets, enriched with discrete variables. In particular, it supports parametric time Petri nets, a formalism shown to be close to PTAs in terms of expressiveness [24, 70]. ROMÉO allows the use of parametric linear expressions in the time

---

[4] https://embedded.eecs.berkeley.edu/research/hytech/.

intervals of the transitions, and the addition of linear constraints on the parameters to restrict their domain. ROMÉO provides a simulator and an integrated model-checker supporting a subset of parametric TCTL (including reachability-synthesis and unavoidability-synthesis), in which "Until" modalities cannot be nested. It also features optimal cost reachability and parameter synthesis for cost-bounded reachability. ROMÉO implements in particular an original algorithm for integer parameter synthesis using a symbolic (continuous) representation [50]. ROMÉO is mainly written in C++, and makes use of the Parma Polyhedra Library [22]. It has been successfully used in a few and diverse case-studies including the analysis of resilience properties in oscillatory biological systems [20]; the synthesis of environment requirements for an aerial video tracking system [61]; and the analysis of operational scenarios modelling in the DGA OMOTESC project [66].

6.3 Spatula

Spatula [52,53] implements the theory of action synthesis outlined in Sect. 5. The tool is written in C++ and employs CUDD library [67] for representing and manipulating BDDs. Spatula accepts models represented as networks of automata written in a simplified C-like input language and pmARCTL as property description language. The result of synthesis for a given property is a BDD that represents all the valuations that make the property true. Spatula can also list all the minimal valuations ($w.r.t.$ bitwise comparison) for the existential part of the logic.

References

1. Aceto, L., Bouyer, P., Burgueño, A., Larsen, K.G.: The power of reachability testing for timed automata. In: Arvind, V., Ramanujam, S. (eds.) FSTTCS 1998. LNCS, vol. 1530, pp. 245–256. Springer, Heidelberg (1998). https://doi.org/10.1007/978-3-540-49382-2_22

2. Alur, R., Courcoubetis, C., Dill, D.L.: Model-checking in dense real-time. Inf. Comput. **104**(1), 2–34 (1993). https://doi.org/10.1006/inco.1993.1024

3. Alur, R., Dill, D.L.: A theory of timed automata. Theoret. Comput. Sci. **126**(2), 183–235 (1994). https://doi.org/10.1016/0304-3975(94)90010-8

4. Alur, R., Henzinger, T.A., Vardi, M.Y.: Parametric real-time reasoning. In: Kosaraju, S.R., Johnson, D.S., Aggarwal, A. (eds.) STOC, pp. 592–601. ACM, New York (1993). https://doi.org/10.1145/167088.167242

5. André, É.: Parametric deadlock-freeness checking timed automata. In: Sampaio, A., Wang, F. (eds.) ICTAC 2016. LNCS, vol. 9965, pp. 469–478. Springer, Cham (2016). https://doi.org/10.1007/978-3-319-46750-4_27

6. André, É.: A benchmark library for parametric timed model checking. In: Artho, C., Ölveczky, P.C. (eds.) FTSCS 2018. CCIS, vol. 1008, pp. 75–83. Springer, Cham (2019). https://doi.org/10.1007/978-3-030-12988-0_5

7. André, É.: What's decidable about parametric timed automata? Int. J. Softw. Tools Technol. Transf. **21**(2), 203–219 (2019). https://doi.org/10.1007/s10009-017-0467-0

8. André, É., Bloemen, V., Petrucci, L., van de Pol, J.: Minimal-time synthesis for parametric timed automata. In: Vojnar, T., Zhang, L. (eds.) TACAS 2019. LNCS, vol. 11428, pp. 211–228. Springer, Cham (2019). https://doi.org/10.1007/978-3-030-17465-1_12

9. André, É., Chatain, T., Encrenaz, E., Fribourg, L.: An inverse method for parametric timed automata. Int. J. Found. Comput. Sci. **20**(5), 819–836 (2009). https://doi.org/10.1142/S0129054109006905

10. André, É., Fribourg, L., Kühne, U., Soulat, R.: IMITATOR 2.5: a tool for analyzing robustness in scheduling problems. In: Giannakopoulou, D., Méry, D. (eds.) FM 2012. LNCS, vol. 7436, pp. 33–36. Springer, Heidelberg (2012). https://doi.org/10.1007/978-3-642-32759-9_6

11. André, É., Hasuo, I., Waga, M.: Offline timed pattern matching under uncertainty. In: Lin, A.W., Sun, J. (eds.) ICECCS, pp. 10–20. IEEE CPS (2018). https://doi.org/10.1109/ICECCS2018.2018.00010

12. André, É., Lime, D.: Liveness in L/U-parametric timed automata. In: Legay, A., Schneider, K. (eds.) ACSD, pp. 9–18. IEEE (2017). https://doi.org/10.1109/ACSD.2017.19

13. André, É., Lime, D., Ramparison, M.: TCTL model checking lower/upper-bound parametric timed automata without invariants. In: Jansen, D.N., Prabhakar, P. (eds.) FORMATS 2018. LNCS, vol. 11022, pp. 37–52. Springer, Cham (2018). https://doi.org/10.1007/978-3-030-00151-3_3

14. André, É., Lime, D., Roux, O.H.: Decision problems for parametric timed automata. In: Ogata, K., Lawford, M., Liu, S. (eds.) ICFEM 2016. LNCS, vol. 10009, pp. 400–416. Springer, Cham (2016). https://doi.org/10.1007/978-3-319-47846-3_25

15. André, É., Lipari, G., Nguyen, H.G., Sun, Y.: Reachability preservation based parameter synthesis for timed automata. In: Havelund, K., Holzmann, G., Joshi, R. (eds.) NFM 2015. LNCS, vol. 9058, pp. 50–65. Springer, Cham (2015). https://doi.org/10.1007/978-3-319-17524-9_5

16. André, É., Liu, Y., Sun, J., Dong, J.S., Lin, S.-W.: PSyHCoS: parameter synthesis for hierarchical concurrent real-time systems. In: Sharygina, N., Veith, H. (eds.) CAV 2013. LNCS, vol. 8044, pp. 984–989. Springer, Heidelberg (2013). https://doi.org/10.1007/978-3-642-39799-8_70

17. André, É., Markey, N.: Language preservation problems in parametric timed automata. In: Sankaranarayanan, S., Vicario, E. (eds.) FORMATS 2015. LNCS, vol. 9268, pp. 27–43. Springer, Cham (2015). https://doi.org/10.1007/978-3-319-22975-1_3

18. André, É., Nguyen, H.G., Petrucci, L., Sun, J.: Parametric model checking timed automata under non-zenoness assumption. In: Barrett, C., Davies, M., Kahsai, T. (eds.) NFM 2017. LNCS, vol. 10227, pp. 35–51. Springer, Cham (2017). https://doi.org/10.1007/978-3-319-57288-8_3

19. André, É., Soulat, R.: The Inverse Method. FOCUS Series in Computer Engineering and Information Technology, ISTE Ltd and Wiley, 176 p. (2013)

20. Andreychenko, A., Magnin, M., Inoue, K.: Analyzing resilience properties in oscillatory biological systems using parametric model checking. Biosystems **149**, 50–58 (2016). https://doi.org/10.1016/j.biosystems.2016.09.002. Selected Papers from the Computational Methods in Systems Biology 2015 Conference

21. Aştefănoaei, L., Bensalem, S., Bozga, M., Cheng, C.-H., Ruess, H.: Compositional parameter synthesis. In: Fitzgerald, J., Heitmeyer, C., Gnesi, S., Philippou, A. (eds.) FM 2016. LNCS, vol. 9995, pp. 60–68. Springer, Cham (2016). https://doi.org/10.1007/978-3-319-48989-6_4

22. Bagnara, R., Hill, P.M., Zaffanella, E.: The parma polyhedra library: toward a complete set of numerical abstractions for the analysis and verification of hardware and software systems. Sci. Comput. Program. **72**(1–2), 3–21 (2008). https://doi.org/10.1016/j.scico.2007.08.001
23. Beneš, N., Bezděk, P., Larsen, K.G., Srba, J.: Language emptiness of continuous-time parametric timed automata. In: Halldórsson, M.M., Iwama, K., Kobayashi, N., Speckmann, B. (eds.) ICALP 2015. LNCS, vol. 9135, pp. 69–81. Springer, Heidelberg (2015). https://doi.org/10.1007/978-3-662-47666-6_6
24. Bérard, B., Cassez, F., Haddad, S., Lime, D., Roux, O.H.: Comparison of the expressiveness of timed automata and time petri nets. In: Pettersson, P., Yi, W. (eds.) FORMATS 2005. LNCS, vol. 3829, pp. 211–225. Springer, Heidelberg (2005). https://doi.org/10.1007/11603009_17
25. Bouyer, P., Markey, N., Sankur, O.: Robustness in timed automata. In: Abdulla, P.A., Potapov, I. (eds.) RP 2013. LNCS, vol. 8169, pp. 1–18. Springer, Heidelberg (2013). https://doi.org/10.1007/978-3-642-41036-9_1
26. Bozzelli, L., La Torre, S.: Decision problems for lower/upper bound parametric timed automata. Formal Methods Syst. Design **35**(2), 121–151 (2009). https://doi.org/10.1007/s10703-009-0074-0
27. Bryant, R.E.: Graph-based algorithms for Boolean function manipulation. IEEE Trans. Comput. **35**(8), 677–691 (1986). https://doi.org/10.1109/TC.1986.1676819
28. Bundala, D., Ouaknine, J.: Advances in parametric real-time reasoning. In: Csuhaj-Varjú, E., Dietzfelbinger, M., Ésik, Z. (eds.) MFCS 2014. LNCS, vol. 8634, pp. 123–134. Springer, Heidelberg (2014). https://doi.org/10.1007/978-3-662-44522-8_11
29. Burch, J.R., Clarke, E.M., McMillan, K.L., Dill, D.L., Hwang, L.J.: Symbolic model checking: 10^{20} states and beyond. In: LICS, pp. 428–439. IEEE Computer Society (1990). https://doi.org/10.1109/LICS.1990.113767
30. Chevallier, R., Encrenaz-Tiphène, E., Fribourg, L., Xu, W.: Timed verification of the generic architecture of a memory circuit using parametric timed automata. Formal Methods Syst. Des. **34**(1), 59–81 (2009). https://doi.org/10.1007/s10703-008-0061-x
31. David, N.: Discrete parameters in Petri nets. Ph.D. thesis. University of Nantes, France (2017)
32. David, N., Jard, C., Lime, D., Roux, O.H.: Discrete parameters in Petri nets. In: Devillers, R., Valmari, A. (eds.) PETRI NETS 2015. LNCS, vol. 9115, pp. 137–156. Springer, Cham (2015). https://doi.org/10.1007/978-3-319-19488-2_7
33. David, N., Jard, C., Lime, D., Roux, O.H.: Coverability synthesis in parametric Petri nets. In: Meyer, R., Nestmann, U. (eds.) CONCUR. LIPIcs, Dagstuhl Publishing (2017). https://doi.org/10.4230/LIPIcs.CONCUR.2017.14
34. Delahaye, B.: Consistency for parametric interval Markov chains. In: André, É., Frehse, G. (eds.) SynCoP. OASICS, vol. 44, pp. 17–32. Schloss Dagstuhl - Leibniz-Zentrum für Informatik (2015). https://doi.org/10.4230/OASIcs.SynCoP.2015.17
35. Delahaye, B., Larsen, K.G., Legay, A., Pedersen, M.L., Wasowski, A.: Consistency and refinement for interval Markov chains. J. Log. Algebr. Program. **81**(3), 209–226 (2012). https://doi.org/10.1016/j.jlap.2011.10.003
36. Delahaye, B., Lime, D., Petrucci, L.: Parameter synthesis for parametric interval Markov chains. In: Jobstmann, B., Leino, K.R.M. (eds.) VMCAI 2016. LNCS, vol. 9583, pp. 372–390. Springer, Heidelberg (2016). https://doi.org/10.1007/978-3-662-49122-5_18
37. Demri, S.: On selective unboundedness of VASS. J. Comput. Syst. Sci. **79**(5), 689–713 (2013). https://doi.org/10.1016/j.jcss.2013.01.014

38. Di Giampaolo, B., La Torre, S., Napoli, M.: Parametric metric interval temporal logic. Theoret. Comput. Sci. **564**, 131–148 (2015). https://doi.org/10.1016/j.tcs. 2014.11.019
39. Doyen, L.: Robust parametric reachability for timed automata. Inf. Process. Lett. **102**(5), 208–213 (2007). https://doi.org/10.1016/j.ipl.2006.11.018
40. Fanchon, L., Jacquemard, F.: Formal timing analysis of mixed music scores. In: ICMC. Michigan Publishing, August 2013
41. Frehse, G.: PHAVer: algorithmic verification of hybrid systems past HyTech. Int. J. Softw. Tools Technol. Transf. **10**(3), 263–279 (2008). https://doi.org/10.1007/s10009-007-0062-x
42. Frehse, G., et al.: SpaceEx: scalable verification of hybrid systems. In: Gopalakrishnan, G., Qadeer, S. (eds.) CAV 2011. LNCS, vol. 6806, pp. 379–395. Springer, Heidelberg (2011). https://doi.org/10.1007/978-3-642-22110-1_30
43. Fribourg, L., Lesens, D., Moro, P., Soulat, R.: Robustness analysis for scheduling problems using the inverse method. In: Reynolds, M., Terenziani, P., Moszkowski, B. (eds.) TIME, pp. 73–80. IEEE Computer Society Press, September 2012. https://doi.org/10.1109/TIME.2012.10. http://www.lsv.ens-cachan.fr/Publis/PAPERS/PDF/FLMS-time12.pdf
44. Geeraerts, G., Heußner, A., Praveen, M., Raskin, J.: ω-Petri nets: algorithms and complexity. Fundamenta Informaticae **137**(1), 29–60 (2015)
45. Glabbeek, R.J.: The linear time - branching time spectrum. In: Baeten, J.C.M., Klop, J.W. (eds.) CONCUR 1990. LNCS, vol. 458, pp. 278–297. Springer, Heidelberg (1990). https://doi.org/10.1007/BFb0039066
46. Henzinger, T.A., Ho, P.H., Wong-Toi, H.: HyTech: a model checker for hybrid systems. Int. J. Softw. Tools Technol. Transf. **1**(1–2), 110–122 (1997). https://doi.org/10.1007/s100090050008
47. Hoare, C.: Communicating Sequential Processes. International Series in Computer Science. Prentice-Hall, Upper Saddle River (1985)
48. Hune, T., Romijn, J., Stoelinga, M., Vaandrager, F.W.: Linear parametric model checking of timed automata. J. Log. Algebr. Program. **52–53**, 183–220 (2002). https://doi.org/10.1016/S1567-8326(02)00037-1
49. Huth, M., Ryan, M.: Logic in Computer Science: Modelling and Reasoning about Systems. Cambridge University Press, Cambridge (2004)
50. Jovanović, A., Lime, D., Roux, O.H.: Integer parameter synthesis for real-time systems. IEEE Trans. Softw. Eng. **41**(5), 445–461 (2015). https://doi.org/10.1109/TSE.2014.2357445
51. Karp, R.M., Miller, R.E.: Parallel program schemata. J. Comput. Syst. Sci. **3**(2), 147–195 (1969). https://doi.org/10.1016/S0022-0000(69)80011-5
52. Knapik, M.: https://michalknapik.github.io/spatula
53. Knapik, M., Meski, A., Penczek, W.: Action synthesis for branching time logic: theory and applications. ACM Trans. Embed. Comput. **14**(4), 64 (2015). https://doi.org/10.1145/2746337
54. Knapik, M., Penczek, W.: Bounded model checking for parametric timed automata. Trans. Petri Nets Other Models Concurr. **5**, 141–159 (2012). https://doi.org/10.1007/978-3-642-29072-5_6
55. Knapik, M., Penczek, W.: Fixed-point methods in parametric model checking. In: Angelov, P., et al. (eds.) Intelligent Systems'2014. AISC, vol. 322, pp. 231–242. Springer, Cham (2015). https://doi.org/10.1007/978-3-319-11313-5_22

56. Li, J., Sun, J., Gao, B., André, É.: Classification-based parameter synthesis for parametric timed automata. In: Duan, Z., Ong, L. (eds.) ICFEM 2017. LNCS, vol. 10610, pp. 243–261. Springer, Cham (2017). https://doi.org/10.1007/978-3-319-68690-5_15

57. Lime, D., Roux, O.H., Seidner, C., Traonouez, L.-M.: Romeo: a parametric model-checker for Petri nets with stopwatches. In: Kowalewski, S., Philippou, A. (eds.) TACAS 2009. LNCS, vol. 5505, pp. 54–57. Springer, Heidelberg (2009). https://doi.org/10.1007/978-3-642-00768-2_6

58. Luthmann, L., Stephan, A., Bürdek, J., Lochau, M.: Modeling and testing product lines with unbounded parametric real-time constraints. In: Cohen, M.B., et al. (eds.) SPLC, vol. A, pp. 104–113. ACM (2017). https://doi.org/10.1145/3106195.3106204

59. Miller, J.S.: Decidability and complexity results for timed automata and semi-linear hybrid automata. In: Lynch, N., Krogh, B.H. (eds.) HSCC 2000. LNCS, vol. 1790, pp. 296–310. Springer, Heidelberg (2000). https://doi.org/10.1007/3-540-46430-1_26

60. Minsky, M.L.: Computation: Finite and Infinite Machines. Prentice-Hall Inc., Upper Saddle River (1967)

61. Parquier, B., et al.: Applying parametric model-checking techniques for reusing real-time critical systems. In: Artho, C., Ölveczky, P.C. (eds.) FTSCS 2016. CCIS, vol. 694, pp. 129–144. Springer, Cham (2017). https://doi.org/10.1007/978-3-319-53946-1_8

62. Pecheur, C., Raimondi, F.: Symbolic model checking of logics with actions. In: Edelkamp, S., Lomuscio, A. (eds.) MoChArt 2006. LNCS (LNAI), vol. 4428, pp. 113–128. Springer, Heidelberg (2007). https://doi.org/10.1007/978-3-540-74128-2_8

63. Petrucci, L., van de Pol, J.: Parameter synthesis algorithms for parametric interval Markov chains. In: Baier, C., Caires, L. (eds.) FORTE 2018. LNCS, vol. 10854, pp. 121–140. Springer, Cham (2018). https://doi.org/10.1007/978-3-319-92612-4_7

64. Raimondi, F., Lomuscio, A.: Automatic verification of multi-agent systems by model checking via ordered binary decision diagrams. J. Appl. Log. 5(2), 235–251 (2007). https://doi.org/10.1016/j.jal.2005.12.010

65. Sankur, O.: Symbolic quantitative robustness analysis of timed automata. In: Baier, C., Tinelli, C. (eds.) TACAS 2015. LNCS, vol. 9035, pp. 484–498. Springer, Heidelberg (2015). https://doi.org/10.1007/978-3-662-46681-0_48

66. Seidner, C.: Vérification des EFFBDs: model checking en Ingénierie Système. (EFFBDs verification: model checking in systems engineering). Ph.D. thesis. University of Nantes, France (2009). https://tel.archives-ouvertes.fr/tel-00440677

67. Somenzi, F.: CUDD: CU decision diagram package - release 2.5.0. https://github.com/ivmai/cudd

68. Sun, J., Liu, Y., Dong, J.S., Liu, Y., Shi, L., André, É.: Modeling and verifying hierarchical real-time systems using stateful timed CSP. ACM Trans. Softw. Eng. Methodol. 22(1), 3:1–3:29 (2013). https://doi.org/10.1145/2430536.2430537

69. Sun, Y., André, É., Lipari, G.: Verification of two real-time systems using parametric timed automata. In: Quinton, S., Vardanega, T. (eds.) WATERS, July 2015

70. Traonouez, L.M., Lime, D., Roux, O.H.: Parametric model-checking of stopwatch Petri nets. J. Univ. Comput. Sci. 15(17), 3273–3304 (2009). https://doi.org/10.3217/jucs-015-17-3273

71. Valk, R., Jantzen, M.: The residue of vector sets with applications to decidability problems in Petri nets. Acta Informatica 21(6), 643–674 (1985). https://doi.org/10.1007/BF00289715

Integrated Simulation of Domain-Specific Modeling Languages with Petri Net-Based Transformational Semantics

David Mosteller, Michael Haustermann^(⊠), Daniel Moldt, and Dennis Schmitz

Faculty of Mathematics, Informatics and Natural Sciences,
Department of Informatics, University of Hamburg, Hamburg, Germany
haustermann@informatik.uni-hamburg.de
https://www.inf.uni-hamburg.de/inst/ab/art.html

Abstract. The development of domain specific models requires appropriate tool support for modeling and execution. Meta-modeling facilitates solutions for the generation of modeling tools from abstract language specifications. The RMT approach (RENEW Meta-Modeling and Transformation) applies transformational semantics using Petri net formalisms as target languages in order to produce quick results for the development of modeling techniques. The problem with transformational approaches is that the inspection of the system during execution is not possible in the original representation.

   We present a concept for providing simulation feedback for domain specific modeling languages (DSML) that are developed with the RMT approach on the basis of meta-models and transformational semantics using Petri nets. Details of the application of this new approach are illustrated by some well-known constructs of the Business Process Model and Notation (BPMN).

Keywords: Meta-modeling · BPMN · Petri nets · Reference Nets · Simulation · Graphical feedback

1 Introduction

The construction of abstract models is an essential part of software and systems engineering. Meta-modeling provides a conceptual base for developing modeling languages that are tailored to satisfy the demands of specific application domains. Tools may be generated from the language specifications to support the modeling process.

   The definition of semantics for domain-specific modeling languages (DSML) and the provision of appropriate execution tools is one of the key challenges in model-driven tool development. One way to define semantics for a DSML is to transform it into another language with well-defined semantics and existing tool support. This is especially suited to produce quick results for rapid prototyping. Bryant et al. identified "the mapping of execution results (e.g., error messages,

© Springer-Verlag GmbH Germany, part of Springer Nature 2019
M. Koutny et al. (Eds.): ToPNoC XIV, LNCS 11790, pp. 101–125, 2019.
https://doi.org/10.1007/978-3-662-60651-3_4

debugging traces) back into the DSML in a meaningful manner, such that the domain expert using the modeling language understands the result" [5, p. 228] as one of the challenges for the translation semantics approach. Concerning the user experience, meaningful visual representation of the domain concepts is vital for the communication between different stakeholders, especially for the domain experts that are often non-software engineers [1, p. 233]. The representation of DSML in execution is still considered a challenge in tool generation in general [28, p. 196].

We present a concept for the rapid prototyping and integrated simulation of DSML within the RENEW simulation environment. This concept integrates the simulation feedback from the executed language into the graphical layer of the original DSML. The focus of this contribution is to equip the language developer with the means necessary to develop iteratively and prototypically the visually animated DSML without being dependent on a difficult implementation or graph transformation language. With our contribution, we combine and advance two branches of our current research: first, the development of domain-specific modeling languages using the RENEW Meta-Modeling and Transformation (RMT) framework [31] and secondly, the provisioning and coupling of multiple modeling perspectives during execution within RENEW [29]. The first contribution does not consider the simulation in the source language; it presents the transformation of the source language into Reference Nets without the support for graphical simulation feedback. The second contribution does not use meta-modeling, but presents an approach for coupling and simulating multiple formalisms synchronously. The contribution mentions the execution of finite automata with the possibility of highlighting the active state and state transitions. However, the graphical simulation feedback was hard-coded. In this contribution, we consider the model-based customization of the visual behavior of a simulated DSML. This paper is an extended version of a workshop contribution [32].

The approach provided by the RMT framework supports the rapid prototypical development of domain-specific modeling languages with Petri net-based semantics. The RMT approach is based on the idea of providing transformational semantics for the modeling language in development (source language) through a mapping of its constructs to a target language. The latter is implemented using net components [6, Chapter 5], which are reusable and parameterizable code templates – quite comparable to code macros – modeled as Petri nets. The benefit of this approach consists in leveraging the intuitive notation of Petri nets in order to fulfill the task of specifying an explicit (operational) semantics for the target language. This, however, comes at the cost of limited applicability in the sense that the approach becomes less suited for modeling environments that do not conform to the state-based, process-oriented characteristics of Petri nets. This said, we acknowledge the limitations of our approach and develop a solution for the integrated simulation of executable DSML that aligns with the fundamental idea of the RMT approach and its focus on rapid prototypical development of DSML.

We choose Reference Nets [23] as a target language, but are not restricted to this formalism. Reference Nets combine concepts of object-oriented programming and Petri net theory. They are well-suited as a target language because of their concise syntax and broad applicability. With the presented solution Reference Nets provide the operational semantics for the target language. The simulation events are reflected in the source language during execution.

Tool support for our approach comes from RENEW, which provides a flexible modeling editor and simulation environment for Petri net formalisms. RENEW has a comprehensive development support for the construction of Reference Net-based systems [7,24].

Various approaches to defining the semantics of DSML with the purpose of simulation exist. In Sect. 2, we start with a comprehensive review of the research on graphical feedback in executable DSML engineering and compare our approach with related work. In this contribution, we extend the RMT framework (as presented in Sect. 3) with a direct simulation of the DSML's original representation and discuss the integration into the approach. The presented concept for simulation visualization is based on the highlighting of model constructs. This is achieved by reflecting simulation events of the underlying executed Petri nets (target language) into the DSML (source language). Several types of mappings are evaluated in Sect. 4 regarding their expressiveness and features for modeling. A major challenge for the provision of direct simulation support is the integration into model-driven approaches in the sense that the DSML developer can specify the desired representation of the executed models in a model-driven fashion. The concept presented here includes tools that enable DSML developers to create the necessary artifacts and configurations to manage this task. Section 5 introduces the current implementation that processes these artifacts and configurations to initialize a simulation of the DSML model with graphical feedback. In Sect. 6, we discuss multiple alternatives to provide support for DSML developers to specify the desired representation of the executed models in the RMT approach. As a part of our contribution, a generic compiler is implemented in RENEW. This is used in the previously mentioned processing of artifacts and configurations. On this basis, the generated technique may be executed within RENEW's simulation engine in its original representation, as presented in Sect. 7. We summarize our results in Sect. 8.

2   Related Work

A vital issue of providing graphical feedback for the execution of DSML is the definition of the language semantics. In their analysis on *Challenges and Directions in Formalizing the Semantics of Modeling Languages* [5] Bryant et al. describe categories of approaches to the definition of semantics of modeling languages: first, the category of rewriting grammars that apply a set of graph transformation rules; second, approaches based on metalanguages that include the operational specifications in the meta-model; third, the transformational approaches, which transfer semantics by mapping the concepts to a target language.

Various approaches to the development of visual languages exist but the notion of graphical feedback or semantics of DSML for simulation purposes differs greatly. The demand for graphical feedback in executable DSML comprises, for instance, visualization of simulation states, feedback from a model checking counterexample, or smooth visual animation of graphical objects. Although the requirements are diverse, the following related work is presented respecting the categories of Bryant et al.: graph-grammar, metalanguages and transformational.

Graph grammars are widely applied to define the semantics of DSML. With the work from Biermann et al. graph rewriting rules are not only applied to provide operational semantics during execution but also to define the behavior of visual editors [2]. With AToM^3 de Lara et al. [26] maintain an approach that relies on graph rewriting systems. Rules may be edited graphically, allowing a low-level entry for the language designer comparable to our rapid prototyping approach. In AToM^3 the graph rewriting rules may be applied in step-by-step, continuous or animation mode [26, p. 199]. To visualize the behavior of the models, they are translated into Java and executed by an external application.

Several approaches aim at providing interactive visual behavior for domain-specific modeling tools by using metalanguages. GEMOC Studio is a comprehensive framework for the development of executable DSML (xDSML) based on the EMF. "The framework provides a generic interface to plug in different execution engines associated to their specific metalanguages used to define the discrete-event operational semantics of DSMLs" [3, p. 84]. The animation component, Sirius animator [18], facilitates reacting to simulation events (execution steps) in order to trigger visual effects in the xDSML. This permits, for instance, the adaption of graphical attributes, such as border width, colors or icon images through style specifications. With integration into the Eclipse debug UI it supports interactive (omniscient) debugging of models including the possibility to move forward and backward in the simulation trace [4]. The tool suite offers four execution engines that are integrated into the framework and conform to the event API. With GEMOC Studio the operational semantics are defined by the application of metalanguages, while we follow a transformational approach.

Hegedüs et al. [15] present a systematic approach for replaying information captured from execution traces into the source model. Graph transformation rules are applied to relate the trace model to the dynamic model of a DSML based on their respective meta-model. The trace information is evaluated externally to the original execution, which conceptually distinguishes the approach taken by the authors from the integrated approach, which is presented in this contribution.

Kindler has a comparable approach to ours using annotations with the ePNK framework to provide simulation capabilities for a meta-model-based modeling application [20]. The simulation in ePNK is realized in the form of different handlers that manipulate the annotations representing the simulation state. A presentation handler implements the representation of these annotations as labels or overlays in the graphical editor.

Cramer and Kastens [11] apply the simulation specification language DSIM as a metalanguage to define the semantics of DSML. The visualization engine interpolates the intermediate states between discrete simulation steps to display a smooth animation of the execution.

Sedrakyan et al. present model-driven graphical feedback in a model-to-code transformation to validate semantic conformance especially for novice modelers [36]. Figure decorations visualize inconsistencies, and error messages provide additional information. They focus on feedback for errors that occur in the compiled system rather than a complete interactive inspection of the execution.

Rybicki et al. [35] use model-to-model transformations to map high-level models to low-level executable code or models. With tracing capability added to their approach, they keep track of the relations between the transformed models. The tracing information is used to propagate runtime information during simulation to the high-level models, which permits the inspection of a simulation in the original representation. However, they do not cover the customization of the representation of the models at runtime and the highlighted constructs.

Petri nets are often used as the target language in transformational approaches in combination with high-level modeling languages. Research questions in the context of transformations to Petri nets are often discussed for specific languages. An overview of semantics for business process languages is, for example, provided by Lohmann et al. [27]. The utilization of semantic components resulting in a 1:n mapping is related to static hierarchy concepts, such as the concepts for Coloured Petri Nets [17]. The research results in this area can be transferred to our approach. Object-oriented Petri nets may serve as a target language, where the encapsulation of properties is desired [34]. Additionally, the capabilities of Reference Nets regarding dynamic hierarchies can be helpful for the development of more complex semantics.

With the Access/CPN tool Westergaard enables domain-specific visualizations of Coloured Petri Nets by reflection of simulation events [38]. It even becomes possible to embed simulations into externally executed programs and vice versa. However, as the Access/CPN tool makes no assumptions about the external application, the domain-specific visualizations need to be provided individually. While RENEW potentially offers an interface for accessing simulation entities through generated Java net stubs, the focus in this contribution is on model-based graphical simulation feedback.

To this end, we apply declarative inscriptions on semantic components to establish the link between simulation and graphical framework. The approach presented in this contribution is unique in the sense that it uses Petri nets as the target language, to provide a solid formal base, in combination with a focus on the representation customization of the executed models.

3   The RMT Framework

The RMT framework (RENEW Meta-Modeling and Transformation) [31] is a model-driven framework for the agile development of DSML. It follows concepts from software language engineering (SLE [21]) and enables a short development cycle to be appropriately applied in prototyping environments. The RMT framework is particularly well-suited to develop languages with simulation feedback due to its lightweight approach to SLE and the tight integration with the

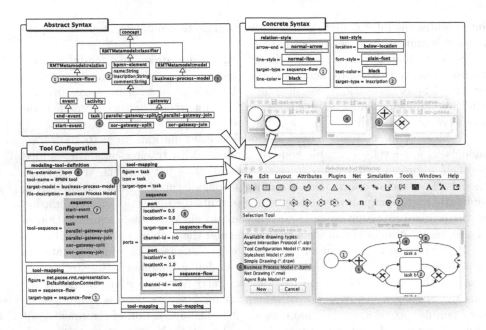

Fig. 1. Excerpt from the RMT models for a BPMN prototype with the generated tool. (Color figure online)

extensible RENEW simulation engine, which supports the propagation of simulation events. Other frameworks for model-driven engineering (MDE), such as the Eclipse Modeling Framework (EMF [13]), could also benefit from the proposed solution for provisioning of simulation feedback. However, this would require integrating an equally adequate simulation engine into the MDE framework.

With the RMT framework, the specification of a language and a corresponding modeling tool may be derived from a set of models, defined by the developer of a modeling technique. A meta-model defines the structure (abstract syntax) of the language, the concepts of its application domain, and their relations. The visual instances (concrete syntax) of the defined concepts and relations are provided using graphical components from RENEW's modeling constructs repertoire. They are configurable by style sheets and complemented with icons and tool configuration model to facilitate the generation of a modeling tool that nicely integrates into the RENEW development environment.

In this contribution, we use a BPMN (Business Process Model and Notation) prototype as our running example to present how the RMT approach is extended with graphical feedback for simulations. Figure 1 displays a selected excerpt from the models required for the BPMN prototype, together with the tool that is generated from these models. The parts of the figure are linked with colored circles and numbers. The meta-model (upper left) defines the concepts for classifiers (4, 5) and relations (1) of the modeling language and the corresponding model (3). Annotations are realized as attributes of these concepts (2). The concrete syntax (upper right) is provided using graphical components, which are created with

RENEW, as it is depicted for the task and the parallel gateway (4, 5). Icon images for the toolbar can be generated from these graphical components. The representation of the inscription annotation (2) and the sequence flow relation (1) is configured with style sheets.

The tool configuration model (lower left) facilitates the connection between abstract and concrete syntax and defines additional tool related settings. The concepts of the meta-model are associated with the previously defined graphical components (4), with custom implementations or default figure classes that are customizable by style sheets (1), and the icons. Connection points for constructs are specified with ports (8). The general tool configuration contains the definition of a file description, an extension (6), and the ordering of tool buttons (7).

With these models, a tool is generated as a RENEW plug-in, as shown at the bottom right side of Fig. 1. A complete example of the RMT models and additional information about the framework and the approach can be found in [31].

4   Net Component-Based Semantics

Our main goal is the provision of an easily applicable approach and easily usable tools for language developers, as well as users. The difficult part is often the definition of the semantics. We propose to apply a mapping of DSML constructs to Petri net constructs, in order to provide operational semantics for the DSML. Within this kind of scenario, graph transformation languages commonly facilitate the semantic mapping either by declarative inscriptions, imperative expressions, or a combination of both [22]. Either way, the developer of a DSML needs to learn the respective language in advance and finally develop the rules for the semantic mappings. With the RMT approach Petri net constructs can simply be modeled with the RENEW editor, in a similar way this is done for the graphical components (cf. Fig. 1). The net elements are annotated with attributes and arranged together in one environment, in order to form a net component. This has the advantage that partly completed mappings become immediately executable and testable, which supports the prototypical development process.

On the downside, this has an impact on the number of mapped constructs in the source and target language. Different types of mappings are possible (1:1, 1:n, n:1, n:m). A general solution that covers semantics for possibly any language would probably require a n:m mapping. In this contribution we mainly consider languages that are more abstract than Reference Nets, consequently utilizing 1:n mappings. Using net components, which group multiple Reference Net constructs into one, the mapping is reduced to cardinalities of 1:1. A 1:1 mapping restricts the expressiveness but there are options to overcome some of the restrictions. We discuss these further on in this section.

For a direct mapping approach, one of the main questions is how the components are bordered and how they are connected. This depends mainly on the handling of relations. A simple way of connecting two net components is using an arc between them, which requires the connection elements in the net components to be of opposing types. Regarding expressiveness, this approach may be

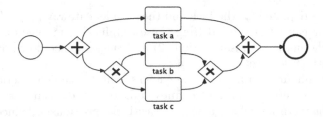

Fig. 2. A BPMN process containing three tasks.

Table 1. Mapping of BPMN and Petri net constructs [12, p. 7].

BPMN	Reference Net	BPMN	Reference Net

sufficient for simple P/T net semantics, but in more complicated scenarios that use colored Petri nets, it becomes necessary to maintain control over the data transported through the relations. One way to overcome this issue is to fuse the connection elements of the net components so that the connecting arc is part of one of the net components, which permits adding inscriptions to that arc.

4.1   BPMN Semantics

Consider the BPMN process displayed in Fig. 2, that shows three tasks, where task a is being executed in parallel to task b and task c, which are mutually exclusively alternative.

There are many ways of defining the semantic mapping. Regarding BPMN, the specification itself [33] describes informal semantics for the execution of BPMN processes based on tokens. Therefore, the Petri net semantics – at least for a subset of BPMN constructs – is straight-forward and may be implemented using a mapping to Petri nets that was proposed by Dijkman et al. [12, p. 7] as displayed in Table 1. Covering the full BPMN standard with Petri net implementations is a challenge of its own that we do not try to resolve in the context of this work. Even this small selected subset of BPMN constructs leaves enough room

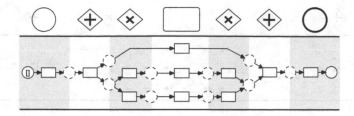

Fig. 3. Place bordered decomposition of the BPMN process.

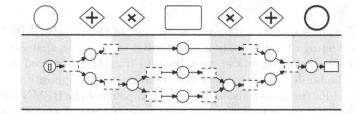

Fig. 4. Transition bordered decomposition of the BPMN process.

for interpretation and discussion about general concepts of Petri net mappings using semantic components.

Each of the Petri nets from Figs. 3 and 4 implements the BPMN process using a slightly different semantics. The vertical lines represent graphical cuts (not Petri net cuts) through the process, which indicate the fusion points between components (highlighted by dashed lines). The first Petri net implements the original mapping from Dijkman et al. [12]. It uses place bordered components and fuses the elements along the places. The sequence flow relations are mapped to places, which dissolve in the fusion points and have no individual semantics. The second net is very similar but uses transition bordered components. These components apply a mapping of sequence flow relations to transitions. Note how the semantics of the individual components slightly varies. For instance, the end event terminates with a marked place when applying the mapping from Dijkman et al. The BPMN standard prescribes that each token must eventually be removed from the process [33, p. 25], so the variant from Fig. 4 is more in conformance with the BPMN standard regarding this aspect. On the other hand, the second variant defines a task as a place surrounded by two transitions (incoming and outgoing), so its character is more like a state rather than an event or process. The BPMN execution semantics of a task is much more complex than this abstract interpretation, but it begins with a state (Inactive) and ends with a state (Closed) [33, p. 428].

There are many possible variants of the semantic mapping from Table 1. Each of the semantic components can be refined with additional net structure, as long as the bordering remains unaltered. The same applies to the relations, in case it becomes desirable to attach semantics to the relations in a similar way, this is done for the constructs.

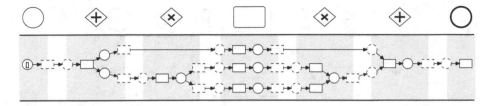

Fig. 5. A Petri net mapping of the BPMN process that consists of alternating semantic components.

Figure 5 again shows an alternative semantics of the presented BPMN process. The bordering of components is alternating in the sense that each BPMN construct is place bordered on the incoming and transition bordered on the outgoing side. The bordering of sequence flow relations is the opposite way around so that it provides the proper fusion points to complement the BPMN constructs. Following the BPMN standard, the outgoing sequence flows of a conditional gateway hold the condition inscriptions. With this mapping, it is possible to specify the conditions on the sequence flow. These could be transformed to guard expressions on the transitions of the mapped sequence flow, which are merged with the outgoing transitions of the conditional gateway in Fig. 5. This allows the mapping of the inscription to remain within the locality of the mapped construct. However, as we do not discuss inscriptions in detail in this contribution, this argument can be omitted. The alternating semantics has a different advantage for defining the highlighting in Sect. 6. Each component locally encapsulates its states and behavior. This property will be used to define two highlighting states "active" and "enabled" for the simulation of BPMN models.

4.2   Reference Nets as a Target Language

The (Java) Reference Net formalism[1] [23] is a high-level Petri net formalism that combines the nets-within-nets paradigm [37] with synchronous channels [9] and a Java inscription language. This formalism makes it possible to build complex systems using dynamic net hierarchies. The nets-within-nets concept is implemented using a reference semantics so that tokens can be references to nets. With the Java inscription language, it is possible to use Java objects as tokens and execute Java code during the firing of transitions. The synchronous channels enable the synchronous firing of multiple transitions distributed among multiple nets and a bidirectional exchange of information. An introduction to Reference Nets is available in the RENEW manual [25], the formal definition can be found in [23] (in German).

In comparison to the previously introduced bordering semantics with Reference Nets, additional variants become possible, such as the connection via virtual

---

[1] The first paragraph originates from one of our previously published contributions [30].

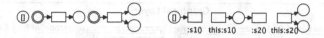

Fig. 6. Connection semantics of Reference Nets.

places or synchronous channels. Displayed on the left side in Fig. 6 are virtual places.

The double-edged place figures are virtual copies of their respective counterpart in the sense that the two places represent the same semantic place but with multiple graphical representations. These could be utilized to implement the place fusion. The synchronous channels displayed to the right are even more powerful. The synchronization of two transition instances supports the bidirectional exchange of information through unification of the channels. Besides supporting the possibility to move information along the edges, synchronous channels provide facilities to define interfaces to query and modify data objects. With syntactical constructs for net instantiation Reference Nets provide the capabilities of modeling dynamic hierarchies. In the context of BPMN, this is useful to implement hierarchies, such as pools and sub-processes. Corradini et al. [10] have identified that existing formalizations of BPMN do not sufficiently account for collaboration across multiple instances, especially regarding the exchange of data and data-based decisions. The authors provide a token-based operational semantics that focusses on these subjects and develop a visual editor that supports animation. According to this model, the synchronous channels of Reference Nets can be used for the formalization of data exchange in collaborative processes. It is something we have applied for quite some time [8]. The focus of this contribution is on the rapid prototypical development of DSML. The results are not yet transferred to BPMN.

Providing a complete formalization of BPMN is a challenge of its own. Van Gorp and Dijkman [14] and Kheldoun et al. [19] aim at covering large subsets of BPMN but still do not cover the full standard. Kheldoun et al. use high-level Petri nets in a similar fashion to our approach for the provision of a formal semantics but without execution support in favor of a verification focus.

5   Graphical Simulation Feedback for DSML

Up to now, there was no (sufficient) user feedback during the execution of models generated by the modeling technique. In this paper, we address the extension of our framework to enable simulation feedback from the underlying Petri net. The main idea is that within the domain models, the internal state (resp. marking) of the Petri net is reflected directly in the domain of the generated modeling technique. This allows for adaptive feedback individually depending on the transformational semantics for each generated modeling technique.

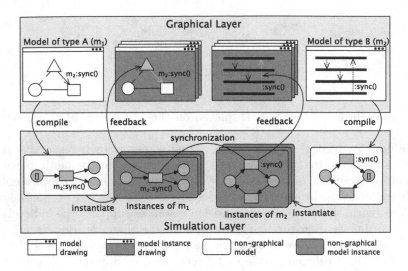

Fig. 7. Conceptual model of the model synchronization from [29, p. 8].

A conceptual image from the simulation of two modeling techniques that interact with each other is displayed in Fig. 7. The image originates from [29, p. 8], where we presented a concept for multi-formalism simulation with the synchronization of multiple modeling techniques based on Reference Nets. The presented solution sketched the idea of providing feedback into a DSML, but the realization was specifically for a finite automata modeling tool. With our current work, this idea is generalized to facilitate feedback for principally any DSML that is developed with the RMT approach, using model-driven development. This opens up the possibility to develop and research different simulation semantics or modes of simulation for these DSML.

The technical realization of our approach relies on RENEW as a simulation backend and the graphical infrastructure provided by the RMT framework. For the provision of graphical feedback in the execution of DSML, the framework is extended with a re-engineered transformation process. This process enables the backpropagation of events from the Petri net simulation and generic classes for graphical instance models that can change the representation depending on the simulation events and state.

The Petri net model displayed in Fig. 8 provides an overview of our solution to implementing graphical feedback for executable DSML. It represents the technical implementation of the compile and feedback steps from the abstract model depicted in Fig. 7. This involves the *Transformation* process (compile), the RENEW *Simulator* and the *Graphical Representation* (feedback) of the DSML model and model instances. A *graphical DSML model* created by a domain practitioner using a domain-specific tool (initially marked place in the *Graphical Representation* box) is the starting point for the simulation. When the simulation is initialized, executable Petri net models are generated through transformation,

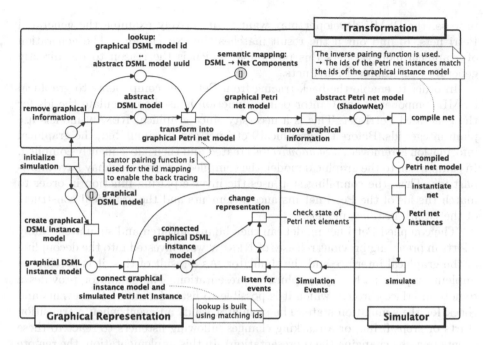

Fig. 8. Transformation and simulation process implementing the graphical feedback.

and the graphical representation classes are instantiated and linked with the simulated Petri nets. The connection between the graphical and simulated elements is established by matching their respective ids.

A prerequisite for the transformation is the provision of a semantic mapping like the one described in Sect. 4 (initially marked place *semantic mapping: DSML → Net Components*). The mapping has to be provided by the DSML developer, in the form of a referrer to the net components repository. Without additional configuration, the constructs are mapped to correspondingly named net components. This is a convenience feature to allow early testing with minimal configuration overhead. Many of the optional configuration parameters follow the *convention over configuration* principle. Alternatively, a semantic model can be used to configure, which construct is mapped to which net component (analogous to the tool configuration or stylesheet model described in Sect. 3).

The traceability of the constructs to properly facilitate the relation between the simulated elements and the graphical representations is a critical aspect of the transformation process. The first step converts the graphical representation of the DSML model into an abstract representation while a lookup is provided that enables the discovery of graphical elements for the abstract ones. Since the ids used in the graphical models are not persistent, the abstract DSML models use UUIDs.

Both, the abstract model and the lookup are combined with the provided semantic mapping to create a graphical Petri net model. During the development process, the DSML developer may want to iteratively examine the generated Petri nets, to find out if the result matches the expectations. The generation of graphical models offers this possibility, which is the reason to not directly generate non-graphical Petri nets.

In order to enable the back tracing from Petri net components to graphical DSML components, the Cantor pairing function is used to calculate the ids for the generated Petri net. This is a necessity since graphical RENEW models use plain integer ids. Before the graphical Petri net can be simulated, the graphical information is removed, a *ShadowNet* is created and then subsequently compiled. In comparison to the graphical model, the compiled net elements may have multipart ids. Thus, the compiling step uses the inverse pairing function in order to match the ids of the Petri net instance's elements and the graphical constructs of the DSML instance models.

The compiled Petri net model can then be instantiated and simulated. Many efforts in preparing previous releases of RENEW were invested into the decoupling of the graphical interface and the simulator. As a result of this, it is possible to implement principally any graphical representation. The simulator provides a remote interface through which it is possible to listen for simulation events and check for the simulation state. The simulator sends an event when a transition starts or stops firing, or a marking changes, allowing listeners to react to these events (e.g., by changing the representation). In this implementation, the remote interface is used by the components of the graphical representation to enable the graphical feedback for the simulation.

While the graphical DSML model is transformed into an executable Petri net, a graphical DSML instance model is also created. As soon as the Petri net instance is available from the simulator, it is connected with the graphical instance model. In this step, the graphical DSML constructs are connected to the simulated Petri net elements based on matching ids. The graphical DSML instance model, in turn, listens for simulation events that are related to the connected Petri net instance. When such an event occurs, the representation of the graphical DSML model is changed accordingly. The state-dependent representations may be arbitrarily implemented in Java by customizing the classes for the instance model. However, this requires knowledge about programming with Java and the details of the representation classes. In the following section, we propose three variants for the declarative description of graphical representations for the simulation of DSML.

6   Model-Based Configuration of the Representation

Two tasks summarize the specification of the representation for DSML models during execution, based on the state of the underlying simulated Petri net in a declarative fashion: The language developer has to specify the connection between the simulation state and the representation, in order to specify on which simulation event the representation of the constructs should be changed. Furthermore, the visualization of the highlighted constructs has to be defined, i.e., how the representation should be altered or extended to represent a specific simulation state.

6.1   Trade-Off Between Expressiveness and Simplicity

We identified some requirements that need to be fulfilled in order to adequately support the developer of a modeling language in these two tasks. Generally, there are two requirements: expressiveness and simplicity. First, the developer must be able to implement the desired result, i.e., has to be able to specify the exact representation of the executed model. Secondly, the implementation and configuration overhead should be minimal. Multiple factors have an impact on the simplicity. The connection between simulator and representation should be specifiable without knowledge of the internals of the simulator and optimally without programming skills. It should be possible to use the same tools as used for the representation of the static constructs to provide representations for highlighted constructs. Often the highlighted representation only minimally differs from the original representation (e.g., by having a different color). It should be possible to specify these slight variations without the requirement to provide multiple copies of the same figure. This is especially important for Petri net components with a high degree of concurrency. These may result in a large number of global states and thus a large number of different representations.

There is a trade-off between expressiveness and simplicity. Based on the identified requirements, we present three variants for realizing model-based simulation highlighting, each with a different focus regarding expressiveness and simplicity. The specification of the connection between simulation and representation may be achieved with an annotation language for the semantic components, which can be used for all of the three variants. For the provision of the altered representations, we propose the strategies simple, style sheet-based, and graphical component-based. With the simple highlighting method, the generic highlighting mechanism of RENEW is used. The style sheet-based mechanism uses style sheets analogously to the specification of the concrete syntax for the DSML constructs. With the annotated template-based highlighting, the representation of the highlighted constructs is provided by drawings that are created with the graphical editor of RENEW.

These three variants are exemplarily illustrated for the parallel gateway in Table 2. The table shows some representations that can be achieved with the different highlighting variants. The semantic component of the parallel gateway may have four different states that are depicted in the first row of the table. With the true concurrency simulation of RENEW there are actually more states, but these are omitted here and can be handled analogously (with the consideration of the state of a firing transition new states of the component emerge). The second row of the table contains the highlighted constructs for the simple highlighting strategy. Exemplary highlighting that can be achieved using the style sheet-based strategy is depicted in the third row. The fourth row shows state illustrations that can be produced with the graphical component-based strategy.

Table 2. Highlighting variants and example representations.

	activatable	activeboth	active1	active2
Semantic Component State				
simple				
stylesheet-based				
graphical component-based				

Before the three variants for *defining the visualization of the highlighted constructs* are described in detail, the basics for *specifying the connection between the simulation state and the representation* are presented in the following section.

6.2 Relating Simulation State and Representation

To obtain graphical feedback in the simulated DSML, one task defines the connection between the simulation events and the graphical representation. Since the underlying executable is a Petri net, the simulation events are net events. In the graphical simulation environment of RENEW it is possible to respond to mainly two types of events: the firing of a transition (with start and end of the firing) and the change of a marking of a place. Additionally, it is possible to check the state of a single net element such as the marking of a place and whether a transition is enabled or firing.

The graphical elements in the representation of the running model are linked to the simulator via the ids of the net elements in the sense that a net element in the simulation (a place or a transition) has exactly one connected graphical component that observes the net element and listens to its events. One graphical component may observe multiple net elements (this is a result of the 1:n mapping).

Depending on the DSML and the semantics, there are multiple possibilities to link simulation events and representations. An essential question in this context is, whether the classifiers and relations reflect the state of the places or the activities of the transitions. This depends mainly on the character of the DSML. For a language focusing on the behavior of a system (such as activity diagrams), it probably makes sense to target the transition events. A classifier, for example,

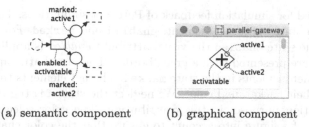

(a) semantic component   (b) graphical component

Fig. 9. Artifacts for simple highlighting. (Color figure online)

could be highlighted when one of the transitions in the corresponding semantic component is firing. A concentration on the place markings is more useful for a language with a strong state focus (such as state diagrams) where a classifier would be highlighted when a place in the semantic component is marked. Many languages (including BPMN) have a hybrid character, requiring different behaviors for different constructs.

Our approach facilitates the implementation of these hybrid languages in a simple way. The DSML developer can define the highlighting behavior for each DSML construct by using an annotation language that we present exemplarily for the parallel gateway. Figure 9a depicts the semantic component of the parallel gateway (as shown in Table 2) with annotations. For illustration purposes in this contribution, the connections between the annotation texts and the corresponding net elements (places and transitions) are presented as dotted lines. In our implementation in the RENEW environment, these annotations are net element inscriptions, which can be added with a specific text tool. These inscriptions resp. annotation texts are stored as attributes of each net element. The graphical interface can retrieve these attributes during simulation in order to determine the highlighting of the corresponding DSML's graphical components. An annotation contains two parts divided by a colon. The first part represents the simulation state or a simulation event for the particular net element, and the second part is a state concerning the whole BPMN component. In this contribution we name the first part *element state/event*, as it describes the state/event of the single element (a place or transition), and the second part *component state/event*, as it describes the state/event of the entire construct. For example, the annotation `marked:active1` in Fig. 9a indicates that the parallel gateway is in the state `active1` whenever the upper place contains a token. These component related states are not disjoint because the upper and the lower place may be marked at the same time, which results in the construct being in the state `active1` and `active2`. These states can now be used to specify the representation of the highlighted components.

6.3   Simple Highlighting

The simple highlighting strategy uses RENEW's generic highlighting mechanism where a color change highlights constructs and parts of constructs. This is already applied for simulation feedback of Petri net simulations. The mechanism reacts to two states of a net element: enabled and marked. For net elements that are in the state enabled the visual attribute *enabled* is applied. Thus, the net elements are presented with a green border. The visual attribute *highlighted* is applied to net elements in the state marked. The net elements therefor receive a change of their background color. We neglect the state firing of a transition for simplification purposes in this contribution. The color selection takes the original color of a figure into account to ensure that the color change is noticeable. This highlighting mechanism is also suitable for figures that are utilized in the RMT approach. For example, the highlighted representations of the parallel gateway construct, which can be achieved with the simple highlighting strategy, are included in the second row of Table 2.

Figure 9 shows the artifacts that are required to achieve the result in Table 2. In this case, the annotations correspond to the component states defined within the semantic component (activatable, active1 and active2). Together with the concrete syntax the highlighting information can be provided with a drawing that is created with RENEW – the graphical component. The graphical component is extended with annotations for the graphical elements (these annotations can be created analogously to the annotations for net elements as described in Sect. 6.2). The annotation activatable is connected to the diamond element of the gateway construct. This results in a green coloring of the border when the gateway component is in the state activatable, because of the usage of RENEW's generic mechanism the style class *enabled* is applied. Analogously, the two circles representing the ports receive a gray background in the states active1 or active2.

Of the three strategies, this is the easiest to implement for the DSML developer, but in return, it is limited in the sense that it is not possible to customize the highlighted representation.

6.4   Stylesheet-Based Highlighting

The problem with the simple highlighting strategy is that highlighting is limited to a change of colors, which are not selectable. The style sheet-based highlighting makes it possible to create custom style classes and, e.g., specify the highlighting color and other style parameters (such as line style/shape, arrow shape). Many attributes are predefined. With this strategy, representations like the ones depicted in the third row of Table 2 become possible.

Figure 10 depicts the artifacts required to achieve this variant of highlighting. Similar to the simple highlighting the DSML developer has to annotate the graphical component in order to specify which part of the figure to style (see Fig. 10b). However, the annotation syntax is extended with boolean expressions over the component states (active1 OR active2) and a style class that will be applied to the respective figure (e.g., bgc-green). In this way, the DSML developer is free to create custom styles according to her/his requirements. The definitions for the custom styles are depicted in Fig. 10c.

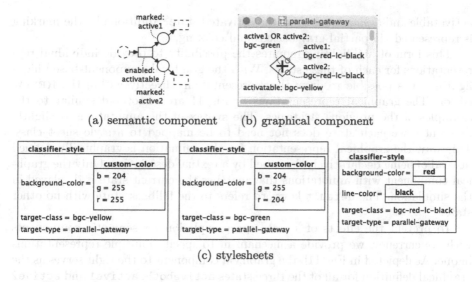

(a) semantic component   (b) graphical component

(c) stylesheets

Fig. 10. Artifacts for stylesheet-based highlighting. (Color figure online)

This variant provides more flexibility than the simple highlighting since the graphical representations may be customized with style sheets, but it requires more effort since these style sheets need to be explicitly defined. The style sheets-based highlighting is especially useful if the representation of the static models is already defined with style sheets. The amount of flexibility is still limited since the possibilities for customization are restricted to RENEW's predefined visual attributes.

6.5   Graphical Component-Based Highlighting

Sometimes the requirements for the highlighted representations exceed the possibilities of predefined style attributes, such as in the last row of Table 2 where the highlighted constructs are extended with additional graphical objects or images. In this example the constructs are amended with a *pause* symbol to represent the

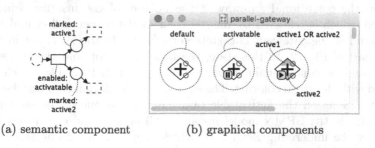

(a) semantic component   (b) graphical components

Fig. 11. Artifacts for graphical component-based highlighting.

activatable and a *play* symbol for the activated state. Additionally, the marking is represented via partial gray background coloring.

This form of highlighting requires the possibility to provide individual representations for each of these states. With the graphical component-based highlighting, it is possible to draw the representations directly within the RENEW editor. The graphical representations in Fig. 11 are annotated similar to the examples of the previous strategies. The syntax of the annotations is slightly different since each state does not need to be mapped to a style sheet class. This way, the graphical representation for the animation is graphically defined instead of being declaratively described by a textual description. Only the graphical component with annotations that matches the current state is displayed in the simulation. The `default` keyword refers to the fallback state with no other state.

To prevent the effects of a state space explosion for semantic components with concurrency, we provide a mechanism to specify multiple representations in one. As depicted in Fig. 11b the graphical component to the right serves as the graphical definition for all of the three states `activeboth`, `active1` and `active2` from Table 2.

Compared to the other two variants, this is the most flexible one when it comes to visualization. However, providing only the graphical component-based highlighting is not optimal because it requires a lot of modeling effort for a simple style or color change. In combination, the three presented highlighting strategies provide the possibility of quickly implementing graphical feedback for a DSML by selecting the most suitable variant depending on the requirements.

7   Simulation of the BPMN Example

Figure 12 shows a snapshot from the simulation of a BPMN model. The topmost part shows RENEW's main window with context menus, editor tool bars and the status bar. The two overlapping windows in the middle are the template (white background) and an instance (light blue background) of a BPMN process. The template drawing was modeled using the constructs from the BPMN toolbar.

The window on the right side contains an instance of this model and was created from that template. It represents the simulation state using the simple highlighting strategy. The simulation is paused in a state where the sequence flow and the conditional gateway at the bottom of the instance window are activated, which is reflected by the red color of the sequence flow and the gray background of the gateway. This state corresponds to the Petri net in the lowermost part of Fig. 12. In this state, the task at the top and the two sequence flows behind the conditional gateway (in conflict) are activatable, which is again reflected with the green coloring. The subsequent simulation step may be invoked by right-clicking on the activatable task figure or on one of the sequence flow connections in the BPMN model instance. All actions and executions are performed by the underlying Petri nets, which therefore determine the semantics of the domain specific language model while the interaction of the user is performed through the BPMN model. The behavior may be customized by providing

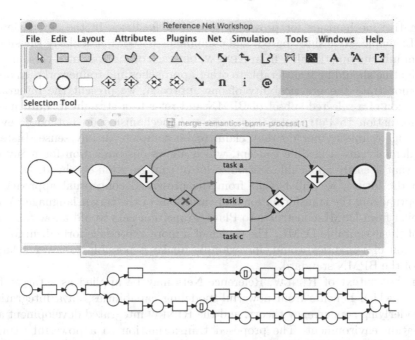

Fig. 12. Running BPMN simulation and the corresponding Petri net. (Color figure online)

alternative net components that may contain colored tokens, variables, inscriptions, synchronous channels, etc. The GUI interaction is provided with the RMT integration.

The Petri net in the lowermost part of the figure is the representation of the simulated net instance, which was generated using the semantic mapping from Sect. 4 (cf. Fig. 5). For the presentation in this paper, the Petri net model was created by hand. The generated Petri net that performs the simulation has no visual representation at all. This is a design decision in order to maintain the ability to execute these models without graphical feedback in server mode, which is essential to building large scale applications from net models. In all, this facilitates the provision of graphical feedback in the BPMN model by reflecting simulation events from the simulated (invisible) Petri net to the above layer.

8 Conclusion

In this contribution, we present a concept for providing simulation feedback for DSML that are developed with the RMT approach on the basis of meta-models and transformational semantics using Petri nets. Based on our new feature for direct visual feedback in the simulated domain specific model, we provide an improved experimentation environment for interactively experiencing with the behavior of newly designed domain specific languages without additional work on animating the models. The central part of our contribution consists of three alternative mechanisms for providing graphical feedback in the simulation of DSML: simple, style sheet-based, and component-based. Particularly the simple variant inherits functionality from RENEW, but each of the presented concepts for highlighting should be transferable to other approaches and frameworks. In order to demonstrate the practicability of our approach, we present the integrated simulation of a selected subset of BPMN and refer to a straight-forward model transformation to Petri nets. The presented mechanisms do not cover every modeling technique, nor do they claim to be complete in any sense. Instead, they demonstrate a flexible concept, which allows customization for many use cases that is easily applicable without a lot of configuration overhead.

In the future, we will benefit from the presented conceptual approach by conceptualizing the transformation and variations of the target language. A 1:n mapping from DSML constructs to Place/Transition nets would allow the analysis of the executable DSML. The choice of a more expressive formalism for the definition of semantics would, for example, make it possible to cover a larger part of the BPMN standard.

In the context of RENEW, Reference Nets may be applied as a target formalism, which benefits from powerful modeling capabilities, Java integration, the underlying concurrency theory, and the RENEW integrated development and simulation environment. The proposed transformation to a powerful (Turing complete) formalism is attractive on the one hand because the mentioned advantages of this formalism may be exploited. On the other hand, the possibilities to perform formal analysis are restricted due to the complexity of the formalism. The flexibility concerning the formalisms opens up the possibility of applying a whole array of methods from low-level analysis – e.g., using RENEW's integration of LoLA[^2] [16] – to normal software engineering validation like unit testing [39].

A graphical representation of analysis results that originate from the underlying Petri net in the DSML requires a connection to the applied analysis tools. When the graphical representation changes based on results of analysis tools instead of the simulation events, the visualization of structural properties, such as invariants, conflicts, concurrency, mutual exclusion, siphons, or traps becomes possible. In the future we plan to extend RENEW's capabilities for the analysis of domain-specific models and improved tool connectivity.

References

1. Abrahão, S., et al.: User experience for model-driven engineering: challenges and future directions. In: 20th ACM/IEEE International Conference on Model Driven Engineering Languages and Systems, MODELS 2017, Austin, TX, USA, 17–22 September 2017, pp. 229–236. IEEE Computer Society (2017). https://doi.org/10.1109/MODELS.2017.5
2. Biermann, E., Ehrig, K., Ermel, C., Hurrelmann, J.: Generation of simulation views for domain specific modeling languages based on the Eclipse modeling framework. In: 2009 IEEE/ACM International Conference on Automated Software Engineering, pp. 625–629, November 2009. https://doi.org/10.1109/ASE.2009.46

[^2]: LoLA: Low-Level Analyzer: http://www.service-technology.org/lola/.

3. Bousse, E., Degueule, T., Vojtisek, D., Mayerhofer, T., DeAntoni, J., Combemale, B.: Execution framework of the GEMOC studio (tool demo). In: van der Storm, T., Balland, E., Varró, D. (eds.) Proceedings of the 2016 ACM SIGPLAN International Conference on Software Language Engineering, Amsterdam, The Netherlands, 31 October–1 November 2016, pp. 84–89. ACM (2016), http://dl.acm.org/citation.cfm?id=2997384

4. Bousse, E., Leroy, D., Combemale, B., Wimmer, M., Baudry, B.: Omniscient debugging for executable DSLs. J. Syst. Softw. **137**, 261–288 (2017). https://hal.inria.fr/hal-01662336

5. Bryant, B.R., Gray, J., Mernik, M., Clarke, P.J., France, R.B., Karsai, G.: Challenges and directions in formalizing the semantics of modeling languages. Comput. Sci. Inf. Syst. **8**(2), 225–253 (2011). https://doi.org/10.2298/CSIS110114012B

6. Cabac, L.: Modeling Petri Net-Based Multi-Agent Applications, Agent Technology - Theory and Applications, vol. 5. Logos Verlag, Berlin (2010). http://www.logos-verlag.de/cgi-bin/engbuchmid?isbn=2673&lng=eng&id=. http://www.sub.uni-hamburg.de/opus/volltexte/2010/4666/

7. Cabac, L., Haustermann, M., Mosteller, D.: Renew 2.5 – towards a comprehensive integrated development environment for Petri net-based applications. In: Kordon, F., Moldt, D. (eds.) PETRI NETS 2016. LNCS, vol. 9698, pp. 101–112. Springer, Cham (2016). https://doi.org/10.1007/978-3-319-39086-4_7

8. Cabac, L., Haustermann, M., Mosteller, D.: Software development with Petri nets and agents: approach, frameworks and tool set. Sci. Comput. Program. **157**, 56–70 (2018). https://doi.org/10.1016/j.scico.2017.12.003

9. Christensen, S., Damgaard Hansen, N.: Coloured Petri nets extended with channels for synchronous communication. In: Valette, R. (ed.) ICATPN 1994. LNCS, vol. 815, pp. 159–178. Springer, Heidelberg (1994). https://doi.org/10.1007/3-540-58152-9_10

10. Corradini, F., Muzi, C., Re, B., Rossi, L., Tiezzi, F.: Animating multiple instances in BPMN collaborations: from formal semantics to tool support. In: Weske, M., Montali, M., Weber, I., vom Brocke, J. (eds.) BPM 2018. LNCS, vol. 11080, pp. 83–101. Springer, Cham (2018). https://doi.org/10.1007/978-3-319-98648-7_6

11. Cramer, B., Kastens, U.: Animation automatically generated from simulation specifications. In: 2009 IEEE Symposium on Visual Languages and Human-Centric Computing (VL/HCC), pp. 157–164, September 2009. https://doi.org/10.1109/VLHCC.2009.5295274

12. Dijkman, R.M., Dumas, M., Ouyang, C.: Semantics and analysis of business process models in BPMN. Inf. Softw. Technol. **50**(12), 1281–1294 (2008). https://doi.org/10.1016/j.infsof.2008.02.006

13. Eclipse Foundation Inc.: Eclipse Modeling Framework (EMF) (2018). https://www.eclipse.org/modeling/emf/. Accessed 24 May 2018

14. Gorp, P.V., Dijkman, R.M.: A visual token-based formalization of BPMN 2.0 based on in-place transformations. Inf. Softw. Technol. **55**(2), 365–394 (2013). https://doi.org/10.1016/j.infsof.2012.08.014

15. Hegedüs, Áb., Ráth, I., Varró, D.: Replaying execution trace models for dynamic modeling languages. Period. Polytech. Electr. Eng. Comput. Sci. **56**(3), 71–82 (2012). https://pp.bme.hu/eecs/article/view/7078

16. Hewelt, M., Wagner, T., Cabac, L.: Integrating verification into the PAOSE approach. In: Duvigneau, M., Moldt, D., Hiraishi, K. (eds.) Petri Nets and Software Engineering. International Workshop PNSE 2011, Newcastle upon Tyne, UK, June 2011. Proceedings. CEUR Workshop Proceedings, vol. 723, pp. 124–135. CEUR-WS.org, June 2011. http://ceur-ws.org/Vol-723/

17. Huber, P., Jensen, K., Shapiro, R.M.: Hierarchies in coloured Petri nets. In: Rozenberg, G. (ed.) ICATPN 1989. LNCS, vol. 483, pp. 313–341. Springer, Heidelberg (1991). https://doi.org/10.1007/3-540-53863-1_30

18. The GEMOC Initiative: The GEMOC initiative - breathe life into your designer! model simulation, animation and debugging with sirius animator, part of the GEMOC studio. http://gemoc.org/breathe-life-into-your-designer. Accessed 9 Oct 2018

19. Kheldoun, A., Barkaoui, K., Ioualalen, M.: Formal verification of complex business processes based on high-level Petri nets. Inf. Sci. **385**, 39–54 (2017). https://doi.org/10.1016/j.ins.2016.12.044

20. Kindler, E.: ePNK applications and annotations: a simulator for YAWL nets. In: Khomenko, V., Roux, O.H. (eds.) PETRI NETS 2018. LNCS, vol. 10877, pp. 339–350. Springer, Cham (2018). https://doi.org/10.1007/978-3-319-91268-4_17

21. Kleppe, A.: Software Language Engineering: Creating Domain-Specific Languages Using Metamodels. Pearson Education, London (2008). https://dl.acm.org/citation.cfm?id=1496375

22. Kolovos, D.S., Paige, R.F., Polack, F.A.C.: The epsilon transformation language. In: Vallecillo, A., Gray, J., Pierantonio, A. (eds.) ICMT 2008. LNCS, vol. 5063, pp. 46–60. Springer, Heidelberg (2008). https://doi.org/10.1007/978-3-540-69927-9_4

23. Kummer, O.: Referenznetze. Logos Verlag, Berlin (2002). http://www.logos-verlag.de/cgi-bin/engbuchmid?isbn=0035&lng=eng&id=

24. Kummer, O., Wienberg, F., Duvigneau, M., Cabac, L., Haustermann, M., Mosteller, D.: Renew - the Reference Net Workshop, June 2016. http://www.renew.de/. release 2.5

25. Kummer, O., Wienberg, F., Duvigneau, M., Cabac, L., Haustermann, M., Mosteller, D.: Renew - User Guide (Release 2.5). University of Hamburg, Faculty of Informatics, Theoretical Foundations Group, Hamburg, June 2016

26. de Lara, J., Vangheluwe, H., Alfonseca, M.: Meta-modelling and graph grammars for multi-paradigm modelling in atom3. Softw. Syst. Model. **3**(3), 194–209 (2004). https://doi.org/10.1007/s10270-003-0047-5

27. Lohmann, N., Verbeek, E., Dijkman, R.M.: Petri net transformations for business processes - a survey. Trans. Petri Nets Other Models Concurr. **2**, 46–63 (2009)

28. Mayerhofer, T., Combemale, B.: The tool generation challenge for executable domain-specific modeling languages. In: Seidl, M., Zschaler, S. (eds.) STAF 2017. LNCS, vol. 10748, pp. 193–199. Springer, Cham (2018). https://doi.org/10.1007/978-3-319-74730-9_18

29. Möller, P., Haustermann, M., Mosteller, D., Schmitz, D.: Simulating multiple formalisms concurrently based on reference nets. In: Moldt, D., Cabac, L., Rölke, H. (eds.) Petri Nets and Software Engineering. International Workshop, PNSE 2017, Zaragoza, Spain, 25–26 June 2017. Proceedings. CEUR Workshop Proceedings, vol. 1846, pp. 137–156. CEUR-WS.org (2017). http://CEUR-WS.org/Vol-1846/

30. Möller, P., Haustermann, M., Mosteller, D., Schmitz, D.: Model synchronization and concurrent simulation of multiple formalisms based on reference nets. Trans. Petri Nets Other Models Concurr. XIII **13**, 93–115 (2018). https://doi.org/10.1007/978-3-662-58381-4_5

31. Mosteller, D., Cabac, L., Haustermann, M.: Integrating Petri net semantics in a model-driven approach: the Renew meta-modeling and transformation framework. Trans. Petri Nets Other Models Concurr. XI **11**, 92–113 (2016). https://doi.org/10.1007/978-3-662-53401-4_5

32. Mosteller, D., Haustermann, M., Moldt, D., Schmitz, D.: Graphical simulation feedback in Petri net-based domain-specific languages within a meta-modeling environment. In: Moldt, D., Kindler, E., Rölke, H. (eds.) Petri Nets and Software Engineering. International Workshop, PNSE 2018, Bratislava, Slovakia, 25–26 June 2018. Proceedings. CEUR Workshop Proceedings, vol. 2138, pp. 56–75. CEUR-WS.org (2018). http://ceur-ws.org/Vol-2138/
33. OMG: Object Management Group: Business Process Model and Notation (BPMN) - Version 2.0.2 (2013). http://www.omg.org/spec/BPMN/2.0.2
34. Pedro, L., Lucio, L., Buchs, D.: System prototype and verification using metamodel-based transformations. IEEE Distrib. Syst. Online **8**(4) (2007). https://doi.org/10.1109/MDSO.2007.22
35. Rybicki, F., Smyth, S., Motika, C., Schulz-Rosengarten, A., von Hanxleden, R.: Interactive model-based compilation continued – incremental hardware synthesis for SCCharts. In: Margaria, T., Steffen, B. (eds.) ISoLA 2016. LNCS, vol. 9953, pp. 150–170. Springer, Cham (2016). https://doi.org/10.1007/978-3-319-47169-3_12
36. Sedrakyan, G., Snoeck, M.: Enriching model execution with feedback to support testing of semantic conformance between models and requirements - design and evaluation of feedback automation architecture. In: Calabrò, A., Lonetti, F., Marchetti, E. (eds.) Proceedings of the International Workshop on domAin specific Model-based AppRoaches to vErificaTion and validaTiOn, AMARETTO@MODELSWARD 2016, Rome, Italy, 19–21 February 2016, pp. 14–22. SciTePress (2016). https://doi.org/10.5220/0005841800140022
37. Valk, R.: Petri nets as token objects. In: Desel, J., Silva, M. (eds.) ICATPN 1998. LNCS, vol. 1420, pp. 1–24. Springer, Heidelberg (1998). https://doi.org/10.1007/3-540-69108-1_1
38. Westergaard, M.: Access/CPN 2.0: a high-level interface to coloured petri net models. In: Kristensen, L.M., Petrucci, L. (eds.) PETRI NETS 2011. LNCS, vol. 6709, pp. 328–337. Springer, Heidelberg (2011). https://doi.org/10.1007/978-3-642-21834-7_19
39. Wincierz, M.: A tool chain for test-driven development of reference net software components in the context of CAPA agents. In: Moldt, D., Cabac, L., Rölke, H. (eds.) Petri Nets and Software Engineering. International Workshop, PNSE 2017, Zaragoza, Spain, 25–26 June 2017. Proceedings. CEUR Workshop Proceedings, vol. 1846, pp. 197–214. CEUR-WS.org (2017). http://CEUR-WS.org/Vol-1846/

Formal Modelling and Incremental Verification of the MQTT IoT Protocol

Alejandro Rodríguez[(✉)], Lars Michael Kristensen, and Adrian Rutle

Department of Computing, Mathematics, and Physics,
Western Norway University of Applied Sciences, Bergen, Norway
{arte,lmkr,aru}@hvl.no

Abstract. Machine to Machine (M2M) communication and Internet of Things (IoT) are becoming still more pervasive with the increase of communicating devices used in cyber-physical environments. A prominent approach to communication between distributed devices in highly dynamic IoT environments is the use of publish-subscribe protocols such as the Message Queuing Telemetry Transport (MQTT) protocol. MQTT is designed to be light-weight while still being resilient to connectivity loss and component failures. We have developed a Coloured Petri Net model of the MQTT protocol logic using CPN Tools. The model covers all three quality of service levels provided by MQTT (at most once, at least once, and exactly once). For the verification of the protocol model, we show how an incremental model checking approach can be used to reduce the effect of the state explosion problem. This is done by exploiting that the MQTT protocol operates in phases comprised of connect, subscribe, publish, unsubscribe, and disconnect.

1 Introduction

Publish-subscribe messaging systems support data-centric communication and have been widely used in enterprise networks and applications. A main reason for this is that a software system architecture based on publish-subscribe messaging provides better support for scalability and adaptability than the traditional client-server architecture used in distributed systems. The interaction and exchange of messages between clients based on the publish-subscribe paradigm are based on middleware usually referred to as a *broker* (or a bus) that manages *topics*. The broker provides space decoupling [9] allowing a client acting as a publisher on a given topic to send messages to other clients acting as subscribers to the topic without the need to know the identity of the receiving clients. The broker also provides synchronisation decoupling in that clients can exchange messages without being executing at the same time. Furthermore, the processing in the broker can be parallelized and handled using event-driven techniques.

The loose coupling and support for asynchronous point-to-multipoint messaging, make the publish-subscribe paradigm attractive also in the context of Internet of Things (IoT) which has experienced significant growth in applications and adoptability in recent years [17]. The IoT paradigm blends the virtual

© Springer-Verlag GmbH Germany, part of Springer Nature 2019
M. Koutny et al. (Eds.): ToPNoC XIV, LNCS 11790, pp. 126–145, 2019.
https://doi.org/10.1007/978-3-662-60651-3_5

and the physical worlds by bringing different concepts and technical components together: pervasive networks, miniaturisation of devices, mobile communication, and new ecosystems [6]. Moreover, the implementation of a connected product typically requires the combination of multiple software and hardware components distributed in a multi-layer stack of IoT technologies.

MQTT [3] is a publish-subscribe messaging protocol for IoT designed with the aim of being light-weight and easy to implement. These characteristics make it a suitable candidate for constrained environments such as Machine-to-Machine communication (M2M) and IoT contexts where a small memory footprint is required and where network bandwidth is often a scarce resource. Even though MQTT has been designed to be easy to implement, it still contains relatively complex protocol logic for handling connections, subscriptions, and the various quality of service levels related to message delivery. Furthermore, MQTT is expected to play a key role in future IoT applications and will be implemented for a wide range of platforms and in a broad range of programming languages making interoperability a key issue. This, combined with the fact that MQTT is only backed by an (ambiguous) natural language specification, motivated us to develop a formal and executable specification of the MQTT protocol.

We have used Coloured Petri Nets (CPNs) [12] for the modelling and verification of the MQTT specification. The main reason is that CPNs have been successfully applied in earlier work to build formal specifications of communication protocols [8], data networks [5], and embedded systems [1]. To ensure the proper operations of the constructed CPN model, we have validated the CPN model using simulation and verified an elaborate set of behavioural properties of the constructed model using model checking and state space exploration. In the course of our work on the MQTT specification [3] and the development of the CPN model, we have identified a number of issues related in particular to the implementation of the quality of service levels. These issues are a potential source of interoperability problems between implementations. For the construction of the model we have applied some general modelling patterns for CPN models of publish-subscribe protocols. Compared to earlier work on modelling and verification of publish-subscribe protocols [4,10,18] (which we discuss in more details towards the end of this paper) our work specifically targets MQTT, and we consider a more extensive set of behavioural properties.

The rest of this paper is organised as follows. In Sect. 2 we present the MQTT protocol context and give a high-level overview of the constructed CPN model. Section 3 details selected parts of the CPN model of the MQTT protocol. In Sect. 4 we present our experimental results on using simulation and model checking to validate and verify central properties of MQTT and the CPN model. Finally, in Sect. 5 we sum up the conclusions, discuss related work, and outlines directions for future work. Due to space limitations, we cannot present the complete CPN model of the MQTT protocol. The constructed CPN model is available via [15]. The reader is assumed to be familiar with the basic concepts of Petri Nets and High-level Petri Nets [12].

2   MQTT Protocol and CPN Model Overview

MQTT [3] runs over the TCP/IP protocol or other transport protocols that
provide ordered, lossless and bidirectional connections. MQTT applies topic-
based filtering of messages with a topic being part of each published message.
An MQTT client can subscribe to a topic to receive messages, publish on a
topic, and clients can subscribe to as many topics as they are interested in. As
described in [14], an MQTT client can operate as a publisher or subscribe and we
use the term client to generally refer to a publisher or a subscriber. The MQTT
broker [14] is the core of any publish/subscribe protocol and is responsible for
keeping track of subscriptions, receiving and filtering messages, deciding to which
clients they will be dispatched, and sending them to all subscribed clients. There
are no restrictions in terms of hardware to run as an MQTT client, and any
device equipped with an MQTT library and connected to an MQTT broker can
operate as a client.

2.1   Modelling Roles and Messages

Figure 1 shows the top-level module of the CPN MQTT model which consists of
two *substitution transitions* (drawn as rectangles with double-lined borders) representing the Clients and the Broker roles of MQTT. Substitution transitions constitute the basic syntactical structuring mechanism of CPNs and each of the substitution transitions has an associated *module* that models the detailed behaviour
of the clients and the broker, respectively. The name of the (sub)module associated with a substitution transition is written in the rectangular tag positioned
next to the transition.

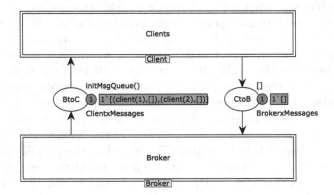

Fig. 1. The top-level module of the MQTT CPN model.

The complete CPN model of the MQTT protocol consists of 24 modules
organised into six hierarchical levels. We have constructed a parametric CPN
model which makes it easy to change the number of clients and topics without

making changes to the net-structure. This makes it possible to investigate different configuration of MQTT and it is a main benefit provided by CPNs in comparison to ordinary Petri Nets.

The two substitution transitions in Fig. 1 are connected via directed arcs to the two places CtoB and BtoC. The clients and the broker interact by producing and consuming tokens on the places. Figure 2 shows the central data type definitions used for the colour sets of the places CtoB and BtoC and the modelling of clients and messages. The colour sets QoS is used for modelling the three quality of service levels supported by MQTT (see below), and the colour set PID is used for modelling the packet identifiers which plays a central role in the MQTT protocol logic. We have abstracted from the actual payload of the published messages. The reason for this is that the message transport structure and the protocol logic of MQTT is agnostic to the payload contained, i.e., the actual content that will be sent in the messages. For similar reasons, we also abstract from the hierarchical structuring of topics.

```
val T = 5;   (* number of topics *)
val C = 2;   (* number of clients *)

colset Client = index client with 1..C;
colset Topic  = index topic with 1..T;
colset QoS    = index QoS with 0..2; (* quality of service *)
colset PID    = INT;                 (* packet identifiers *)

colset TopicxPID      = product Topic * PID;
colset TopicxQoSxPID  = product Topic * QoS * PID;

colset Message = union CONNECT + CONNACK +
   SUBSCRIBE : TopicxQoSxPID + UNSUBSCRIBE : TopicxPID +
   SUBACK    : TopicxQoSxPID + UNSUBACK     : TopicxPID
   PUBLISH   : TopicxQoSxPID +
   PUBACK    : TopicxPID     + PUBREC       : TopicxPID +
   PUBREL    : TopicxPID     + PUBCOMP      : TopicxPID +
   DISCONNECT;

colset Messages = list Message;

colset ClientxMessage       = product Client * Message;
colset BrokerxMessages      = list ClientxMessage;

colset ClientxMessageQueue = product Client * Messages;
colset ClientxMessages     = list ClientxMessageQueue;
```

Fig. 2. Client and message colour set definitions (Color figure online)

The places CtoB and BtoC are designed to behave as queues. The queue mechanism offers some advantages that the MQTT specification implicitly indicates. The purpose of this is to ensure the ordered message distribution as assumed from the transport service on top of which MQTT operates. Even so, the CtoB and BtoC places are slightly different; while CtoB is modelled as a single queue that the broker manages to consume messages, BtoC is designed to maintain an incoming queue of messages for each client. This construction assures that all clients will have their own queue, individually respecting the ordered reception of messages. The function initMsgQueue() initialises the queues according to the number of clients specified by the symbolic constant C. The BrokerxMessages colour set for the CtoB place used at the bottom of Fig. 2 consists of a list of ClientxMessage which are pairs of Client and Messages.

We represent all the messages that the clients and the broker can use by means of the Message colour set. We use the terms packet and message indistinguishably when we refer to control packets. The control information used depends on the messages considered. As an example, a Connect message (packet) does not contain control information, but a Publish message requires a specific Topic, QoS, and PID. The Topic and QoS colour sets are both indexed types containing values (topic(1), topic(2) ... topic(T) depending on the constant T, and QoS(0), QoS(1) and QoS(2), respectively. The ClientxMessages colour set for the BtoC place encapsulates all the queues (each one declared as a pair of Client and Messages in the ClientxMessageQueue colour set) in one single queue. This modelling pattern allows us to deal with the distribution of multiple messages in a single step in the broker side which in turn simplifies the modelling of the broker and reduces the number of reachable states of the model.

2.2 Quality of Service

The MQTT protocol delivers application messages according to the three Quality of Service (QoS) levels defined in [3]. The QoS levels are motivated by the different needs that IoT applications may have in terms of reliable delivery of messages. It should be noted that even if MQTT has been designed to operate over a transport service with lossless and ordered delivery, then message reliability still must be addressed as logical transport connections may be lost.

The delivery protocol is symmetric, and the clients and the broker can each take the role of either a sender or a receiver. The delivery protocol is concerned solely with the delivery of an application message from a single sender to a single receiver. When the broker is delivering an application message to more than one client, each client is treated independently. The QoS level used to deliver an outbound message from the broker could differ from the QoS level designated in the inbound message. Therefore, we need to distinguish two different parts of delivering a message: a client that publishes to the broker and the broker that forwards the message to the subscribing clients. The three MQTT QoS levels for message delivery are:

At most once: (QoS level 0): The message is delivered according to the capabilities of the underlying network. No response is sent by the receiver and no retry is performed by the sender. The message arrives at the receiver either once or not at all. An application of this QoS level is in environments where sensors are constantly sending data and it does not matter if an individual reading is lost as the next one will be published soon after.

At least once (QoS level 1): Where messages are assured to arrive, but duplicates can occur. It fits adequately for situations where delivery assurance is required but duplication will not cause inconsistencies. An application of this are idempotent operations on actuators, such as closing a valve or turning on a motor.

Exactly once (QoS level 2): Where messages are assured to arrive exactly once. This is for use when neither loss nor duplication of messages are acceptable. This level could be used, for example, with billing systems where duplicate or lost messages could lead to incorrect charges being applied.

When a client subscribes to a specific topic with a certain QoS level it means that the client is determining the maximum QoS that can be expected for that topic. When the broker transfers the message to a subscribing client it uses the QoS of the subscription made by the client. Hence QoS guarantees can get downgraded for a particular receiving client if subscribed with a lower QoS. This means that if a receiver is subscribed to a topic with a QoS level 0, no matter if a sender publishes in this topic with a QoS level 2, then the receiver will proceed with its QoS level 0.

3 Modelling the Protocol Roles and Their Interaction

We now consider the different phases and the client-broker interaction in the MQTT protocol, and show how we have modelled the MQTT protocol logic using CPNs. MQTT defines five main operations: connect, subscribe, publish, unsubscribe, and disconnect. Such operations, except the connect which must be the first one for the clients, are independent of each other and can be triggered in parallel by either the clients or the broker. The model is organized following a modelling pattern that ensures modularity and therefore, encapsulation of the protocol logic and behaviour of such operations. This offers advantages both for readability and understandability of the model and also, for making easier to detect and fix errors during the incremental verification.

3.1 Interaction Overview

In order to show how the clients and the broker interact, we describe the different actions that clients may carry out by considering an example. Figure 3 shows a sequence diagram for a scenario where two clients connect, perform subscribe, publish and unsubscribe, and finally disconnect from the broker. The numbers on each step of the communication define the interaction of the protocol as follows:

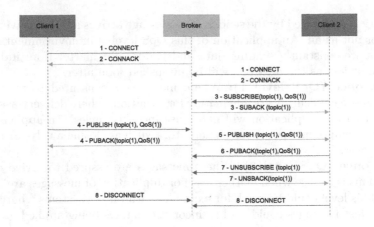

Fig. 3. Message sequence diagram illustrating the MQTT phases.

1. Client 1 and Client 2 request a connection to the Broker.
2. The Broker sends back a connection acknowledgement to confirm the establishment of the connection.
3. Client 2 subscribes to topic 1 with a QoS level 1, and the Broker confirms the subscription with a subscribe acknowledgement message.
4. Client 1 publishes on topic 1 with a QoS level 1. The Broker responds with a corresponding publish acknowledgement.
5. The Broker transmits the publish message to Client 2 which is subscribed to the topic.
6. Client 2 gets the published message, and sends a publish acknowledgement back as a confirmation to the Broker that it has received the message.
7. Client 2 unsubscribes to topic 1, and the Broker responds with an unsubscribe acknowledgement.
8. Client 1 and Client 2 disconnect.

3.2   Client and Broker State Modelling

The colour sets defined for modelling the client state are shown in Fig. 4. The place Clients (top-left place in Fig. 5) uses a token for each client to store their respective state during the communication. This is a modelling pattern that allows not only to parameterize the model so we can change the number of clients without modifying the structure, but also to maintain all the clients independently in only one place and with a proper data structure that encapsulates all the information required. The states of the clients are represented by the `ClientxState` colour set which is a product of `Client` and `ClientState`. The record colour set `ClientState` is used to represent the state of a client which consists of a list of `TopicxQoS`, a `State`, and a `PID`. Using this, a client stores the topics it is subscribed to, and the quality of service level of each subscription.

```
colset State = with READY | DISC | CON | WAIT;

colset TopicxQoS     = product Topic * QoS;
colset ListTopicxQoS = list TopicxQoS;

colset ClientState  = record topics : ListTopicxQoS *
                             state  : State *
                             pid    : PID;

colset ClientxState = product Client * ClientState;
```

Fig. 4. Colour set definitions used for modelling client state. (Color figure online)

The State colour set is an enumeration type containing the values READY (for the initial state), WAIT (when the client is waiting to be connected), CON (when the client is connected), and DISC (for when the client has disconnected).

Below we present selected parts of the model by first presenting a high-level view of the clients and broker sides, and then illustrating how the model captures the execution scenario described in Sect. 3.1 where two clients connects, one subscribes to a topic, and the other client publishes on this topic. The unsubscribe and the disconnection phases are not detailed due to space limitations.

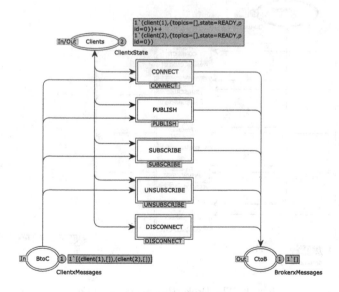

Fig. 5. ClientProcessing submodule.

3.3   Client Modelling

The ClientProcessing submodule in Fig. 5 models all the operations that a client can carry out. Clients can behave as senders and receivers, and the five substitution transitions CONNECT, PUBLISH, SUBSCRIBE, UNSUBSCRIBE and DISCONNECT has been constructed to capture both behaviours.

The socket place Clients stores the information of all the clients that are created at the beginning of the execution of the model. In this scenario there are two clients, and the value of the tokens representing the state of the two clients is provided in the green rectangle (the marking of the place) next to the Clients place. The BtoC and CtoB port places are associated with the socket places already shown in Fig. 2.

3.4   Broker Modelling

We have modelled the broker similarly as we have done for clients. This can be seen from Fig. 6 which shows the BrokerProcessing submodule. The Connected-Clients place keeps the information of all clients as perceived by the broker. This place is designed as a central storage, and it is used by the broker to distribute the messages over the network. The broker behaviour is different from that of the clients, since it will have to manage all the requests and generate responses for several clients at the same time.

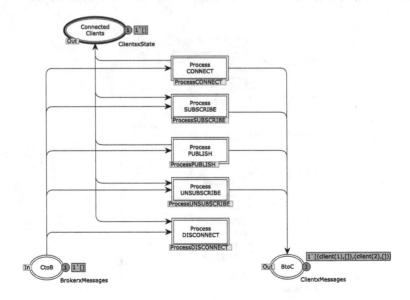

Fig. 6. The BrokerProcessing module.

3.5   Connection Phase

The first step for a client to be part of the message exchange is to connect to the broker. A client will send a CONNECT request, and the broker will respond with a CONNACK message to complete the connection establishment. Figure 7 shows the CONNECT submodule in a marking where client(1) has sent a CONNECT request and it is waiting (state = WAIT) for the broker acknowledgement processing to finish such that the connection state can be set to CON.

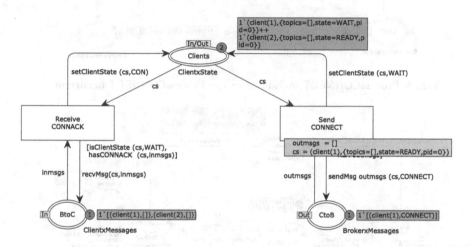

Fig. 7. CONNECT module after the sendCONNECT occurrence.

The broker will receive the CONNECT request. The broker will register the client in the place ConnectedClients and send back the acknowledgement. Figure 8 shows the situation where client(1) is connected in the broker side and the CONNACK response has been sent back to the client. The function connectClient() used on the arc from the ProcessCONNECT transition to the ConnectedClients place will record the connected client on the broker side. The last step of the connection establishment will occur again in the clients side, where the transition ReceiveCONNACK (in Fig. 7) will be enabled, meaning that the confirmation for the connection of client(1) can proceed.

3.6   Subscription Phase

Starting from the point where both clients are connected (i.e., for both clients, the state is CON as shown at the top of Fig. 9), client(2) will send a SUBSCRIBE request to topic(1) with QoS(1). The place PendingAcks represents a queue that each client maintains to store the PIDs that are waiting to be acknowledged. In this example, the message has assigned a PID = 0, and client(2) is waiting for an acknowledgement to this subscription with a PID = 0. When a client receives a SUBACK (subscribe acknowledgement) it will check that the packet identifier (0

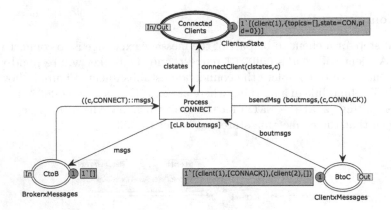

Fig. 8. ProcessCONNECT module after the ProcessCONNECT occurrence.

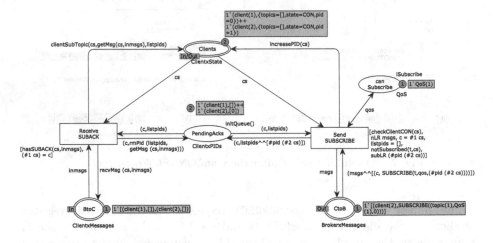

Fig. 9. SUBSCRIBE module after the SUBSCRIBE occurrence.

in this case) is the same to ensure that the correct packet is being received. At the bottom right side of the Fig. 9, the message has been sent to the broker.

We show now the situation where the SUBSCRIBE request has been processed by the broker as represented in Fig. 10. The function brokerSubscribeUpdate() manages the subscription process, so if the client is subscribing to a new topic, it will be added to the client state stored in the broker. If the client is already subscribed to this topic it will update it. In the example, one can see that client(1) keeps the same state, but client(2) has appended this new topic to its list. The corresponding SUBACK message has been sent to client(2) (with the PID set to 0) to confirm the subscription. Next, client(2) will detect that the response has arrived and it will check that the packet identifiers correspond to each other.

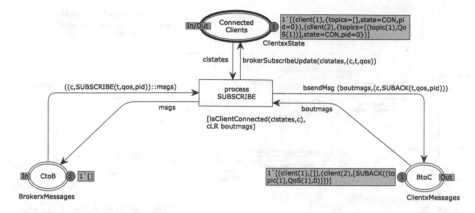

Fig. 10. ProcessSUBSCRIBE module after occurrence of ProcessSUBSCRIBE.

3.7 Publishing Phase

The publishing process in the considered scenario requires two steps to be completed. First a client sends a PUBLISH in a specific topic, with a specific QoS, which is received by broker. The broker will answer back with the corresponding acknowledgement, depending on the quality of service previously set. Second, the broker, that stores information for all clients, will propagate the PUBLISH sent by the client to any clients subscribing to that topic. We have modelled the clients and broker sides using different submodules depending on the quality of service that is being applied for sending and receiving. In our example, client(1) will publish in topic(1) with a QoS(1). This means that the broker will acknowledge back with a PUBACK to client(1), and will create a PUBLISH message for client(2), which is subscribed to this topic with a QoS(1). In this case, there is no downgrading for the client(2), so the publication process will be similar to step 1, i.e, client(2) will send back a PUBACK to the broker.

Figure 11 shows the situation in the model where client(1) has sent a PUBLISH with a QoS(1) for the topic(1). Similar to the subscription process, the place CtoB holds the message that the broker will receive, and the place Publishing keeps the information (PID and topic in this case) of the packet that needs to be acknowledged. The transition TimeOut models the behaviour for the re-transmission of packets. Quality of service level 1 assures that the message will be received at least once. The TimeOut transition will be enabled to re-send the message until the client has received the acknowledgement from the broker.

The Broker module models the logic for both receiver and sender behaviours. Figure 12 shows a marking corresponding to the state where the broker has processed the PUBLISH request made by client(1), and it has generated both the answer to this client and the PUBLISH message for client(2) (in this case, only one client is subscribed to the topic). The port place BPID (Broker PID), at top right of Fig. 12, will hold a packet identifier for each message that the broker re-publishes to the clients. The port place Publishing keeps information

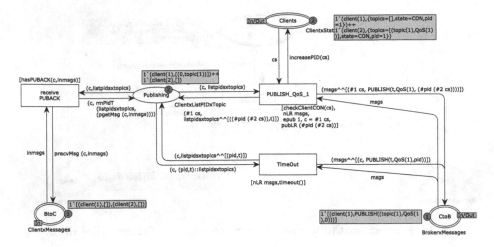

Fig. 11. PUBLISH_QoS_1 module after the PUBLISH_QoS_1 occurrence.

for all the clients that will acknowledge back the publish messages transmitted by the broker. Again, a TimeOut is modelled which, in this case, creates PUBLISH messages for all the clients subscribed to the topic in question. In the BtoC place (bottom right of Fig. 12), one can see that both messages have been sent, one for the original sender client(1) (PUBACK packet), and one for the only receiver client(2) (PUBLISH packet).

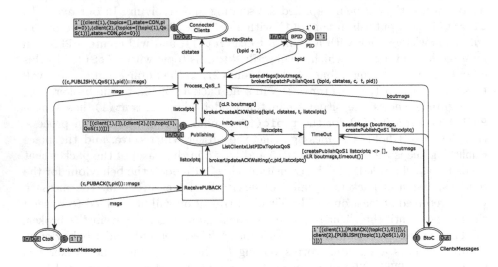

Fig. 12. Process_QoS_1 module after the Process_QoS_1 occurrence.

To finish the process, client(2) will notice that there has been a message published in topic(1). Since client(2) is subscribed to this topic with QoS(1), it must send a PUBACK acknowledgement to the broker to confirm that it has received the published message. Figure 13 shows the Receive_QoS_1 submodule in the clients side. The transition Receive_QoS_1 has been fired meaning that client(2) has received the publish message from the broker, and has sent the corresponding PUBACK. When the broker detects the incoming PUBACK message, it will check if there is some confirmation pending in the Publishing place (in Fig. 12 where client(2) is waiting for a PID = 0 in topic(1) with QoS(1).

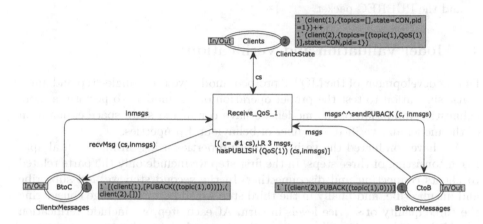

Fig. 13. Receive module after the transition Receive_QoS_1 occurrence

3.8 Findings

In the course of constructing the CPN model based on the informal MQTT specification, we encountered several parts that were vaguely defined and which could lead developers to obtain different implementations. The most significant issues are detailed below.

- There is a gap in the specification related to the MQTT protocol being described to run over TCP/IP, or other transport protocols that provide ordered, lossless and bidirectional connections. The QoS level 0 description establishes that message loss can occur, but the specification is not clear as to whether this is related to termination of TCP connections and/or clients connecting and disconnecting from the broker.
- It is specified that the receiver (assuming the broker role) is not required to complete delivery of the application message before sending the PUBACK (for QoS1) or PUBREC or PUBCOMP (for QoS2) and the decision of handling this is up to the implementer. For instance, in the case of QoS level 2, the specification provides two alternatives with respect to forward the publish request to the subscribers: (1) The broker will forward the messages when it receives the PUBLISH from the original sender; or (2) The broker will

forward the messages after the reception of the PUBREL from the original sender. Even it is assured that either choice does not modify the behaviour of the QoS level 2, this could lead to different implementation decisions and therefore consequent interoperability problems.

- The documentation specifies that when the original sender receives the PUB-ACK packet (with a QoS level 1), ownership of the application message is transferred to the receiver. It is unclear how to determine that the original sender has received the PUBACK packet. The same applies for QoS level 2 and the PUBREC packet.

4   Model Validation and Verification

During development of the MQTT protocol model we used single-step and automatic simulation to test the proper operation of the model. To perform a more exhaustive validation of the model, we have conducted state space exploration of the model and verified a number of behavioural properties.

We have conducted the verification of properties using an incremental approach consisting of three steps. In the first step we include only the parts related to clients connecting and disconnecting. In the second step we add subscribe and unsubscribe, and finally in the third step we add data exchange considering the three quality of service levels in turn. At each step, we include verification of additional properties. The main motivation underlying this incremental approach is to be able to control the effect of the state explosion problem. Errors in the model will often manifest themselves in small configurations and leading to a very large state space. Hence, by incrementally adding the protocol features, we can mitigate the effect of this phenomenon. We identified several modelling errors in the course of conducting this incremental model validation based on the phases of the MQTT protocol.

In addition, we have developed a mechanism to be able to explore different scenarios and check the behavioural properties against them fully automatically. This has been done by providing the model with a set of parameterized options, which we can easily change. This feature allows us to first modify add or remove new configurations, and secondly to run them automatically. For each new modification in the parameters, we always run the six incremental executions and check the behavioural properties. Among others, one can quickly change the number of clients, the roles that such clients can perform (either subscriber, publisher, or both), switch between acyclic or cyclic version (where reconnection of clients is allowed) or enable/disable the possibility to retransmit packets (by means of timeouts).

To obtain a finite state space, we have to limit the number of clients and topics, and also bound the packet identifiers. It can be observed that there is no interaction between clients and brokers across topics as the protocol treats each topic in isolation. Executing the protocol with multiple topics is equivalent to running multiple instances of the protocol in parallel. We therefore only consider a single topic for the model validation. Initially, we consider two clients. The packet identifiers are incremented throughput the execution of the different phases of the protocol (connect, subscribe, data exchange, unsubscribe, and disconnect). This means that we cannot use a single global bound on the packet identifiers as a client could reach this bound, e.g., already during the publish phase and hence the global bound would prevent (block) a subsequent unsubscribe to take place. We therefore introduce a local upper bound on packet identifiers for each phase. This local bound expresses that the given phase may use packet identifiers up to this local bound. Note that the use of bounds does not guarantee that the client uses packet identifiers up to bound. It is the guard on the transitions sending packets from the clients that ensures that these local bounds are enforced. Finally, we enforce an upper bound on the number of messages that can be in the message queues on the places CtoB and BtoC.

Below we describe each step of the model validation and the behavioural properties verified. The properties verified in each step include the properties from the previous step. We summarise the experimental results at the end. For the actual checking of properties, we have used the state and action-oriented variant of CTL supported by the ASK-CTL library of CPN Tools.

Step 1 – Connect and Disconnect. In the first step, we consider only the part of the model related to clients connecting and disconnecting to the broker. We consider the following behavioural properties:

S1-P1-ConsistentConnect. The clients and the broker have a consistent view of the connection state. This means that if the clients side is in a connect state, then also the broker has the client recorded as connected.

S1-P2-ClientsCanConnect. For each client, there exists a reachable state in which the client is connected to the broker.

S1-P3-ConsistentTermination. In each terminal state (dead marking), clients are in a disconnect state, the broker has recorded the clients as disconnected, no clients are recorded as subscribed on both clients and broker sides, and there are no outstanding messages in the message buffers.

S1-P4-PossibleTermination. The protocol can always be terminated, i.e., a terminal state (dead marking) can always be reached.

The two properties S1-P3 and S1-P4 imply partial correctness of the protocol as it states that the protocol can always be terminated, and if it terminates, then it will be in a correct state. The state space obtained in this step is acyclic when we do not allow reconnections. This together with S1-P3 implies the stronger property of eventual correct termination. This is, however, more a property of how the model has been constructed as in a real implementation there is nothing forcing a client to disconnect.

Step 2 – Subscribe and Unsubscribe. In the second step, we add the ability for the clients to subscribe and unsubscribe (in addition to connect and disconnect from step 1). For subscribe and unsubscribe we additionally consider the following properties:

S2-P1-CanSubscribe. For each of the clients, there exists states in which both the clients and the broker sides consider the client to be subscribed.

S2-P2-ConsistentSubscription. If the broker side considers the client to be subscribed, then the clients side considers the client to be subscribed.

S2-P3-EventualSubscribed. If the client sends a subscribe message, then eventually both the clients and the broker sides will consider the client to be subscribed.

S2-P4-CanUnsubscribe. For each client there exists executions in which the client sends an unsubscribe message.

S2-P5-EventualUnsubscribed. If the client sends an unsubscribe message, then eventually both the clients and the broker sides considers the client to be unsubscribed.

It should be noted that for property S2-P2, the antecedent of the implication deliberately refers to the broker side. This is because the broker side unsubscribes the client upon reception of the unsubscribe message, whereas the client side does not remove the topic from the set of subscribed topics until the subscribe acknowledgement message is received from the broker. Hence, during unsubscribe, we may have the situation that the broker has unsubscribed the client, but the subscribe acknowledgement has not yet been received on the client side.

Step 3 – Publish and QoS Levels. In this step we also consider publication of data for each of the three quality of service levels. As we do not model the concrete data contained in the messages, we use the packet identifiers attached to the message published to identity the packets being sent and received by the clients. In order to reduce the effect of state explosion, we verify properties for each QoS level in isolation. To make it simpler to check properties related to data being sent, we record for each client the packet identifiers of messages sent. For all three service levels, we consider the following properties:

S3-P1-PublishConnect. A client only publishes a message if it is in a connected state.

S3-P2-CanPublish. For each client there exists executions in which the client publishes a message.

S3-P3-CanReceive. For each client there exists executions in which the client receives a message.

S3-P4-Publish. Any data (packet identifiers) received on the client side must also have been sent on the client side.

S3-P5-ReceiveSubscribed. A client only receives data if it is subscribed to the topic, i.e., the client side considers the client to be subscribed.

It should be noted that it is possible for a client to publish to a topic without being subscribed. The only requirement is that the client is connected to the broker. What data can correctly be received depends on the quality of service level considered. We therefore have one of the following three properties depending on the quality of service considered.

S3-P6-Publish-QoS0. The data (packet identifiers) received by the subscribing clients must be a subset of the data (packet identifiers) sent by the clients.

S3-P7-Publish-QoS1. The data sent on the client side must be a subset of the multi-set of packets received by the subscribing clients.

S3-P8-Publish-QoS2. The data received by each client is identical to the packet identifiers sent by the clients.

To check the above properties related to data received, we accumulate the packet identifiers received such that they can be compared to the packet identifiers sent. To simplify the verification of data received, we force (using priorities) both clients to be subscribed before data exchange takes places since otherwise the data that can be received depends on the time at which the clients were subscribed and unsubscribed.

Table 1 summarises the validation statistics where each configuration (scenario) is represented by a row comprised of Clients, Roles and Version. We report the size of the state space (number of states/number of arcs) and the number of dead markings (written below the state space size). We do not show the dead markings for the cyclic configurations as they are always 0. The columns S3.1, S3.2 and S3.3 correspond to the results considering QoS level 0, QoS level 1 and QoS level 2, respectively. Cells containing a hyphen represent configurations where the state space exploration and model checking did not complete within 12 h which we used as a cut-off point.

Table 1. Summary of configurations and experimental results for model validation

N° Clients	Roles	Version	State space (states/arcs) Number of dead markings				
			Step 1	Step 2	Step 3		
					S3.1	S3.2	S3.3
2	1 sub/1 pub	Acyclic	35/48 1	258/480 4	622/1074 21	1312/2616 21	3234/6394 21
	2 sub-pub		35/48 1	1849/4120 16	4282/8840 70	11462/23934 70	43791/85682 76
	1 sub/1 pub	Cyclic	24/38	271/547	1149/2265	2376/5045	5996/12267
	2 sub-pub		24/38	2954/6798	8138/17714	20362/43572	79913/159254
3	2 sub/1 pub	Acyclic	163/292 1	9529/25408 16	31765/76848 165	103176/262254 165	–
	1 sub/2 pub		163/292 1	1262/2862 4	10360/21604 90	46721/120321 90	–
	2 sub/1 pub	Cyclic	84/175	12650/35875	87450/235887	254095/679920	–
	1 sub/2 pub		84/175	1057/2662	23817/59342	101794/279871	–

5   Conclusions and Related Work

We have presented a formal CPN model based on the most recent specification of the MQTT protocol (version 3.1.1 [3]). The constructed CPN model represents a formal and executable specification of the MQTT protocol. While performing an exhaustive review of the MQTT specification to develop the model, we found several issues that might lead to not interoperable implementations. Consequently, this may add extra complexity for interoperability in the heterogeneous ecosystem that surrounds the application of a protocol such as MQTT.

The model has been built using a set of general CPN modelling patterns ensuring modular organisation of the protocol roles and protocol processing logic. Furthermore, we incorporated parameterization that makes it easy to change, among others, the number of clients and topics without having to make changes in the CPN model structure. In addition, we have applied modelling patterns related to the input and output message queues of the clients (publishers and subscribers) and brokers. These modelling patterns apply generally for modelling distributed systems that include one-to-one and one-to-many communication.

For the validation of the model, we have conducted simulation and state space exploration in order to verify an extensive list of behavioural properties and thereby validate the correctness of the model. In particular, our modelling approach makes it possible to apply an incremental verification technique where the functionality of the protocol is gradually introduced and properties are verified in each incremental step. A main advantage of the modelling patterns used for communication and message queues is that they avoid intermediate states and hence contributes to making state space exploration feasible.

There exists previous work on modelling and validation of the MQTT protocol. In [11], the authors uses the UPPAAL SMC model checker [7] to evaluate different quantitative and qualitative (safety, liveness and reachability) properties against a formal model of the MQTT protocol defined with probabilistic timed automata. Compared to their work, we have verified a larger set of behavioural properties using the incremental approach adding more operations in each step. In [13], tests are conducted over three industrial implementations of MQTT against a subset of the requirements specified in the MQTT version 3.1.1 standard using the TTCN-3 test specification language. In comparison to our work, test-based approaches do not cover all the possible executions but only randomly generated scenarios. With the exploration of state spaces, we considered all the possible cases. In [2], the authors first define a formal model of MQTT based on timed message-passing process algebra, and they conduct analysis of the three QoS levels. In contrast, our work is not limited to the publishing/subscribing process, but considers all operations of the MQTT specification.

We are planning to extend the features supported by the model in order to be able to simulate more sophisticated scenarios. For instance, we will allow the model to deal with persistence of data, so clients can receive the messages on reconnections lost suddenly in the middle of some operation. Furthermore, we plan to improve the mechanism to simulate loss of packets as an extension of the timeout system already implemented. In addition to aiding in the development of compatible MQTT implementations, the CPN MQTT model may also be used as basis for testing of MQTT implementations. As part of future work, we plan to explore model-based testing of MQTT protocol implementations following the approach presented in [16].

References

1. Adamski, M.A., Karatkevich, A., Wegrzyn, M.: Design of Embedded Control Systems, vol. 267. Springer, Boston (2005). https://doi.org/10.1007/0-387-28327-7
2. Aziz, B.: A formal model and analysis of an IoT protocol. Ad Hoc Netw. **36**, 49–57 (2016)
3. Banks, A., Gupta, R.: MQTT Version 3.1.1. OASIS Standard, 29 (2014). http:// docs.oasis-open.org/mqtt/mqtt/v3.1.1/mqtt-v3.1.1.html
4. Baresi, L., Ghezzi, C., Mottola, L.: On accurate automatic verification of publish-subscribe architectures. In: Proceedings of the 29th International Conference on Software Engineering, pp. 199–208. IEEE Computer Society (2007)
5. Billington, J., Diaz, M.: Application of Petri Nets to Communication Networks: Advances in Petri Nets, vol. 1605. Springer, Heidelberg (1999). https://doi.org/10.1007/BFb0097770
6. Chen, S., Xu, H., Liu, D., Hu, B., Wang, H.: A vision of IoT: applications, challenges, and opportunities with China perspective. IEEE Internet Things J. **1**(4), 349–359 (2014)
7. David, A., Larsen, K.G., Legay, A., Mikučionis, M., Poulsen, D.B.: Uppaal SMC tutorial. Int. J. Softw. Tools Technol. Transf. **17**(4), 397–415 (2015)
8. Desel, J., Reisig, W., Rozenberg, G. (eds.): Lectures on Concurrency and Petri Nets, Advances in Petri Nets. LNCS, vol. 3018. Springer, Heidelberg (2004). https://doi.org/10.1007/b98282
9. Eugster, P.T., Felber, P.A., Guerraoui, R., Kermarrec, A.-M.: The many faces of publish/subscribe. ACM Comput. Surv. (CSUR) **35**(2), 114–131 (2003)
10. Garlan, D., Khersonsky, S., Kim, J.S.: Model checking publish-subscribe systems. In: Ball, T., Rajamani, S.K. (eds.) SPIN 2003. LNCS, vol. 2648, pp. 166–180. Springer, Heidelberg (2003). https://doi.org/10.1007/3-540-44829-2_11
11. Houimli, M., Kahloul, L., Benaoun, S.: Formal specification, verification and evaluation of the MQTT protocol in the Internet of Things. In: Mathematics and Information Technology, pp. 214–221. IEEE (2017)
12. Jensen, K., Kristensen, L.: Coloured Petri nets: a graphical language for modelling and validation of concurrent systems. Commun. ACM **58**(6), 61–70 (2015)
13. Mladenov, K.: Formal verification of the implementation of the MQTT protocol in IoT devices. Master thesis, University of Amsterdam (2017)
14. MQTT essentials part 3: Client, broker and connection establishment. https:// www.hivemq.com/blog/mqtt-essentials-part2-publish-subscribe
15. Rodriguez, A., Kristensen, L.M., Rutle, A.: Complete CPN model of the MQTT Protocol. via Dropbox. http://www.goo.gl/6FPVUq
16. Wang, R., Kristensen, L., Meling, H., Stolz, V.: Application of model-based testing on a quorum-based distributed storage. In: Proceedings of PNSE 2017, volume 1846 of CEUR Workshop Proceedings, pp. 177–196 (2017)
17. Wortmann, F., Flüchter, K.: Internet of things. Bus. Inf. Syst. Eng. **57**(3), 221–224 (2015)
18. Zanolin, L., Ghezzi, C., Baresi, L.: An approach to model and validate publish/-subscribe architectures. Proc. SAVCBS **3**, 35–41 (2003)

Kleene Theorems for Free Choice Automata over Distributed Alphabets

Ramchandra Phawade$^{(\boxtimes)}$

Indian Institute of Technology Dharwad, Dharwad, India
prb@iitdh.ac.in

Abstract. We provided (PNSE'2014) expressions for free choice nets having *distributed choice property* which makes the nets *direct product representable*. In a recent work (PNSE'2016), we gave equivalent syntax for a larger class of free choice nets obtained by dropping distributed choice property.

In both these works, the classes of free choice nets were restricted by a *product condition* on the set of final markings. In this paper we do away with this restriction and give expressions for the resultant classes of nets which correspond to *free choice synchronous products and Zielonka automata*. For free choice nets with distributed choice property, we give an alternative characterization using properties checkable in polynomial time.

Free choice nets we consider are 1-bounded, S-coverable, and are labelled with distributed alphabets, where S-components of the associated S-cover respect the given alphabet distribution.

Keywords: Kleene theorems · Petri nets · Distributed automata

1 Introduction

There are several different notions of acceptance to define languages for labelled place transition Petri nets, depending on restrictions on labelling and *final* markings [13]. The language of a place transition net with an initial marking and a finite set of final markings, is called L-type language [8]. One goal of this work is to give syntax of expressions for L-type languages for various subclasses of 1-bounded, free choice nets labelled with distributed alphabets. One advantage of using distributed alphabet is that we can see free choice nets as products of automata [12], enabling us to write expressions for the nets using components. This also enables us to compare expressiveness of nets and products of automata. Three kinds of formulations of automata over distributed alphabets, in the increasing order of expressiveness: direct products, synchronous products, and asynchronous products are described in [12]. In the present paper[1], we

---

[1] A preliminary version of this paper appeared at 14th PNSE workshop, held at Bratislava [16].

© Springer-Verlag GmbH Germany, part of Springer Nature 2019
M. Koutny et al. (Eds.): ToPNoC XIV, LNCS 11790, pp. 146–171, 2019.
https://doi.org/10.1007/978-3-662-60651-3_6

present a hirearchy of 1-bounded free choice nets like automata over distributed alphabets, and also introduce a fourth product automata in the current hierarchy which is utilized to get the syntax. In this hierarchy, there are four kinds of free choice nets labelled over distributed alphabets. Two out of these four classes were introduced earlier [15, 18, 19]. Two new classes of systems are given in this work. To understand the complete hierarchy and their relations to other formalisms like expressions and automata over distributed alphabets we invite the reader to read these earlier works [15, 18, 19].

We use product automata to get expressions for the Free choice nets, and give correspondences for all these three formalisms for various classes. This kind of correspondence has been used in concurrent code generation for discrete event systems [7].

We construct expressions for L-type languages of free choice nets via *free choice Zielonka automata*.

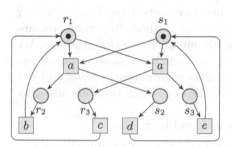

Fig. 1. Free choice net without distributed choice

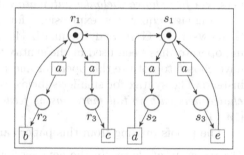

Fig. 2. S-cover of the net in Fig. 1

Consider the net N of Fig. 1 with $G = \{\{r_1, s_1\}, \{r_2, s_2\}\}$ as its set of final markings, with its decomposition into finite state machines in Fig. 2. Because this net is decomposable into state machines [3, 6], its markings can be written in tuple form, where each s-component has a place in the tuple: for example $G = \{(r_1, s_1), (r_2, s_2)\}$. For the final marking (r_1, s_1), its language can be expressed by $fsync((ab+ac)^*, (ad+ae)^*)$ [18, 19]. Similarly, for the final marking (r_2, s_2) the language equivalent expression can be given by $fsync((ab + ac)^*a, (ad + ae)^*a)$. In general, if the places involved in the final markings form a *product* [18], then its language is specified by taking product of component expressions, using *free choice Zielonka automata with product-acceptance* [15] as intermediary. Even though r_1 and s_2 participate in final markings, marking $\{r_1, s_2\}$ does not belong to G, hence set G do not form a product. The language L of net system (N, G) can be described by, $fsync((ab+ac)^*, (ad+ae)^*)+fsync((ab+ac)^*a, (ad+ae)^*a)$. The key idea is ability to express the language of a net as the union of languages of nets complying with the product condition on final set of markings. This closure under union may not be always possible for restricted classes of languages defined over a distributed alphabet. For example, the union of *direct product* languages $L_1 = \{ca, cb\}$ and $L_2 = \{caa, cbb\}$ defined over $\Sigma_1 = \{c, a\}$ and $\Sigma_2 = \{c, b\}$ respectively, is not expressible as a *direct product language*. But this language is accepted by *synchronous products*: the direct products extended with subset-acceptance [12].

For the restricted class of direct product representable free choice nets, with its set of final markings having product condition-we gave expressions via product systems with matchings (matched states of product system correspond to places of a cluster in net) and product-acceptance [18,19]. As a second goal, we develop syntax for free choice nets with distributed choice, now extended with subset-acceptance. For a net in this class also, its language can be expressed as the union of languages accepted by product system with matchings and product-acceptance. This union is accepted by product systems with matching and extended with subset-acceptance (*free choice synchronous products*). As a third contribution, we develop an alternate characterization of this class of nets, via *free choice Zielonka automata with product-moves*.

Language equivalent expressions for 1-bounded nets have been given by Grabowski [5], Garg and Ragunath [4] and other authors [9], where renaming operator has been used in the syntax to disambiguate synchronizations. We have chosen to not use this operator and to exploit the S-decompositions of nets instead. The syntax for smaller subclasses of nets such *marked graphs* and *free choice nets with initial markings as feedback vertex set* has been given earlier [10,14].

The proofs missing from this paper can be found here [17].

Organization of Paper. In the next section, we begin with preliminaries on distributed alphabets and nets. In Sect. 3 we define product systems with globals and subset-acceptance, and show that their languages can be expressed as the union of languages accepted by product systems with globals and product-acceptance. These product systems are used as intermediary to get expressions for nets and vice versa. The following section relates these product systems to nets. In Sect. 5 we develop syntax of expressions for product systems with subset acceptance, and next section establishes the correspondence between various classes of product systems and expressions. In the last section we conclude, with an overview of established correspondences between all three formalisms.

2   Preliminaries

$\mathbb{N}$ denotes the set of natural numbers including 0. Let Σ be a finite alphabet and Σ^* be the set of all finite words over the alphabet Σ, including the empty word ε. A language over an alphabet Σ is a subset $L \subseteq \Sigma^*$. The projection of a word $w \in \Sigma^*$ to a set $\Delta \subseteq \Sigma$, denoted as $w\!\downarrow_\Delta$, is defined by: $\varepsilon\!\downarrow_\Delta = \varepsilon$ and

$$(a\sigma)\!\downarrow_\Delta = \begin{cases} a(\sigma\!\downarrow_\Delta) & \text{if } a \in \Delta, \\ \sigma\!\downarrow_\Delta & \text{if } a \notin \Delta. \end{cases}$$

Given languages $L_1, L_2, \ldots, L_m$, their synchronized shuffle $L = L_1 \| \ldots \| L_m$ is defined as: $w \in L$ iff for all $i \in \{1, \ldots, m\}, w\!\downarrow_{\Sigma_i} \in L_i$.

Definition 1 (Distributed Alphabet). *Let Loc denote the set* $\{1, 2, \ldots, k\}$. *A distribution of* Σ *over Loc is a tuple of nonempty sets* $(\Sigma_1, \Sigma_2, \ldots, \Sigma_k)$ *with* $\Sigma = \bigcup_{1 \leq i \leq k} \Sigma_i$. *For each action* $a \in \Sigma$, *its locations are the set* $loc(a) = \{i \mid a \in \Sigma_i\}$. *Actions* $a \in \Sigma$ *such that* $|loc(a)| = 1$ *are called local, otherwise they are called global.*

A global action is global in the locations in which it occurs. For a set S let $\wp(S)$ denote the set of all its subsets. For singleton sets like $\{p\}$, sometimes we may write it as p.

We will sometimes write p instead of the singleton $\{p\}$.

Let $I = \{i_1, \ldots, i_m\} \subseteq \{1, \ldots, k\}$ be a set of indices with $1 \leq i_1 < i_2 < \ldots < i_m \leq k$, and let $Z_1, \ldots, Z_k$ be finite sets. Then $\Pi_{i \in I} Z_i = \{(z_{i_1}, \ldots, z_{i_m}) \mid z_{i_j} \in Z_{i_j}, \forall j \in \{1, \ldots, m\}\}$.

Let $Z = \Pi_{i \in Loc} Z_i$ and $z = (z_1, \ldots, z_k) \in Z$. Then restriction of z to I is the subset of its components taken in the order given by I i.e., $z{\downarrow}I = (z_{i_1}, \ldots, z_{i_m})$. And its generalization $Z{\downarrow}I = \{(z_{i_1}, \ldots, z_{i_m}) \mid \exists z \in Z \text{ with } z{\downarrow}I = (z_{i_1}, \ldots, z_{i_m})\}$.

2.1   Nets

Definition 2. *A labelled net N is a tuple (S, T, F, λ), where S is a finite set of places, T is a finite set (disjoint from S) of transitions labelled by the function $\lambda : T \to \Sigma$ and $F \subseteq (T \times S) \cup (S \times T)$ is the flow relation.*

Elements of $S \cup T$ are called nodes of N. Given a node z of net N, set $^{\bullet}z = \{x \mid (x, z) \in F\}$ is called pre-set of z and $z^{\bullet} = \{x \mid (z, x) \in F\}$ is called post-set of z. Given a set Z of nodes of N, let $^{\bullet}Z = \bigcup_{z \in Z} {}^{\bullet}z$ and $Z^{\bullet} = \bigcup_{z \in Z} z^{\bullet}$.

We only consider nets in which every transition has nonempty pre- and post-set. For each action a in Σ let $T_a = \{t \mid t \in T \text{ and } \lambda(t) = a\}$.

A path of net N is a nonempty sequence $x_1 \ldots x_n$ of nodes of N where $(x_i, x_{i+1}) \in F$ for all i in $\{1, \ldots, n-1\}$. We say that this path leads from node x_1 to x_n. Net N is said to be connected if for any two nodes x and y there exists a path leading x to y or from y to x. The net is strongly connected if for any two nodes x and y there exists a path leading from x to y and a path from y to x.

A net is called an S-net [3] if for any transition t we have $|^{\bullet}t| = 1 = |t^{\bullet}|$.

A marking of a net N is mapping $M : S \to \mathbb{N}$. At marking M, a place p is said to be marked if $M(p) \geq 1$, and is said to be unmarked if $M(p) = 0$.

Definition 3. *A labelled net system is a tuple $(N, M_0, \mathcal{G})$ where $N = (S, T, F, \lambda)$ is a labelled net; M_0 an initial marking; and a finite set of final markings $\mathcal{G}$.*

A transition t is enabled at a marking M if all places in its pre-set are marked by M. In such a case, t can be fired or occurs at M, to produce the new marking M' which is defined as : for each place p in S, $M'(p) = M(p) + F(t, p) - F(p, t)$, where $F(x, y) = 1$ if $(x, y) \in F$ and 0 otherwise. We write this as $M \xrightarrow{t} M'$ or $M \xrightarrow{\lambda(t)} M'$.

For some markings $M_0, M_1, \ldots, M_n$ if we have $M_0 \xrightarrow{t_1} M_1 \xrightarrow{t_2} \ldots \xrightarrow{t_n} M_n$, then the sequence $\sigma = t_1 t_2 \ldots t_n$ is called occurrence or firing sequence. We write

$M_0 \xrightarrow{\sigma} M_n$ and call M_n the marking reached by σ. This includes an empty transition sequence ε. For each marking M we have $M \xrightarrow{\varepsilon} M$. We write $M \xrightarrow{*} M'$ and call M' reachable from M if it is reached by some occurrence sequence σ from M.

A net system $(N, M_0, \mathcal{G})$ is called 1-bounded if for every place p of the net and every reachable marking M, we have $M(p) \leq 1$. Any marking M of a 1-bounded net can be alternately represented by the subset of places which are marked at M. In this paper, we consider only 1-bounded nets.

We say a net system (N, M_0) is live if, for every reachable marking M and every transition t, there exists a marking M' reachable from M which enables t.

Definition 4. *For a labelled net system* $(N, M_0, \mathcal{G})$, *its language is defined as* $Lang(N, M_0, \mathcal{G}) = \{\lambda(\sigma) \in \Sigma^* \mid \sigma \in T^* \text{ and } M_0 \xrightarrow{\sigma} M, \text{ for some } M \in \mathcal{G}\}$.

Net Systems and Its Components. First we define subnet of a net.

Let X be a set of nodes of net $N = (S, T, F)$. Then the triple $N' = (S \cap X, T \cap X, F \cap (X \times X))$ is a subnet of net N. Flow relation $F \cap (X \times X)$ is said to be induced by nodes X; and N' is said to be a subnet of N generated by nodes X of N.

We follow the convention that if N' is a subnet of N and z is a node of N' then $^\bullet z$ and $z^\bullet$ denote the pre-set and post-set taken in N, i.e., $^\bullet z = \{x \mid (x, z) \in F\}$ and $z^\bullet = \{x \mid (z, x) \in F\}$.

Definition 5. *Subnet* N' *is called a* **component** *of* N *if,*

- *For each place* s *of* X, $^\bullet s, s^\bullet \subseteq X$,
- N' *is an S-net,*
- N' *is connected.*

A set $\mathcal{C}$ *of components of net* N *is called* **S-cover** *for* N, *if every place of the net belongs to some component of* $\mathcal{C}$.

Our notion of component does not require strong connectedness and so it is different from notion of S-component in [3], and therefore our notion of S-cover also differs from theirs.

A net is covered by components or S-coverable if it has an S-cover.

Fix a distribution $(\Sigma_1, \Sigma_2, \ldots, \Sigma_k)$ of Σ. We define s-decomposition [6] of a net into sequential components. Note that S-decomposition given here is for labelled nets unlike [3,6] and is different from [15,16,18] also, as it takes into account the initial marking of the net.

Definition 6. *A labelled net system* $(N, M_0, \mathcal{G})$ *is called* **S-decomposable** *if, there exists an S-cover* $\mathcal{C}$ *for net* $N = (S, T, F, \lambda)$, *such that for each* $T_i = \bigcup_{a \in \Sigma_i} \lambda^{-1}(a)$, *there exists* $S_i \subseteq S$ *and the subnet generated by* $S_i \cup T_i$ *is a component in* $\mathcal{C}$, *and the initial marking* M_0 *marks only one place of the component.*

Now each S-decomposable net N admits an S-cover, since there exist subsets $S_1, S_2, \ldots, S_k$ of places S, such that $S = S_1 \cup S_2 \cup \ldots S_k$ and ${}^\bullet S_i \cup S_i^\bullet = T_i$, such that the subnet (S_i, T_i, F_i) generated by S_i and T_i is an S-net, where F_i is the induced flow relation from S_i and T_i.

Note that, the initial marking, of a 1-bounded and S-decomposable net system, marks exactly one place in each S-component of the given S-cover $S_1, S_2, \ldots, S_k$. At any reachable markings of such a net, the total number of tokens in an S-component remains constant [3,6]. Therefore, at any reachable marking M, each S-component has only on token, so at that marking only one place of that component is marked. Also, if we collect each place from an S-component we get back the marking of net. Hence, marking M can be written as a k-tuple from its component places $S_1 \times S_2 \times \ldots \times S_k$.

We use a *product* condition [18] on the set of final markings of a net system which is known [12,20] to restrict classes of languages.

Definition 7. *An S-decomposable labelled net system* $(N, M_0, \mathcal{G})$ *is said to have product-acceptance if its set of final markings* $\mathcal{G}$ *satisfies product condition: if* $\langle q_1, q_2, \ldots q_k \rangle \in \mathcal{G}$ *and* $\langle q_1', q_2', \ldots q_k' \rangle \in \mathcal{G}$ *then* $\{q_1, q_1'\} \times \{q_2, q_2'\} \times \ldots \times \{q_k, q_k'\} \subseteq \mathcal{G}$.

Let t be a transition in T_a. Then by S-decomposability a pre-place and a post-place of t belongs to each S_i for all i in $loc(a)$. Let $t[i]$ denote the tuple $\langle p, a, p' \rangle$ such that $(p, t), (t, p') \in F_i$, and $p, p' \in P_i$ for all i in $loc(a)$.

2.2   Free Choice Nets and Their Properties

Let x be a node of a net N. The cluster of x, denoted by $[x]$, is the minimal set of nodes containing x such that

- if a place $s \in [x]$ then $s^\bullet$ is included in $[x]$, and
- if a transition $t \in [x]$ then ${}^\bullet t$ is included in $[x]$.

For a cluster C, we denote its set of places by S_C, and its set of transitions by T_C.

The set of all a-labelled transitions along with places r_1 and s_1 form a cluster of the net shown in Fig. 3.

Definition 8 (Free choice nets [3]**).** *A cluster C is called free choice (FC) if all transitions in C have the same pre-set. A net is called free choice if all its clusters are free choice.*

In a labelled net N, for a free choice cluster C define the a-labelled transitions $C_a = \{t \in T_C \mid \lambda(t) = a\}$. If the net has an S-decomposition then we associate a post-product $\pi(t) = \Pi_{i \in loc(a)}(t^\bullet \cap S_i)$ with every such transition t. This is well defined since in S-nets, every transition will have at most one post-place in S_i. Let $post(C_a) = \bigcup_{t \in C_a} \pi(t)$. Let $C_a[i] = C_a{}^\bullet \cap S_i$ and $postdecomp(C_a) = \Pi_{i \in loc(a)} C_a[i]$. Clearly $post(C_a) \subseteq postdecomp(C_a)$. Sometimes, we may call

$C_a[i]$ as post-projection of the cluster C with respect to label a and location i. Also, $postdecomp(C_a)$ is called post-decomposition of cluster C with respect to label a.

The following definition from [18,19] is used to get direct product representability.

Definition 9 (distributed choice property). *An S-decomposable free choice net $N = (S, T, F, \lambda)$ is said to have distributed choice property (DCP) if, for all a in Σ and for all clusters C of N, $postdecomp(C_a) \subseteq post(C_a)$.*

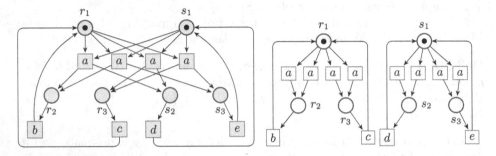

Fig. 3. Labelled free choice net system with distributed choice

Fig. 4. S-cover of the net in Fig. 3

Example 1 (Free choice net system without distributed choice and with product-acceptance). Consider a distributed alphabet $\Sigma = (\Sigma_1 = \{a, b, c\}, \Sigma_2 = \{a, d, e\})$ and the net system N shown in Fig. 1, labelled over Σ. Its (only possible) S-cover having two S-components with sets of places $S_1 = \{r_1, r_2, r_3\}$ and $S_2 = \{s_1, s_2, s_3\}$ respectively, is given in Fig. 2. For the cluster C of r_1, we have the set of a-labelled transitions $C_a = \{t_1, t_2\}$ with $C_a[1] = \{r_2, r_3\}$ and $C_a[2] = \{s_2, s_3\}$. So we get $postdecomp(C_a) = \{(r_2, s_2), (r_2, s_3), (r_3, s_2), (r_3, s_3)\}$.

As $\pi(t_1) = \{(r_2, s_2)\}$ and $\pi(t_2) = \{(r_3, s_3)\}$ so $post(C_a) = \{(r_2, s_2), (r_3, s_3)\}$. Since $postdecomp(C_a) \nsubseteq post(C_a)$, this cluster does not have distributed choice, so the net system does not have it.

With the set of final markings $\{(r_1, s_1), (r_1, s_2), (r_2, s_1), (r_2, s_2)\}$ satisfying product condition, the language L_p accepted by this net system is $r^*[\varepsilon + a + ab + ad]$ where $r = (a(bd + db) + a(ce + ec))$.

Example 2 (Free choice net system without distributed choice and not satisfying product condition of the set of final markings). Consider the net system of Example 1 whose underlying net is shown in Fig. 1. With set of final markings $\{(r_1, s_1), (r_2, s_2)\}$, which do not satisfy product condition, the language L_s accepted by this net system is $r^*[\varepsilon + a]$ where $r = (a(bd + db) + a(ce + ec))$.

Example 3 (A net with distributed choice property and product acceptance condition). Consider the labelled net system $(N, (r_1, s_1), \mathcal{G}))$ of Fig. 3, defined over distributed alphabet $\Sigma = (\Sigma_1 = \{a, b, c\}, \Sigma_2 = \{a, d, e\})$, and where $\mathcal{G} = \{(r_1, s_1), (r_1, s_2), (r_2, s_1), (r_2, s_2)\}$ is the set of final markings satisfying product condition. Its two S-components with sets of places $S_1 = \{r_1, r_2, r_3\}$ and $S_2 = \{s_1, s_2, s_3\}$, are shown in Fig. 4. For cluster C of r_1, we have $C_a = \{t_1, t_2, t_3, t_4\}$, $C_a[1] = \{r_2, r_3\}$ and $C_a[2] = \{s_2, s_3\}$, hence $postdecomp(C_a) = \{(r_2, s_2), (r_2, s_3), (r_3, s_2), (r_3, s_3)\}$. We have $\pi(t_1) = \{(r_1, s_2)\}$, $\pi(t_2) = \{(r_2, s_3)\}$, $\pi(t_3) = \{(r_3, s_2)\}$ and $\pi(t_4) = \{(r_3, s_3)\}$.
So $post(C_a) = \{(r_2, s_2), (r_2, s_3), (r_3, s_2), (r_3, s_3)\}$. Therefore, $postdecomp(C_a) = post(C_a)$. For all other clusters this holds trivially, because each of them have only one transition and only one post-place, hence the net has distributed choice. Language L_3 accepted by the net system is $r^*[\varepsilon + a + a(b + c) + a(d + e)]$ where $r = (a(bd + db) + a(be + eb) + a(cd + dc) + a(ce + ec))$.

Example 4 (A net with distributed choice and subset-acceptance). Consider the net system of Example 3, with the underlying net shown in Fig. 3 with set of final markings $\{(r_1, s_1), (r_2, s_2)\}$. The language L_4 accepted by this net system is $r^*[\varepsilon + a]$ where $r = (a(bd + db) + a(be + eb) + a(cd + dc) + a(ce + ec))$.

3   Product Systems

We define product systems over a fixed distribution $(\Sigma_1, \Sigma_2, \ldots, \Sigma_k)$ of Σ. First we define sequential systems.

Definition 10. *A* **sequential system** *over a set of actions Σ_i is a finite state automaton $A_i = \langle P_i, \rightarrow_i, G_i, p_i^0 \rangle$ where P_i are called* **states***, $G_i \subseteq P_i$ are final states, $p_i^0 \in P_i$ is the initial state, and $\rightarrow_i \subseteq P_i \times \Sigma_i \times P_i$ is a set of* **local moves***.*

For a local move $t = \langle p, a, p' \rangle$ of $\rightarrow_i$ state p is called **pre-state** sometimes denoted by $pre(t)$ and p' is called **post-state** of t, sometimes denoted by $post(t)$. Such a move is sometimes called an a-**move** or an a-**labelled** move.

Let $\rightarrow_a^i$ denote the set of all a-labelled moves in the sequential system A_i. The language of a sequential system is defined as usual.

Definition 11. *Let $A_i = \langle P_i, \rightarrow_i, G_i, p_i^0 \rangle$ be a sequential system over alphabet Σ_i for $1 \leq i \leq k$. A* **product system** *A over the distribution $\Sigma = (\Sigma_1, \ldots, \Sigma_k)$ sometimes denoted by $\langle A_1, \ldots, A_k \rangle$ is a tuple $\langle P, \Rightarrow, R^0, G \rangle$, where:*
$P = \Pi_{i \in Loc} P_i$ is the set of **product states** *of A; $R^0 = (p_1^0, \ldots, p_k^0)$ is the initial product state of A; $G \subseteq \Pi_{i \in Loc} G_i$ is the set of final product states of A; and, $\Rightarrow \subseteq \bigcup_{a \in \Sigma} \Rightarrow_a$, denotes the* **global moves** *of A where $\Rightarrow_a = \Pi_{i \in loc(a)} \rightarrow_a^i$.*

Elements of $\Rightarrow_a$ are sometimes called global a-moves. Any global a-move is *global* within the set of component sequential machines where action a occurs. For a global a-move g, we define its set of pre-states $pre(g)$ as the set of pre-states of all its component a-moves; the set of post-states $post(g)$ as the set of post-states of all its component a-moves; and, use notation $g[i]$ for its i-th component–local a-move–belonging to A_i, for all i in $loc(a)$. We use $R[i]$ for the projection of a product state R in A_i.

3.1 Direct Products

With set of global moves $\Rightarrow = \bigcup_{a \in \Sigma} \Rightarrow_a$ and final states $G = \Pi_{i \in Loc} G_i$ A is called product system with product-acceptance. These systems are called direct products in [12].

With set of global moves $\Rightarrow = \bigcup_{a \in \Sigma} \Rightarrow_a$ and final states $G \subseteq \Pi_{i \in Loc} G_i$ A is called product system with subset-acceptance. These systems are called synchronous products in [12].

The runs of a product system A over some word w are described by associating product states with prefixes of w: the empty word is assigned initial product state R^0, and for every prefix va of w, if R is the product state reached after v and Q is reached after va where, for all $j \in loc(a), \langle R[j], a, Q[j] \rangle \in \rightarrow_j$, and for all $j \notin loc(a), R[j] = Q[j]$. A run of a product system over word w is said to be accepting if the product state reached after w is in G. We define the language $Lang(A)$ of product system A, as the set of words on which the product system has an accepting run. The set of languages accepted by direct (resp. synchronous) products is called direct (resp. synchronous) product languages.

We use a characterization from [12] of languages accepted by direct products.

Proposition 1. *Let L be a language defined over distributed alphabet Σ. The language L is a direct product language iff $L = \{w \in \Sigma^* \mid \forall i \in \{1, \ldots, k\}, \exists u_i \in L$ such that $w \downarrow_{\Sigma_i} = u_i \downarrow_{\Sigma_i}\}$.*

If $L = Lang(A)$ for direct product $A = \langle A_1, \ldots, A_k \rangle$ defined over distributed alphabet Σ then $L = Lang(A_1) \| \ldots \| Lang(A_k)$.

We also use a characterization of synchronous product languages [12].

Proposition 2. *A language over distributed alphabet Σ is accepted by a product system with subset-acceptance if and only if it can be expressed as a finite union of direct product languages.*

The following property of direct products from [18] clubs together the places of product system which correspond to places of a cluster in the net.

Definition 12 (PS-matchings). *For global $a \in \Sigma$, matching(a) is a subset of tuples $\Pi_{i \in loc(a)} P_i$ such that for all i in $loc(a)$, projection of these tuples is the set of all pre-states of a-moves in $\rightarrow_i^a$, and if a state $p \in P_i$ appears in one tuple, it does not appear in another tuple. We say a product state R is in matching(a) if its projection $R \downarrow loc(a)$ is in the matching.*

A product system is said to have matching of labels if for all global $a \in \Sigma$, there is a suitable matching(a). Such a system is denoted by PS-matchings.

We have *PS-matchings* with product-acceptance, if the set of final product states of it is a product of final states of component machines, or **PS-matchings with subset-acceptance**, if the set of final product states is a subset of product of final states of individual components.

A run of *PS-matchings* A is said to be consistent with a matching of labels [18] if for all global actions a and every prefix of the run $R^0 \overset{v}{\Rightarrow} R \overset{a}{\Rightarrow} Q$, the pre-states $R \downarrow loc(a)$ are in the matching.

Consistency of matchings is a behavioural property and to check if a *PS-matchings* A has it and can be done in PSPACE [18,19].

The following property from [18] is used to capture free choice property.

Definition 13 (conflict-equivalent matchings for *PS-matchings*). *In a product system, we say the local move $\langle p, a, q_1 \rangle \in \rightarrow_i$ is conflict-equivalent to the local move $\langle p', a, q_1' \rangle \in \rightarrow_j$, if for every other local move $\langle p, b, q_2 \rangle \in \rightarrow_i$, there is a local move $\langle p', b, q_2' \rangle \in \rightarrow_j$ and, conversely, for moves from p' there are corresponding outgoing moves from p. For global action a, its matching(a) is called conflict-equivalent matching, if whenever p, p' are related by the matching(a), their outgoing local a-moves are conflict-equivalent.*

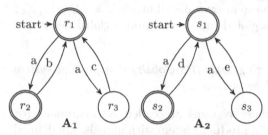

Figure 5, shows a product system over a distributed alphabet $\Sigma = (\Sigma_1 = \{a, b, c\}, \Sigma_2 = \{a, d, e\})$. It has two components A_1, and A_2 with final states $G_1 = \{r_1, r_2\}$ and, $G_2 = \{s_1, s_2\}$, respectively.

Fig. 5. Product system (A_1, A_2)

Example 5 (Product system with matchings). Consider product system of Fig. 5 and relation $matching(a) = \{(r_1, s_1)\}$ relation. This matching is conflict-equivalent and the system is consistent with this matching relation.

We have a *PS-matchings* $\mathcal{A} = (A_1, A_2)$ with product acceptance condition, if its set of final states is $G_1 \times G_2$. With the set of final states as $\{(r_1, s_1), (r_2, s_2)\} \subseteq G_1 \times G_2$, we have a *PS-matchings* $\mathcal{B} = (A_1, A_2)$ having subset-acceptance.

Lemma 1 presents a language not accepted by any direct product.

Lemma 1. *The language $L_4 = \{(abd + adb + abe + aeb + ace + aec + acd + adc)^*(\varepsilon + a)\}$ from Example 4 is not accepted by any direct product.*

We know that the class of synchronous product languages is strictly larger than the class of direct product languages [12]. With the matching relations this relationship is preserved. The *PS-matchings* $\mathcal{B}$ of Example 5 accepts language L_4 which by Lemma 1, is not accepted by any direct product. Hence, the class of languages accepted by *PS-matchings* with subset-acceptance condition, is strictly larger, than the class of languages accepted by *PS-matchings* with product-acceptance.

However, using Proposition 2 we have the following characterization of *PS-matchings* with subset-acceptance.

Corollary 1. *A language L is accepted by a product system with the subset-acceptance and, having conflict-equivalent and consistent matchings if and only if L can be expressed as a finite union of languages accepted by product system with product-acceptance and, having conflict-equivalent and consistent matchings.*

Lemma 2 presents a language not accepted by any synchronous product.

Lemma 2. *The language $L_s = \{abd, adb, ace, aec\}^*(\varepsilon + a)$ of Example 2 is not a synchronous product language.*

So we have language L_s which is not accepted by any *PS-matchings* with subset-acceptance. This motivates the bigger class of automata over distributed alphabets, which we discuss next.

3.2 Product Systems with Globals

Let $A = \langle A_1, \dots, A_k \rangle$ be a product system over distribution $\Sigma = (\Sigma_1, \dots, \Sigma_k)$ and, let *globals(a)* be a subset of its global moves $\Rightarrow_a$, and a-global denote an element of *globals(a)*.

Definition 14. *A product system with globals (PS-globals) is a product system with relations globals(a), for each global action a in Σ.*

With subset-acceptance condition these systems are called Asynchronous (or Zielonka) automaton [12,20]. Runs of a product system with globals, are defined in the same way as for the direct products, with an additional requirement of $\Pi_{j\in loc(a)}(\langle R[j], a, Q[j] \rangle) \in globals(a)$, to be satisfied when $R \xrightarrow{a} Q$ is to be taken. With abuse of notation sometimes we use $pre(a)$ to denote the set $\{R \mid \exists Q, R \xrightarrow{a} Q\}$.

The following property from [15], of product systems with globals, relates to free choice property of nets.

Definition 15 (same source property). *A product system with globals have same source property if, any two global moves share a pre-state then their sets of pre-states are same.*

Example 6 (Product system with globals). Consider the product system of Fig. 5. Let $globals(a) = \{((r_1 \xrightarrow{a} r_2), (s_1 \xrightarrow{a} s_2)), ((r_1 \xrightarrow{a} r_3), (s_1 \xrightarrow{a} s_3))\}$. This system has same source property.

With the given *globals(a)* we have a *PS-globals* $\mathcal{C} = (A_1, A_2)$ with product acceptance condition, if its set of final states is $G_1 \times G_2$. And, for the set of final states $\{(r_1, s_1), (r_2, s_2)\} \subseteq G_1 \times G_2$, we have a *PS-globals* $\mathcal{D} = (A_1, A_2)$ with subset-acceptance.

The language $L_s = \{abd, adb, ace, aec\}^*(\varepsilon + a)\}$ of Lemma 2, is accepted by product system with globals $\mathcal{D}$ of Example 6 with same source property.

Product systems with globals and product-acceptance are not considered in [12]. This class of systems are strictly more expressive, as shown in Lemma 3. This lemma is new and was not present in [16].

Lemma 3. *The language $L_p = \{(abd + adb + ace + aec)^*(\varepsilon + a + ab + ad)\}$ from Example 1 is not accepted by any direct product.*

Proof. Consider a word $w = abe$ not in L_p and, words $u_1 = abd$, $u_2 = ace$ which are in L_p. We have projections, $w{\downarrow}_{\Sigma_1} = ab = u_1{\downarrow}_{\Sigma_1}$, $w{\downarrow}_{\Sigma_2} = ae = u_2{\downarrow}_{\Sigma_2}$. Therefore, by Proposition 1, word w is in L_p, which is a contradiction.   □

We give in Lemma 4, a characterization of class of languages accepted by product systems with globals and having subset-acceptance, in terms of *PS-globals* and product-acceptance.

Lemma 4. *A language is accepted by a PS-globals with subset-acceptance if and only if it can be expressed as a finite union of languages accepted by PS-globals with product-acceptance.*

In the construction, transition structure of local components is preserved, and so are the global moves, hence, we have Corollary 2, which is used to get syntax for product systems with subset-acceptance condition.

Corollary 2. *A language L is accepted by a PS-globals with subset-acceptance and having same source property if and only if L can be expressed as a finite union of languages accepted by PS-globals with product-acceptance and same source property.*

In a product system with globals and having same source property, global moves for an action a can be partitioned into different compartments: two global a-moves belong to same compartment if they have the same set of pre-states. For any a-global g of a same source compartment $\Rightarrow_a^{SS}$, we associate a **target-configuration** $\pi(g) = \Pi_{i \in loc(a)} post(g) \cap P_i$. Let $post(\Rightarrow_a^{SS}) = \{\pi(g) \mid g \in \Rightarrow_a^{SS}\}$. We define $\Rightarrow_a^{SS}[i] = post(\Rightarrow_a^{SS}) \cap P_i$ and $postdecomp(\Rightarrow_a^{SS}) = \Pi_{i \in loc(a)} \Rightarrow_a^{SS}[i]$. We may call $\Rightarrow_a^{SS}[i]$ as **post-projection** and $postdecomp(\Rightarrow_a^{SS})$ as **post-decomposition** of a compartment.

The following property relates to distributed choice property of nets.

Definition 16. *A product system with globals and having same source property, is said to have **product moves property**, if for all a in Σ, and for all same source compartments $\Rightarrow_a^{SS}$ of a-globals, $postdecomp(\Rightarrow_a^{SS}) \subseteq post(\Rightarrow_a^{SS})$.*

Product systems $\mathcal{C}$ and $\mathcal{D}$ of Example 6 do not have product moves property. Product system of Example 7 has product moves property.

Example 7 (Product system with globals and product moves property). Consider product system $\mathcal{A}$ of Example 5 where the set of final states is $G_1 \times G_2$. With $globals(a) = \{ ((r_1 \xrightarrow{a} r_2), (s_1 \xrightarrow{a} s_2)), ((r_1 \xrightarrow{a} r_2), (s_1 \xrightarrow{a} s_3)), ((r_1 \xrightarrow{a} r_3), (s_1 \xrightarrow{a} s_2)) ((r_1 \xrightarrow{a} r_3), (s_1 \xrightarrow{a} s_3)) \}$ relation, we have *PS-globals* $\mathcal{A}'$ which has product moves property. This also has same source property.

Now consider the product system $\mathcal{B}$ of Example 5, with $\{(r_1, s_1), (r_2, s_2)\} \subseteq G_1 \times G_2$ as its final states, and having the $globals(a)$ relation as above, we get a *PS-globals* $\mathcal{B}'$ with subset-acceptance condition, having product moves and same source property.

A product system with globals is said to be live, if for any global move g and any reachable product state R, there exists a product state Q such that g is enabled at Q.

3.3   Relating Product Systems with Matchings and Globals

First we show, in Theorem 1, how to construct a product system with consistent and conflict-equivalent matchings from a *PS-globals* with same source property.

Theorem 1. *Let Σ be a distributed alphabet and A be a product system with globals defined over it. Then we can construct a product system B with matchings, linear in the size of product system A with globals such that,*

1. *if A has same source property then B has conflict-equivalent matchings,*
2. *in addition, if A is live then*
   (a) B is consistent with matchings, and
   (b) Lang(A) = Lang(B).

Now, from a *PS-matchings* with consistent and conflict-equivalent matchings we construct a product system with globals having same source property.

Theorem 2. *Let Σ be a distributed alphabet and let B be a product system with conflict equivalent and consistent matchings. Then for the language of B we can construct a product system A with globals over Σ having same source and product moves property. The constructed product system A with globals is exponential in the size of system B having matching of labels.*

The proofs missing from this section are given in detail here [17].

4   Nets and Product Systems

We first present a generic construction of a 1-bounded S-decomposable labelled net systems, from product systems with globals.

Definition 17 (*PS-globals* to nets). *Given a PS-globals $A = \langle A_1, \ldots, A_k \rangle$ over distribution Σ, a net system $(N = (S, T, F, \lambda), M_0, \mathcal{G})$ is constructed as follows: The set of places is $S = \bigcup_i P_i$, and the set of transitions is $T = \bigcup_{a \in \Sigma} globals(a)$. Define $T_i = \{\lambda^{-1}(a) \mid a \in \Sigma_i\}$. The labelling function λ labels by action a the transitions in globals(a). The flow relation is $F = \{(p, g), (g, q) \mid g \in T_a, g[i] = \langle p, a, q \rangle, \text{ for all } i \in loc(a)\}$, define F_i as its restriction to the transitions T_i for $i \in loc(a)$. See that F is union of all F_is. Let $M_0 = \{p_1^0, \ldots, p_k^0\}$, be the initial product state and $\mathcal{G} = G$ as the set of final global states.*

We get one to one correspondence between reachable states of product system and reachable markings of nets because the set of transitions of resultant net is same as the set of global moves in the product system, and construction preserves pre as well as post places.

Lemma 5. *The constructed net system N from a PS-globals A, as in Definition 17, is S-decomposable and $Lang(N, M_0, \mathcal{G}) = Lang(A)$. The size of constructed net is linear in the size of product system.*

Applying the generic construction above to product systems with same source property, we get a free choice net, because any two global moves having same set of pre-places are put into one cluster.

Theorem 3. *Let $(N, M_0, \mathcal{G})$ be the net system constructed from PS-globals A as in Definition 17.*

- *If A has same source property then N is a free choice net,*
- *In addition if A has product moves property, then N has distributed choice.*

In the construction, if the product system has subset-acceptance then we get a net with a set of final markings, which may not have product condition. Since A has subset-acceptance, we generalize the results obtained in [15].

For the product system $\mathcal{D}$ of Example 6, accepting language L_s we can construct the net system of Example 2.

In the case that the product system A has matchings, transitions of net constructed are reachable global moves of system A [18,19].

Now we describe a linear-size construction of a product system from a net which is S-decomposable.

Definition 18 (nets to *PS-globals*). *Given a 1-bounded labelled and an S-decomposable net system $(N, M_0, \mathcal{G})$, with $N = (S, T, F, \lambda)$ the underlying net and $N_i = (S_i, T_i, F_i)$ the components in the S-cover, for i in $\{1, 2, \ldots, k\}$, we define a product system $A = \langle A_1, \ldots, A_k \rangle$, as follows. Take $P_i = S_i$, and p_i^0 the unique state in $M_0 \cap P_i$. Define local moves $\to_i = \{\langle p, \lambda(t), p' \rangle \mid t \in T_i \text{ and } (p, t), (t, p') \in F_i, \text{ for } p, p' \in P_i\}$. So we get sequential systems $A_i = \langle P_i, \to_i, p_i^0 \rangle$, and the product system $A = \langle A_1, A_2, \ldots, A_k \rangle$ over alphabet Σ. Global moves are $globals(a) = \{\Pi_{i \in loc(a)} t[i] \mid t \in T_a\}$. And, the set of final states is $G = \mathcal{G}$.*

Lemma 6. *From net system N with a final set of markings, the construction of the PS-globals A in Definition 18 above preserves language. The product system A is linear in the size of net, and product system has subset-acceptance.*

For each a-labelled transition of the net we get one global a-move in the product system having same set of pre-places and post-places. And, for each global a-move in product system we have an a-labelled transition in the net having same pre and post-places. We get one to one correspondence between reachable states of product system and reachable markings of the net we started with. Therefore, if we begin with a free choice net, we get same source property in the obtained product system. And, for each transition in the net we have a global transition hence, A has product moves property if the net has distributed choice.

Theorem 4. *Let $(N, M_0, \mathcal{G})$ be a 1-bounded, and an S-decomposable labelled net with a set of final markings $\mathcal{G}$. Then*

- *if N has free choice property, then constructed product system A with globals, has same source property,*
- *in addition, if the net has distributed choice, then A has product moves.*

In construction of Definition 18, we start with a net having distributed choice and a final set of markings then we get product system with matching with subset-acceptance condition. Note that in this case we do not have to construct globals [18, 19].

For the net system of Example 2 and accepting language L_s we can construct the product system $\mathcal{D}$ of Example 6. Therefore, we generalize the results from [18, 19].

Theorem 5. *For a 1-bounded, S-decomposable labelled net having distributed choice and given with a set of final markings. Then one can construct a product system with conflict-equivalent and consistent matchings and having subset-acceptance.*

Given below is the converse result.

Theorem 6. *For a product system with conflict-equivalent, consistent matchings, and subset-acceptance, we get language equivalent free choice net with distributed choice and having a set of final markings.*

5   Expressions

First we define regular expressions and its derivatives.

5.1   Regular Expressions and Their Properties

A regular expression over alphabet Σ_i such that constants 0 and 1 are not in Σ_i is given by:

$$s ::= 0 \mid 1 \mid a \in \Sigma_i \mid s_1 \cdot s_2 \mid s_1 + s_2 \mid s_1^*$$

The language of constant 0 is $\emptyset$ and that of 1 is $\{\varepsilon\}$. For a symbol $a \in \Sigma_i$, its language is $Lang(a) = \{a\}$. For regular expressions $s_1 + s_2, s_1 \cdot s_2$ and s_1^*, its languages are defined inductively as union, concatenation and Kleene star of the component languages respectively.

As a measure of the size of an expression we will use $wd(s)$ for its alphabetic width—the total number of occurrences of letters of Σ in s.

For each regular expression s over Σ_i, let $Lang(s)$ be its language and its initial actions form the set $Init(s) = \{a \mid \exists v \in \Sigma_i^* \text{ and } av \in Lang(s)\}$ which can be defined syntactically. We can syntactically check whether the empty word $\varepsilon \in Lang(s)$.

We use derivatives of regular expressions which are known since the time of Brzozowski [2], Mirkin [11] and Antimirov [1].

Inductively $Der_{aw}(s) = Der_w(Der_a(s))$.

The set of all partial derivatives $Der(s) = \bigcup_{w \in \Sigma_i^*} Der_w(s)$, where $Der_\varepsilon(s) = \{s\}$.

We have derivatives $Der_a(ab + ac) = \{b, c\}$ and $Der_a(a(b + c)) = \{b + c\}$.

A derivative d of s with action $a \in Init(d)$ is called an a-site of s. An expression is said to have **equal choice** if for all a, its a-sites have the same set of initial actions. For a set D of derivatives, we collect all initial actions to form $Init(D)$. Two sets of derivatives have equal choice if their $Init$ sets are same.

As in [18] we put together derivatives which may correspond to the same state in a finite automaton.

Definition 19 ([18]). *Let s be a regular expression and $L = \mathrm{Lang}(s)$. For a set D of a-sites of regular expression s and an action a, we define the relativized language $L_a^D = \{xay \mid xay \in L, \exists d \in Der_x(s) \cap D, \exists d' \in Der_{ay}(d)$ with $\varepsilon \in \mathrm{Lang}(d')\}$, and the prefixes $Pref_a^D(L) = \{x \mid xay \in L_a^D\}$, and the suffixes $Suf_a^D(L) = \{y \mid xay \in L_a^D\}$. We say that the derivatives in set D a-bifurcate L if $L_a^D = Pref_a^D(L)\ a\ Suf_a^D(L)$.*

We use partitions of the a-sites of s into blocks such that each block (that is, element of the partition) a-bifurcates L [18].

For an action a, let $Part_a(s)$ denote such a partition. In addition to thinking of blocks of the partition as places of an automaton, we can think of pairs of blocks and their effects as local moves.

Definition 20 ([15]). *Given an action a, and a set of a-sites B of regular expression s, and a specified set of a-effects $E \subseteq Der_a(B)$, we define the relativized languages*

$$L_a^{(B,E)} = \{xay \in L \mid \exists d \in Der_x(s) \cap B, \exists d' \in Der_a(d) \cap E, \text{ and}$$
$$\exists d'' \in Der_y(d') \text{ with } \varepsilon \in \mathrm{Lang}(d'')\}.$$

We define the prefixes $Pref_a^{(B,E)}(L) = \{x \mid xay \in L_a^{(B,E)}\}$ and the suffixes $Suf_a^{(B,E)}(L) = \{y \mid xay \in L_a^{(B,E)}\}$. We say that a tuple (B,E) a-funnels L if $L_a^{(B,E)} = Pref_a^B(L) \cdot a \cdot Suf_a^{(B,E)}(L)$. In such a pair (B,E), if B is a block in the $Part_a(s)$ and E is a nonempty subset of a-effects of B, then it is called as an a-duct.

For an a-duct (B,E), we define its set of initial actions $Init(B,E)$ as $Init(B,E) = Init(B)$, call B as its pre-block and call E as its post-effect. For all i in $loc(a)$ let a-ducts(s_i) denote the set of all a-ducts of regular expression s_i. For any two a-ducts (B,E) and (B',E') in a-ducts(s_i), define $(B,E) = (B',E')$ if $B = B'$ and $E = E'$. Given an a-duct $d = (B,E)$ its post-effect E is sometimes denoted by $d^{\bullet}$ and its pre-block B can be denoted as $^{\bullet}d$. For a collection of ducts z, the set of all their post-effects (resp. pre-blocks) is denoted as $z^{\bullet}$ (resp. $^{\bullet}z$). In a similar way, we define the set of post-effects of an a-cable D, as $D^{\bullet} = \{D[i]^{\bullet} \mid i \in loc(a)\}$ and its set of pre-blocks as $^{\bullet}D = \{^{\bullet}D[i] \mid i \in loc(a)\}$.

5.2   Connected Expressions over a Distributed Alphabets

The syntax of connected expressions defined over a distribution $(\Sigma_1, \Sigma_2, \ldots, \Sigma_k)$ of alphabet Σ is given below.

$$e ::= 0 | fsync(s_1, s_2, \ldots, s_k), \quad \text{where} \quad s_i \text{ is a regular expression over } \Sigma_i$$

When $e = fsync(s_1, s_2, \ldots, s_k)$ and $I \subseteq Loc$, let the projection $e{\downarrow}I = \Pi_{i \in I} s_i$.

A connected expression $e = fsync(s_1, s_2, \ldots, s_k)$ over Σ, is said to have equal choice if, for all global actions a in Σ and for any i, j in $loc(a)$, any a-site of s_i have same *Init* set as of any a-site of s_j.

For a connected expression defined over distributed alphabet its derivatives and semantics were given in [18], and are given as follows. For the connected expression 0, we have $Lang(0) = \emptyset$. For the connected expression $e = fsync(s_1, \ldots, s_k)$, its language is $Lang(e) = Lang(s_1) \| Lang(s_2) \| \ldots \| Lang(s_k)$.

The definitions of derivatives extended to connected expressions [18] is as follows. The expression 0 has no derivatives on any action. Given an expression $e = fsync(s_1, s_2, \ldots, s_k)$, its derivatives are defined by induction using the derivatives of the s_i on action a:

$$Der_a(e) = \{fsync(r_1, \ldots, r_k) \mid \forall i \in loc(a), r_i \in Der_a(s_i); \text{ otherwise } r_j = s_j\}.$$

5.3   Connected Expressions with Pairings

We recall some properties of connected expressions over a distribution, which were, useful in construction of free choice nets. This property relates to matchings of direct products [18].

Definition 21 ([18]). *Let $e = fsync(s_1, s_2, \ldots, s_k)$ be a connected expression over Σ. For a global action a, pairing(a) is a subset of tuples $\Pi_{i \in loc(a)} Part_a(s_i)$ such that the projection of these tuples includes all the blocks of $Part_a(s_i)$, and if a block of $Part_a(s_j), j \in loc(a)$ appears in one tuple of the pairing, it does not appear in another tuple. (For convenience we also write pairing(a) as a subset of $\Pi_{i \in loc(a)} Der(s_i)$ which respects the partition.) We call pairing(a) equal choice if for every tuple in the pairing, the blocks of derivatives in the tuple have equal choice.*

Derivatives for connected expressions with *pairing* are defined as follows. A derivative $fsync(r_1, \ldots, r_k)$ is in *pairing*(a) if there is a tuple $D \in pairing(a)$ such that $r_i \in D[i]$ for all $i \in loc(a)$. For convenience we may write a derivative as an element of *pairing*(a). Expression e is said to have (equal choice) pairing of actions if for all global actions a, there exists an (equal choice) pairing of a. Expression e is said to be consistent with a pairing of actions if every reachable a-site $d \in Der(e)$ is in $pairing(a)$. Expression e is said to have equal choice property if it has equal choice pairing of actions for all global actions a in Σ.

Given a connected expression e with pairings, checking if it is consistent with pairing of actions can be done in PSPACE [18,19].

5.4   Connected Expression with Cables (*CE-cables*)

We give some properties of connected expressions over a distribution, which extend the notion of pairing, and have been related to product systems with globals [15]. The notion of cables corresponds to notion of globals of product systems, and hence it corresponds to transitions of a net.

Definition 22 ([15]). *Let $e = \mathrm{fsync}(s_1, s_2, \ldots, s_k)$ be a connected expression over Σ. For each action a in Σ, we define a-cables$(e) = \Pi_{i \in loc(a)} a$-ducts$(s_i)$. For an action a, an a-cable is an element of the set a-cables(e). We say that a block B of $\mathrm{Part}_a(s_i)$ appears in an a-cable D if there exists j in $loc(a)$ and there exists $Y \subseteq \mathrm{Der}_a(B)$ such that $D[j] = (B, Y)$, i.e. if B is a pre-block of a component a-duct of D. For any a-cable D, its set of pre-blocks $^{\bullet}D = \cup_{i \in loc(a)} \{B_i \mid B_i$ appears in $D\}$, i.e. the set of pre-blocks of all the of its component a-ducts.*
For expression e, let cables$(a) \subseteq a$-cables(e), such that for all i in $loc(a)$

1. *Each block B in $\mathrm{Part}_a(s_i)$, appears in at least one a-cable of it.*
2. *for all (B, E) and (B', E') in a-ducts(s_i) with $(B, E) \neq (B', E')$, if $B = B' \implies E \cap E' = \emptyset$, i.e. if any two distinct a-ducts of s_i appearing in it have same pre-block then, they must have disjoint post-effects.*

Connected expressions with cables were defined in [15], as follows.
A connected expression with cables (CE-cables) is a connected expression with relations cables(a) of it, for each global action a in Σ.

Derivatives of a connected expression with cables are [15] defined as follows. The *CE-cables* 0 has no derivatives on any action. For expression $e = \mathrm{fsync}(s_1, s_2, \ldots, s_k)$, we define its derivatives on action a, by induction, using a-ducts and the derivatives of s_j as:

$$\mathrm{Der}_a(e) = \{\mathrm{fsync}(r_1, r_2, \ldots, r_k) \mid r_j \in \mathrm{Der}_a(s_j) \text{ if there exists an } a\text{-cable } D$$
in cables(a) such that, for all j in $loc(a)$, s_j is in pre-block B_j and r_j is in X_j of a-duct $D[j] = (B_j, X_j)$ of s_j, otherwise $r_j = s_j\}$.

We use the word derivative for expressions such as $d = \mathrm{fsync}(r_1, \ldots, r_k)$ given above. The reachable derivatives are $\mathrm{Der}(e) = \{d \mid d \in \mathrm{Der}_x(e), x \in \Sigma^*\}$. A *CE-cables* is said to have equal source property if for any pair of two cables sharing a common pre-block have same set of pre-blocks. This property corresponds to same source property of product systems and relates to transitions belonging to same cluster of nets.

Language of e is the set of words over Σ defined using derivatives as below.

$$Lang(e) = \{w \in \Sigma^* \mid \exists e' \in \mathrm{Der}_w(e) \text{ such that } \varepsilon \in Lang(r_i), \text{ where } e'[i] = r_i\}.$$

So we can have next derivative on action a, if it is allowed by the cables(a) relation. The number of derivatives may be exponential in k. Let $\Sigma = (\Sigma_1 = \{a, b, c\}, \Sigma_2 = \{a, d, e\})$ be a distributed alphabet.

Example 8 (CE-pairings and CE-cables). Let $e = \text{fsync}((ab+ac)^*, (ad+ae)^*)$ be a connected expression defined over Σ. Here, $r_1 = (ab+ac)^*$ and $s_1 = (ad+ae)^*$. The set of derivatives of r_1 is $\text{Der}(r_1) = \{r_1, r_2 = br_1, r_3 = cr_1\}$ and for s_1 it is $\text{Der}(s_1) = \{s_1, s_2 = ds_1, s_3 = es_1\}$. We have $a\text{-sites}(r_1) = r_1$ and $\text{Part}_a(r_1) = \{D = r_1\}$. Similarly, $a\text{-sites}(s_1) = r_1$ and $\text{Part}_a(s_1) = \{D' = s_1\}$. The only possible pairing relation is $pairing(a) = \{(D, D')\}$. We have $\text{Der}_a(e) = \{\text{fsync}(r_i, s_j) \mid i, j \in \{2, 3\}\}$. Expression e satisfies equal choice property.

Now we associate a cabling relation with e. The set of a-effects of D is $\text{Der}_a(D) = \{r_2, r_3\}$. The set of a-ducts of r_1 is $\{(D, r_2), (D, r_3), (D, \{r_2, r_3\})\}$. The set of a-effects of D' is $\text{Der}_a(D') = \{s_2, s_3\}$ and the set of a-ducts of component expression s_1 is $\{(D', s_2), (D', s_3), (D', \{s_2, s_3\})\}$. A possible $cables(a)$ relations for expression e is $\{((D, r_2), (D', s_2)), ((D, r_3), (D', s_3))\}$. See that each block in the $\text{Part}_a(r_1)$ and $\text{Part}_a(s_1)$ appears at least once in the $cables(a)$ relation. And two a-ducts of r_1 appearing in this relation, have same pre-block D, so their set of post-effects r_2 and r_3 are disjoint. This condition also holds for a-ducts of s_1.

For both a-cables set of pre-blocks is identical, therefore $cables(a)$ satisfies equal source property. We have $\text{Der}_a(e) = \{\text{fsync}(r_2, s_2), \text{fsync}(r_3, s_3)\}$, but expression $\text{fsync}(r_2, s_3)$ is not in $\text{Der}_a(e)$, because only post-effect of D containing r_2 is the set $\{r_2\}$ and similarly, only post-effect of D' containing s_3 is the set $\{s_3\}$ and there does not exist an a-cable with $(D, \{r_2\})$ and $(D', \{s_3\})$ as its components. We have $\text{Der}(e) = \{e, (r_2, s_2), (r_3, s_3), (r_1, s_2), (r_1, s_3), (r_2, s_1), (r_3, s_1)\}$.

Another such example of connected expression with pairings (resp. cables) is given below.

Example 9 (CE-pairings). Let $e = \text{fsync}((ab+ac)^*a, (ad+ae)^*a)$ be a connected expression defined over Σ. Let $p_1 = (ab+ac)^*a$ with language L_1 and $q_1 = (ad+ae)^*a$ with language L_2. The set of derivatives are $\text{Der}(p_1) = \{p_1, p_2 = bp_1, p_3 = cp_1, p_4 = \varepsilon\}$ and $\text{Der}(q_1) = \{q_1, q_2 = dq_1, q_3 = eq_1, q_4 = \varepsilon\}$. The partitions of a-sites are $\text{Part}_a(p_1) = \{B = \{p_1\}\}$ and $\text{Part}_a(q_1) = \{B' = \{q_1\}\}$. A pairing relation is $pairing(a) = \{(B, B')\}$ and with respect to that $\text{Der}_a(e) = \{\text{fsync}(p_i, q_j) \mid i, j \in \{2, 3, 4\}\}$. Expression e has equal choice property.

Now we associate a cabling relation with e. The set of a-effects of B is $\text{Der}_a(B) = \{p_2, p_3, p_4\}$ and $\text{Der}_a(B') = \{q_2, q_3, q_4\}$. The set of a-ducts for p_1 is $\{(B, \{p_2\}), (B, \{p_3\}), (B, \{p_2, p_4\}), (B, \{p_3, p_4\})(B, \{p_2, p_4, p_3\})\}$ and for q_1 is $\{(B', q_2), (B', q_3), (B', \{q_2, q_4\}), (B', \{q_3, q_4\})(B', \{q_2, q_4, q_3\})\}$. A $cables(a)$ relation is $\{((B, p_2), (B', q_2)), ((B, p_3), (B', q_3)), ((B, p_2), (B', q_2))\}$. Expression e has equal source property and its set of derivatives with respect to letter a is $\text{Der}_a(e) = \{\text{fsync}(p_2, q_2), \text{fsync}(p_3, q_3), \text{fsync}(p_4, q_4)\}$.

Now we give two new properties of connected expressions. In a connected expression with cables and having equal source property, cables for an global action a can be partitioned into different compartments: two a-cables belong to same compartment if they have equal source. To any such a-cable D belonging to an equal source compartment ES_a of a-cables, we can associate a set of its post-blocks listed in some order as $\pi(D) = \Pi_{i \in loc(a)}(D^\bullet \cap \chi_i)$, where

$\chi_i = \{Part_a(s_i) \mid a \in \Sigma\}$. Let $post(ES_a) = \{\pi(D) \mid D \in ES_a\}$. Then we define $ES_a[i] = post(ES_a^i) \cap \chi_i$ and $postdecomp(ES_a) = \Pi_{i \in loc(a)} ES_a[i]$. We call $ES_a[i]$ as **post-projection** and $postdecomp(ES_a)$ as **post-decomposition** of the compartment ES_a.

The following property of connected expressions will later be related to product-moves property of direct products and hence to distributed-choice of nets.

Definition 23 (product-derivatives property). *A connected expression with cables and having equal source property, has **product-derivatives property**, if for all a in Σ, and for all equal source compartments ES_a of a-cables $postdecomp(ES_a) \subseteq post(ES_a)$.*

Example 10. The connected expression $e = fsync((ab+ac)^*, (ad+ae)^*)$ of Example 8 does not have product-derivative property with the given cabling relation $\{((D, r_2), (D', s_2)), ((D, r_3), (D', s_3))\}$. If we associate the cabling relation $\{((D, r_2), (D', s_2)), ((D, r_3), (D', s_3)), ((D, r_2), (D', s_3)), ((D, r_3), (D', s_2))\}$ with e then it has product-derivative property.

Definition 24. *A connected expression e is **action-live** if for all actions a in Σ, from any reachable derivative of e, we can reach an a-derivative of e.*

5.5   Relating Connected-Expressions with Pairings and with Cables

First, we show how connected expressions with equal choice and consistent pairings can be seen as connected expression with cables and having equal source and product-derivatives property.

Theorem 7. *Let Σ be a distributed alphabet and e be a connected expression having equal choice and consistent pairing of actions, defined over Σ. Then for the language of e, we can construct a connected expression e' with cables having equal source and product-derivatives property. The constructed expression e' with cables is exponential in the size of expression e with pairings.*

Now we give a language preserving construction of connected expression with equal choice and consistent pairings from a *CE-cables* having equal source and product-derivatives property.

Theorem 8. *Let Σ be a distributed alphabet and e' be a connected expression with cables defined over it. Then we can construct a connected expression e with pairings linear in the size of e'. And, if e' has equal source then e has equal choice property. In addition, if e' is action-live then if e' has product moves property then e has consistency of pairing. $Lang(e) = Lang(e')$.*

Expressions given in the following subsection correspond to product systems with subset acceptance condition.

5.6   Sum of Connected Expressions (*SCE*)

We give syntax for sum of connected expressions (*SCE*) defined over a distribution $(\Sigma_1, \Sigma_2, \ldots, \Sigma_k)$ of alphabet Σ.

$e ::= 0 | e_1 + \cdots + e_m$,   where e_i is a connected expression (*CE*) over Σ

For an *SCE* e its semantics is given as follows: For the *SCE* 0, we have $Lang(0) = \emptyset$. For the *SCE* $e = e_1 + \cdots + e_m$, its language is given as $Lang(e) = Lang(e_1) \cup Lang(e_2) \cup \cdots \cup Lang(s_m)$. The definitions of derivatives extended to *SCE*s is as given below. The expression 0 has no derivatives on any action. Derivative of an *SCE* with respect to a letter a is defined as: $Der_a(e) = Der_a(e_1) \cup Der_a(e_2) \cup \cdots \cup Der_a(s_m)$. Inductively $Der_{aw}(e) = Der_w(Der_a(e))$. The set of all derivatives $Der(e)$ is union of sets of derivatives of e over all words w in Σ_i^* for all i in $\{1, \ldots, m\}$.

Definition 25 (*SCE* with pairings)). *An SCE $e = e_1 + \cdots + e_m$ where each e_i is a CE-pairings, is called a sum of connected expressions with pairing (SCE-pairings). An SCE-pairings e is said to have equal choice property if each component CE-pairings e_i of the sum also has it.*

Example 11. The expression $e = e_1 + e_2$ where $e_1 = \text{fsync}((ab+ac)^*, (ad+ae)^*)$ is the *CE-pairings* of Example 8 and $e_2 = \text{fsync}((ab + ac)^*a, (ad + ae)^*a)$ is the *CE-pairings* of Example 9 is an *SCE-pairings* with equal choice property, as both e_1 and e_2 have it.

Definition 26 (*SCE* with cables)). *An SCE $e = e_1 + \cdots + e_m$ where each e_i is a CE-cables, then e is called a sum of connected expressions with cables (SCE-cables). An SCE-cables e is said to have **equal source property** if each component CE-cables e_i of the sum also has it. An SCE-cables e has **product-derivatives property** if each component CE-cables e_i of the sum has it.*

Example 12. The expression $e' = e_3 + e_4$ where $e_3 = \text{fsync}((ab+ac)^*, (ad+ae)^*)$ is the *CE-cables* of Example 8 and $e_4 = \text{fsync}((ab + ac)^*a, (ad + ae)^*a)$ is the *CE-cables* of Example 9.

As an example of how derivatives of *SCE-cables* from derivatives of *SCE-pairings* differ even while having identical components, see that $fsync(r_2, s_3) \in Der_a(e)$ (of Example 11) but it does not belong to $Der_a(e')$.

The missing proofs from this section can be found in [17].

6   Connected Expressions and Product Systems

To get a product system with globals having subset-acceptance, from a sum of connected expression, we use an earlier result from [15], where construction of *PS-globals* with product-acceptance was given from a *CE-cables*.

For each set of derivatives (pre-blocks and after-effects), of a component regular expression, we constructed an unique state of a local component, which gives us product moves property in the constructed product system, if we have product-derivatives property for given *CE-cables*.

Lemma 7 (*CE-cables* to *PS-globals* with product-acceptance [15]). *Let e be a CE-cables, defined over a distribution Σ. Then for the language of e, we can compute a PS-globals with product-acceptance linear in the size of expression e. Further, if e had equal-source, then system A has same source property; and, if e had product-derivatives then A has product moves property.*

Using Lemma 7, we get *PS-globals* with subset-acceptance, from sum of connected expressions.

Theorem 9 (*SCE*-cables to *PS-globals* with subset-acceptance). *Let e be an sum of connected expression defined over Σ. Then we can construct a PS-globals A with subset-acceptance for the language of e. If e had equal source property, then has same source property. In addition, if e has product-derivatives property then A has product moves property.*

Proof. Let $e = e_1 + \ldots + e_m$ be an sum of *CE-cables* having equal source and product-derivatives property. Language of expression e is $Lang(e) = Lang(e_1) \cup \ldots \cup Lang(e_m)$.

Using Lemma 7, we construct *PS-globals* A_i with product-acceptance condition, for the language of each *CE-cables* e_i having the same source and product-derivatives property. That is $Lang(A_i) = Lang(e_i)$ for all i in $\{1, \ldots, m\}$.

Using language characterization of *PS-globals* with subset-acceptance conditions given in Corollary 2, we get an *PS-globals* A with subset-acceptance condition, over Σ such that $Lang(A) = Lang(A_1) \cup \ldots \cup Lang(A_m)$. Since Corollary 2, preserves global moves of component *PS-globals* and, as underlying *CE-cables* had equal source and product-derivatives property, we get same source property and product moves property for each of the component A_i. Hence A has both these properties as required.   □

Example 13. For *SCE-cables* e' of Example 12, we can construct *PS-globals* with same source property and subset-acceptance $\mathcal{D}$ of Example 6, accepting language L_s, using Theorem 9.

A language preserving construction of connected expressions having the equal source property, from a *PS-globals* with the same source property and product-acceptance, was given in [15]. Since each local state–either a source state or a target state of local move–is mapped uniquely to a set of derivatives of component regular expression, we have product-derivatives property for the expression, if the product system had product-moves property.

Lemma 8 (*PS-globals* with product-acceptance to *CE*-cables [15]). *Let Σ be a distributed alphabet and, A be a product system with globals and product-acceptance, defined over Σ. For the language of A, we can construct a connected expression e with cables, exponential in the size of the given product system. Furthermore, if product system has same source property then connected expression with cables has equal source property, in addition, if it has product-moves property then connected expression with cables has product-derivatives property.*

Now using Lemma 8, we get a sum of connected expressions for *PS-globals* with subset-acceptance.

Theorem 10 (*PS-globals* with subset-acceptance to *SCE*-cables). *Let A be a product system with globals and subset-acceptance, defined over distribution Σ. For the language of A, we can construct a sum of connected expression e with cables. And, if product system has same source property then e has equal source property. Also, in addition, if A has product moves property then e has product-derivatives property.*

Proof. Let A be an *PS-globals* with subset-acceptance condition and same source property. Using Corollary 2, there exist *PS-globals* $A_1, \ldots, A_m$ with product-acceptance conditions such that $Lang(A) = Lang(A_1) \cup \ldots \cup Lang(A_m)$. Note that each A_i has same source property.

For the language of each *PS-globals* A_i with product-acceptance condition, we can construct *CE-cables* e_i with equal source property, using Lemma 8.

From these we construct a sum of *CE-cables* $e = e_1 + \ldots + e_m$ which has equal source property and language $Lang(e_1) \cup \ldots \cup Lang(e_m)$ which is $Lang(A)$. □

Example 14. For a *PS-globals* with same source property and subset-acceptance $\mathcal{D}$ of Example 6, accepting language L_s, we can construct an *SCE-cables* e' of Example 12 using Theorem 10.

Using equivalence of *PS-matchings* with product-acceptance and *CE-cables*, from [18,19], and Corollary 1 we get language equivalent *SCE-pairings* for *PS-matchings* with subset-acceptance, and vice-versa.

Theorem 11. (*PS-matchings* with subset-acceptance to *SCE*-pairings). *Let A be a product system with conflict-equivalent and consistent matchings, having subset-acceptance. For the language of A, we can construct a sum of connected expression e with equal choice and consistent pairings.*

The converse result follows.

Theorem 12. (*SCE*-pairings to *PS-matchings* with subset-acceptance). *Let Σ be a distributed alphabet and a sum of connected expression e defined over it, with equal choice and consistent pairings. Then for its language we can construct a product system with conflict-equivalent and consistent matchings, having subset-acceptance.*

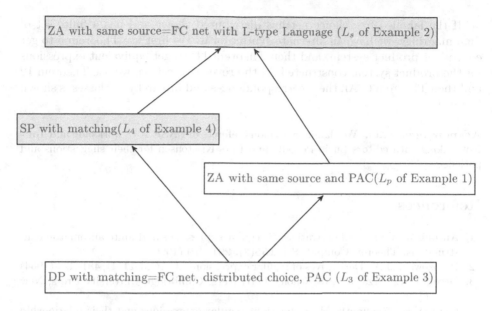

Fig. 6. Free choice net languages and automata over distributed alphabets

7 Conclusion

In this paper, we have given a language (L_s of Example 2) which can be accepted by *free choice Zielonka automata*. This language is not accepted by any synchronous product or direct product, using Lemma 2, proof of which is presented here, and was not given in [16]. We have also given a language (L_4 of Example 4) which can be accepted by a *free choice synchronous product* and not by any direct product. A language which can be accepted by *free choice direct product* (L_3 of Example 3) was given in [18]. With this we have a hierarchy of labelled free choice nets similar to automata over distributed alphabets. In addition we have defined Zielonka automata with product acceptance condition and its free choice restriction. We have given language (L_p of Example 1) of this class. We used this intermediate automata to obtain Kleene theorem for *free choice Zielonka automata*. Lemma 3 shows that this class is strictly more expressive than direct products with matching. In addition, ZA with same source has product moves property (class is not shown in the figure) then it is equivalent to SP with matching.

We give below the summary of correspondences established for the nets, automata over distributed alphabets and expressions. To get an expression, for the language of a labelled 1-bounded and S-coverable free choice net (with or without distributed choice) having a finite set of final markings, we use Theorems 4 and 10. In the reverse direction, we use Theorem 9 to get product system with subset acceptance from expressions and then Theorem 3 to get a language equivalent free choice net system.

If the labelled free choice net has distributed choice and has a finite set of final markings, we have an alternate syntax for it. We first use Theorem 5 to get equivalent product system, and then Theorem 11, to get equivalent expressions for the product system constructed. In the reverse direction, we use Theorem 12 and then Theorem 6. All these correspondences and hirearchy of classes is shown in Fig. 6.

Acknowedgements. We thank anonymous referees of PNSE 2018 workshop and ToP-NoC, along with editors Lucio Pomello and Lars Kristensen for their suggestions and patience.

References

1. Antimirov, V.: Partial derivatives of regular expressions and finite automaton constructions. Theoret. Comput. Sci. **155**(2), 291–319 (1996)
2. Brzozowski, J.A.: Derivatives of regular expressions. J. ACM **11**(4), 481–494 (1964)
3. Desel, J., Esparza, J.: Free Choice Petri Nets. Cambridge University Press, New York (1995)
4. Garg, V.K., Ragunath, M.: Concurrent regular expressions and their relationship to petri nets. Theoret. Comput. Sci. **96**(2), 285–304 (1992)
5. Grabowski, J.: On partial languages. Fundam. Inform. **4**(2), 427–498 (1981)
6. Hack, M.H.T.: Analysis of production schemata by Petri nets. Project Mac Report TR-94, MIT (1972)
7. Iordache, M.V., Antsaklis, P.J.: The ACTS software and its supervisory control framework. In: Proceedings Conference on Decision and Control, CDC, pp. 7238–7243. IEEE (2012)
8. Jantzen, M.: Language theory of Petri nets. In: Brauer, W., Reisig, W., Rozenberg, G. (eds.) ACPN 1986. LNCS, vol. 254, pp. 397–412. Springer, Heidelberg (1987). https://doi.org/10.1007/978-3-540-47919-2_15
9. Lodaya, K.: Product automata and process algebra. In: SEFM. IEEE (2006)
10. Lodaya, K., Mukund, M., Phawade, R.: Kleene theorems for product systems. In: Holzer, M., Kutrib, M., Pighizzini, G. (eds.) DCFS 2011. LNCS, vol. 6808, pp. 235–247. Springer, Heidelberg (2011). https://doi.org/10.1007/978-3-642-22600-7_19
11. Mirkin, B.G.: An algorithm for constructing a base in a language of regular expressions. Eng. Cybern. **5**, 110–116 (1966)
12. Mukund, M.: Automata on distributed alphabets. In: D'Souza, D., Shankar, P. (eds.) Modern Applications of Automata Theory. World Scientific (2011)
13. Petersen, J.L.: Computation sequence sets. J. Comput. Syst. Sci. **13**(1), 1–24 (1976)
14. Phawade, R.: Labelled free choice nets, finite product automata, and expressions. Ph.D. thesis, Homi Bhabha National Institute (2015)
15. Phawade, R.: Kleene theorems for labelled free choice nets without distributed choice. In: Cabac, L., Kristensen, L.M., Rölke, H. (eds.) Proceedings of PNSE. CEUR Workshop Proceedings, vol. 1591, pp. 132–152. CEUR-WS.org (2016)
16. Phawade, R.: Kleene theorems free choice nets labelled with distributed alphabets. In: Daniel Moldt, E.K., Rölke, H. (eds.) Proceedings of PNSE. CEUR Workshop Proceedings, vol. 2138, pp. 77–98. CEUR-WS.org (2018)

17. Phawade, R.: Kleene Theorems for Free Choice Nets Labelled with Distributed Alphabets. arXiv e-prints arXiv:1907.01168, July 2019
18. Phawade, R., Lodaya, K.: Kleene theorems for labelled free choice nets. In: Moldt, D., Rölke, H. (eds.) Proceedings of PNSE. CEUR Workshop Proceedings, vol. 1160, pp. 75–89. CEUR-WS.org (2014)
19. Phawade, R., Lodaya, K.: Kleene theorems for synchronous products with matching. In: Koutny, M., Desel, J., Haddad, S. (eds.) Transactions on Petri Nets and Other Models of Concurrency X. LNCS, vol. 9410, pp. 84–108. Springer, Heidelberg (2015). https://doi.org/10.1007/978-3-662-48650-4_5
20. Zielonka, W.: Notes on finite asynchronous automata. Inform. Theor. Appl. 21(2), 99–135 (1987)

Synthesis of Weighted Marked Graphs from Constrained Labelled Transition Systems: A Geometric Approach

Raymond Devillers[1], Evgeny Erofeev[2], and Thomas Hujsa[3](✉)

[1] Département d'Informatique, Université Libre de Bruxelles, 1050 Brussels, Belgium
rdevil@ulb.ac.be
[2] Department of Computing Science, Carl von Ossietzky Universität Oldenburg,
26111 Oldenburg, Germany
evgeny.erofeev@uni-oldenburg.de
[3] LAAS-CNRS, Université de Toulouse, CNRS, Toulouse, France
thujsa@laas.fr

Abstract. Recent studies investigated the problems of analysing Petri nets and synthesising them from labelled transition systems (LTS) with two labels (transitions) only. In this paper, we extend these works by providing new conditions for the synthesis of Weighted Marked Graphs (WMGs), a well-known and useful class of weighted Petri nets in which each place has at most one input and one output.

Some of these new conditions do not restrict the number of labels; the other ones consider up to 3 labels. Additional constraints are investigated: when the LTS is either finite or infinite, and either cyclic or acyclic. We show that one of these conditions, developed for 3 labels, does not extend to 4 nor to 5 labels. Also, we tackle geometrically the WMG-solvability of finite, acyclic LTS with any number of labels.

Keywords: Weighted Petri net · Marked graph · Synthesis · Labelled transition system · Cycles · Cyclic words · Circular solvability · Theory of regions · Geometric interpretation

1 Introduction

Petri nets form a highly expressive and intuitive operational model of discrete event systems, capturing the mechanisms of synchronisation, conflict and concurrency. Many of their fundamental behavioural properties are decidable, allowing to model and analyse numerous artificial and natural systems. However, most interesting model checking problems are intractable, and the efficiency of synthesis algorithms varies widely depending on the constraints imposed on the desired solution. In this study, we focus on the Petri net synthesis problem from

E. Erofeev—Supported by DFG through grant Be 1267/16-1 ASYST.
T. Hujsa—Supported by the STAE foundation/project DAEDALUS, Toulouse, France.

M. Koutny et al. (Eds.): ToPNoC XIV, LNCS 11790, pp. 172–191, 2019.
https://doi.org/10.1007/978-3-662-60651-3_7

a labelled transition system (LTS), which consists in determining the existence of a Petri net whose reachability graph is isomorphic to the given LTS, and building such a Petri net solution when it exists.

In previous studies on the analysis or synthesis of Petri nets, structural restrictions encompassed *plain* nets (each weight equals 1; also called ordinary nets) [1], *homogeneous* nets (meaning that for each place p, all the output weights of p are equal) [2,3], *free-choice* nets (the net is homogeneous, and any two transitions sharing an input place have the same set of input places) [2,4], choice-free nets (each place has at most one output transition) [5,6], marked graphs (each place has at most one output transition and one input transition) [7–10], join-free nets (each transition has at most one input place) [2,3,11,12], etc.

More recently, another kind of restriction has been considered, limiting the number of different transition labels of the LTS in combination with restrictions on the LTS structure: for the binary case, feasibility of net synthesis from finite linear LTS and LTS with *cycles*[1] has been characterised by rates of labels in the transition system [13,14] and by pseudo-regular expressions [15], giving rise to fast specialised synthesis algorithms; moreover, a complete enumeration of the shapes of synthesisable transition systems is presented in [16].

In this paper, we combine the restriction on the number of labels with the weighted marked graph (WMG) constraint. In addition, we study constraints on the existence of cycles in the LTS: when the LTS is *acyclic*, i.e. it does not contain any *cycle*, and when it is *cyclic*, i.e. it contains at least one cycle. In the latter case, we also study the finite *circular LTS*, meaning strongly connected LTSs that contain a unique cycle: we investigate the *cyclic solvability* of a word w, meaning the existence of a Petri net solution to the finite circular LTS induced by the infinite *cyclic word* w^∞.

An important purpose of studying such constrained LTSs is to better understand the relationship between LTS decompositions and their solvability by Petri nets. Indeed, the unsolvability of simple subgraphs of the given LTS, typically elementary paths (i.e. not containing any node twice) and cycles (i.e. closed paths, whose start and end states are equal), often induces simple conditions of unsolvability for the entire LTS, as highlighted in other works [13,15,17]. Moreover, cycles appear systematically in the reachability graph of live and/or reversible Petri nets [5,18], which are used to model various real-world applications, such as embedded systems [19].

Contributions. In this work, we study further the links between simple LTS structures and the reachability graph of WMGs, as follows.

First, we provide a characterisation of the 2-label (i.e. binary) words being cyclically solvable by a WMG (i.e. WMG-solvable), and extend the analysis to finite cyclic LTSs. We also tackle the case of infinite cyclic LTSs with 2 labels.

---

[1] A set A of k arcs in a LTS G defines a cycle of G if the elements of A can be ordered as a sequence $a_1 \ldots a_k$ such that, for each $i \in \{1, \ldots, k\}$, $a_i = (n_i, \ell_i, n_{i+1})$ and $n_{k+1} = n_1$, i.e. the i-th arc a_i goes from node n_i to node n_{i+1} until the first node n_1 is reached, closing the path. Cycles are also sometimes called circuits, circles and oriented cycles.

Then, when the number of labels is arbitrary, we provide a geometric characterisation of the finite, acyclic, WMG-solvable LTS, as well as a general sufficient condition of WMG-solvability for a cyclic word, using a decomposition into specific cyclically WMG-solvable binary subwords. We prove that this sufficient condition becomes a characterisation of cyclic WMG-solvability for a subclass of the 3-label words. Furthermore, we show, with the help of two counter-examples, that this characterisation does not hold for words with four or five labels.

Comparing with [20], we refine the results and explanations on WMG-solvable, finite, cyclic, binary LTSs by introducing Lemma 1 and upgrading Theorem 2, in Subsect. 3.1. We also provide the new geometric characterisation of WMG-solvability for acyclic LTS with any number of labels, and we sharpen the counter-examples to the characterisation of cyclically solvable ternary words in the cases of four and five labels.

Organisation of the Paper. After recalling classical definitions, notations and properties in Sect. 2, we present the results of WMG-solvability for 2-label words in Sect. 3. Then, in Sect. 4, we propose the geometric characterisation of WMG-solvability for an acyclic LTS with any number of labels. In Sect. 5, we develop the general sufficient condition of circular WMG-solvability for any number of labels. In Sect. 6, we tackle the ternary case and exhibit counter-examples for 4 and 5 labels. Finally, Sect. 7 presents our conclusions and perspectives.

2   Classical Definitions, Notations and Properties

LTSs, Sequences and Reachability. A *labelled transition system with initial state*, abbreviated *LTS*, is a quadruple $TS = (S, \rightarrow, T, \iota)$ where S is the set of *states*, T is the set of *labels*, $\rightarrow \subseteq (S \times T \times S)$ is the *labelled transition relation*, and $\iota \in S$ is the *initial state*.

A label t is *enabled* at $s \in S$, written $s[t\rangle$, if $\exists s' \in S \colon (s, t, s') \in \rightarrow$, in which case s' is said to be *reachable* from s by the firing of t, and we write $s[t\rangle s'$. Generalising to any (firing) sequences $\sigma \in T^*$, $s[\varepsilon\rangle$ and $s[\varepsilon\rangle s$ are always true; and $s[\sigma t\rangle s'$, i.e. σt is *enabled* from state s and leads to s', if there is some s'' with $s[\sigma\rangle s''$ and $s''[t\rangle s'$. A state s' is *reachable* from state s if $\exists \sigma \in T^* \colon s[\sigma\rangle s'$. The set of states reachable from s is denoted by $[s\rangle$.

Petri Nets, Reachability and Languages. A (finite, place-transition) *weighted Petri net*, or *weighted net*, is a tuple $N = (P, T, W)$ where P is a finite set of *places*, T is a finite set of *transitions*, with $P \cap T = \emptyset$, and $W \colon ((P \times T) \cup (T \times P)) \rightarrow \mathbb{N}$ is a *weight* function giving the weight of each arc. A *Petri net system*, or *system*, is a tuple $\mathcal{S} = (N, M_0)$ where N is a net and M_0 is the *initial marking*, which is a mapping $M_0 \colon P \rightarrow \mathbb{N}$ (hence a member of $\mathbb{N}^P$) indicating the initial number of *tokens* in each place. If $W(x, y) > 0$, y is said to be an *output* of x, and x is said to be an *input* of y. The *incidence matrix* C of the net is the integer $P \times T$-matrix with components $C(p, t) = W(t, p) - W(p, t)$.

A transition $t \in T$ is *enabled by* a marking M, denoted by $M[t\rangle$, if for all places $p \in P$, $M(p) \geq W(p, t)$. A place $p \in P$ is *enabled by* a marking M if

$M(p) \geq W(p,t)$ for every output transition t of p, meaning that it is not an obstacle to enabling transitions. If t is enabled at M, then t can *occur* (or *fire*) in M, leading to the marking M' defined by $M'(p) = M(p) - W(p,t) + W(t,p)$; we note $M[t\rangle M'$. A marking M' is *reachable* from M if there is a sequence of firings leading from M to M'; if this sequence of firings defines a sequence of transitions $\sigma \in T^*$, we note $M[\sigma\rangle M'$. The set of markings reachable from M is denoted by $[M\rangle$. The *reachability graph of* $\mathcal{S}$ is the labelled transition system $RG(\mathcal{S})$ with the set of vertices $[M_0\rangle$, the set of labels T, initial state M_0 and transitions $\{(M, t, M') \mid M, M' \in [M_0\rangle \wedge M[t\rangle M'\}$. A system $\mathcal{S}$ is *bounded* if $RG(\mathcal{S})$ is finite.

The language of a Petri net system $\mathcal{S}$ is the set $\mathcal{L}(\mathcal{S}) = \{\sigma \in T^* \mid M_0[\sigma\rangle\}$. These languages are prefix-closed, i.e., if $\sigma = \sigma'\sigma'' \in \mathcal{L}(\mathcal{S})$, then $\sigma' \in \mathcal{L}(\mathcal{S})$. For any language $L \subseteq T^*$, we denote by $PREF(L)$ the language formed by its prefixes.

Vectors. The *support* of a vector is the set of the indices of its non-null components. Consider any net $N = (P, T, W)$ with its incidence matrix C. A *T-vector* is an element of $\mathbb{N}^T$; it is called *prime* if the greatest common divisor of its components is one (i.e. its components do not have a common non-unit factor). A *T-semiflow* ν of the net is a non-null T-vector such that $C \cdot \nu = 0$. A T-semiflow is called *minimal* when it is prime and its support is not a proper superset of the support of any other T-semiflow [5].

The *Parikh vector* $\mathbf{P}(\sigma)$ of a finite sequence σ of transitions is a T-vector counting the number of occurrences of each transition in σ, and the *support* of σ is the support of its Parikh vector, i.e. $supp(\sigma) = supp(\mathbf{P}(\sigma)) = \{t \in T \mid \mathbf{P}(\sigma)(t) > 0\}$.

Strong Connectedness and Cycles in an LTS. The LTS $(S, \rightarrow, T, \iota)$ is said *reversible* if, $\forall s \in [\iota\rangle$, we have $\iota \in [s\rangle$, i.e., it is always possible to go back to the initial state; reversibility implies the strong connectedness of the LTS.

A sequence $s[\sigma\rangle s'$ is called a *cycle*, or more precisely a *cycle at (or around) state* s, if $s = s'$. A non-empty cycle $s[\sigma\rangle s$ is called *small* if there is no non-empty cycle $s'[\sigma'\rangle s'$ in TS with $\mathbf{P}(\sigma') \lneq \mathbf{P}(\sigma)$, meaning that no component of the left vector is greater than the corresponding component of the right vector, and at least one is smaller (the definition of Parikh vectors extending readily to sequences over the set of labels T of the LTS).

A *circular LTS* is a finite, strongly connected LTS that contains a unique cycle; hence, it has the shape of an oriented circle. The circular LTS *induced by* a word $w = w_1 \ldots w_k$ is the LTS with initial state s_0 defined as $s_0[w_1\rangle s_1[w_2\rangle s_2 \ldots [w_k\rangle s_0$.

All notions defined for labelled transition systems apply to Petri nets through their reachability graphs.

Petri Net Subclasses. A net N is *plain* if no arc weight exceeds 1; *pure* if $\forall p \in P: (p^\bullet \cap {}^\bullet p) = \emptyset$, where $p^\bullet = \{t \in T \mid W(p,t) > 0\}$ and ${}^\bullet p = \{t \in T \mid W(t,p) > 0\}$; CF (*choice-free* [5,21]) or ON (place-output-nonbranching [17]) if $\forall p \in P: |p^\bullet| \leq 1$; a WMG (*weighted marked graph* [8]) if $|p^\bullet| \leq 1$ and $|{}^\bullet p| \leq 1$

for all places $p \in P$. The latter form a subclass of the choice-free nets; other subclasses are *marked graphs* [7], which are plain with $|p^\bullet| = 1$ and $|^\bullet p| = 1$ for each place $p \in P$, and *T-systems* [4], which are plain with $|p^\bullet| \leq 1$ and $|^\bullet p| \leq 1$ for each place $p \in P$.

Isomorphism and Solvability. Two LTS $TS_1 = (S_1, \rightarrow_1, T, s_{01})$ and $TS_2 = (S_2, \rightarrow_2, T, s_{02})$ are isomorphic if there is a bijection $\zeta \colon S_1 \rightarrow S_2$ with $\zeta(s_{01}) = s_{02}$ and $(s, t, s') \in \rightarrow_1 \Leftrightarrow (\zeta(s), t, \zeta(s')) \in \rightarrow_2$, for all $s, s' \in S_1$.

If an LTS TS is isomorphic to $RG(\mathcal{S})$ where $\mathcal{S}$ is a system, we say that $\mathcal{S}$ *solves* TS. Solving a word $w = \ell_1 \ldots \ell_k$ amounts to solve the acyclic LTS defined by the single path $\iota[\ell_1\rangle s_1 \ldots [\ell_k\rangle s_k$. A finite word w is *cyclically solvable* if the circular LTS induced by w is solvable. A LTS is WMG-solvable if a WMG solves it.

Other Classical Notions. An LTS $TS = (S, \rightarrow, T, \iota)$ is *fully reachable* if $S = [\iota\rangle$. It is *forward deterministic* if $s[t\rangle s' \wedge s[t\rangle s'' \Rightarrow s' = s''$, and *backward deterministic* if $s'[t\rangle s \wedge s''[t\rangle s \Rightarrow s' = s''$.

A system $\mathcal{S}$ is *forward persistent* if, for any reachable markings M, M_1, M_2, $(M[a\rangle M_1 \wedge M[b\rangle M_2 \wedge a \neq b) \Rightarrow M_1[b\rangle M' \wedge M_2[a\rangle M'$ for a reachable marking M'; it is *backward persistent* if, for any reachable markings M, M_1, M_2, $(M_1[a\rangle M \wedge M_2[b\rangle M \wedge a \neq b) \Rightarrow M'[b\rangle M_1 \wedge M'[a\rangle M_2$ for a reachable marking M'.

Next, we recall classical properties of Petri net reachability graphs.

Proposition 1 (Classical Petri net properties). *If $\mathcal{S}$ is a Petri net system:*

– *$RG(\mathcal{S})$ is a fully reachable LTS.*
– *$RG(\mathcal{S})$ is forward deterministic and backward deterministic.*

For the subclass of WMGs, we have the following dedicated properties, extracted from Proposition 4, Lemma 1, Theorem 2 and Lemma 2 in [10].

Proposition 2 (Properties of WMG). *If $\mathcal{S} = (N, M_0)$ is a WMG system:*

– *It is forward persistent and backward persistent.*
– *If N is connected and has a T-semiflow ν, then there is a unique minimal one π, with support T, and $\nu = k \cdot \pi$ for some positive integer k. Moreover, if there is a non-empty cycle in $RG(\mathcal{S})$, there is one with Parikh vector π in $RG(\mathcal{S})$ around each reachable marking and $RG(\mathcal{S})$ is reversible. If there is no cycle, all the paths starting from some state s and reaching some state s' have the same Parikh vector.*

To simplify our reasoning in the sequel, we introduce the following notation, which captures some of the behavioural properties satisfied by WMG (Propositions 1 and 2). We denote by

– **b** (for basic) the set of properties: forward and backward deterministic, forward and backward persistent, totally reachable;
– **c** (for cyclic) the property: there is a small cycle whose Parikh vector is prime with support T.

A synthesis procedure does not necessarily lead to a connected solution. However, the technique of decomposition into prime factors described in [22,23] can always be applied first, so as to handle connected partial solutions and recombine them afterwards. Hence, in the following, we focus on connected WMGs, without loss of generality. In the next section, we consider the synthesis problem of WMG with exactly two different labels.

3   Synthesis of a WMG from a Cyclic Binary LTS

In this section, we provide conditions for the WMG-solvability of 2-label cyclic LTS. In Subsect. 3.1, we investigate the WMG-solvability of a finite cyclic LTS: first when it is circular, then without this constraint. In Subsect. 3.2, we investigate the WMG-solvability of an infinite cyclic binary LTS.

3.1   WMG-solvable Finite Cyclic Binary LTS

In this subsection, we first consider any circular LTS with only two different labels. Each such LTS is defined by a word $w \in \{a, b\}^*$, corresponding to the labels encountered by firing the circuit once from ι, leading back to ι. Changing the initial state in this LTS amounts to rotate w. Clearly, each such LTS satisfies property **b**, but is not always WMG- (nor even Petri net-) solvable.

The next results consider circuit Petri nets as represented in Fig. 1, where places are named following the direction of the arcs, e.g. $p_{a,b}$ is the output place of a and the input place of b.

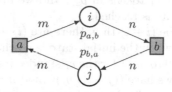

Fig. 1. A generic WMG solving a finite circular LTS induced by a word w over the alphabet $\{a, b\}$, whose initial marking (i, j) depends on the given solvable LTS. We assume that $\mathbf{P}(w) = (n, m)$ is prime.

Theorem 1 (Cyclically WMG-solvable binary words). *Consider a finite binary word w over the alphabet $\{a, b\}$, with $\mathbf{P}(w) = (n, m)$ and $n \leq m$, the case $m \leq n$ being handled symmetrically. Then, w is cyclically solvable if and only if $\gcd(n, m) = 1$ and w is a rotation of the word $w' = ab^{m_0} \ldots ab^{m_{n-1}}$, where the sequence $m_0, \ldots, m_{n-1}$ is the sequence of quotients in the following system of equalities, with $r_0 = 0$:*

$$\begin{cases} r_0 + m = m_0 \cdot n + r_1, & where \ 0 \leq r_1 < n \\ r_1 + m = m_1 \cdot n + r_2, & where \ 0 \leq r_2 < n \\ \ldots \\ r_{n-1} + m = m_{n-1} \cdot n. \end{cases}$$

Moreover, $m + n - 1$ tokens are necessary and sufficient to solve the word cyclically.

Proof. From Proposition 2, for a connected WMG solution to exist, the Parikh vector of the word must be the minimal T-semiflow $\mu = (n, m)$ with support $T = \{a, b\}$, which is prime by definition, thus $gcd(m, n) = 1$. A variant of this problem has been studied in [24], Sect. 6. Basing of this previous study, we highlight the following facts, leading to the claim. If a solution exists, then:

- there exists a WMG solution as pictured in Fig. 1, in which each firing preserves the number of tokens; thus, denoting by $M_s(p)$ the marking of place p at state s, the sum $M_s(p_{a,b}) + M_s(p_{b,a})$ is the same for all states.
- Consider any two different reachable markings M' and M'', then, from the above, $M'(p_{a,b}) \neq M''(p_{a,b})$ and $M'(p_{b,a}) \neq M''(p_{b,a})$.
- $M_s(p_{a,b}) + M_s(p_{b,a}) = m + n - 1$. Indeed, with more tokens, a reachable marking enables both a and b, which is not allowed by the given LTS; with fewer tokens, a deadlock is reached, i.e. a marking that enables no transition.
- For each i, $m_i \in \{\lfloor m/n \rfloor, \lceil m/n \rceil\}$, there are $(m \bmod n)$ b-blocks of size $(\lfloor m/n \rfloor + 1)$, the other ones have size $\lfloor m/n \rfloor$.

Let us start from the state s such that $M_s(p_{a,b}) = 0$ and $M_s(p_{b,a}) = m + n - 1$, with $r_0 = 0$. We denote by r_i the number of tokens in $p_{a,b}$ at the $(i + 1)$-th visited state that enables a. The value m_0 is the maximal number of b's that can be fired after the first a, and then r_1 tokens remain in $p_{a,b}$; hence, there are $m + n - 1 - r_1$ tokens in $p_{b,a}$ (which is at least m) before the second a. After the second a, we have $m + r_1$ tokens in $p_{a,b}$ and we fire m_1 b's. We iterate the process until the initial state is reached.

In the state enabling the $(i+1)$-th a, there are $(i \cdot m) \bmod n$ tokens in $p_{a,b}$, implying that r_n equals 0 when the initial state is reached again. In between, we visited all the values from 0 to $n-1$ for the r_i's: indeed, if $(i \cdot m) \bmod n = (j \cdot m) \bmod n$ for $0 \leq i < j < n$, we have $((j - i) \cdot m) \bmod n = 0$, or $(j - i) \cdot m = k \cdot n$ for some k; but then n must divide $j - i$ since m and n are relatively prime, which is only possible if $i = j$.

Finally, some rotation of w' leads to w and to the associated value of r_0. ⊔

An example is given in Fig. 2, where the elements of the sequence $m_0, \ldots, m_{n-1}$ are put in bold in the system on the left.

Complexity. The number of operations to determine the sequence of m_i's is linear in the smallest weight n, i.e. also in the minimal number of occurrences of a label. In comparison, the previous algorithm of [24] checks a quadratic number of subwords.

The next lemma characterises the set of states reachable in any WMG whose underlying net is the one pictured in Fig. 1 and whose initial marking contains at least $n + m - 1$ tokens.

Lemma 1 (Reachable states w.r.t. the number of tokens). *Let N be a binary WMG as in Fig. 1, such that $\mu = (n, m) \geq 1$ with $\gcd(n, m) = 1$. Then*

$$0 + 21 = \mathbf{2} \cdot 8 + 5$$
$$5 + 21 = \mathbf{3} \cdot 8 + 2$$
$$2 + 21 = \mathbf{2} \cdot 8 + 7$$
$$7 + 21 = \mathbf{3} \cdot 8 + 4$$
$$4 + 21 = \mathbf{3} \cdot 8 + 1$$
$$1 + 21 = \mathbf{2} \cdot 8 + 6$$
$$6 + 21 = \mathbf{3} \cdot 8 + 3$$
$$3 + 21 = \mathbf{3} \cdot 8 + 0.$$

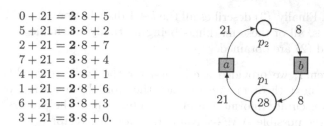

Fig. 2. This system solves the word $w = ab^2ab^3ab^2ab^3ab^3ab^2ab^3ab^3$ cyclically.

for each positive integer $k \geq m+n-1$, and each marking M_0 for N, the following properties are equivalent:
(1) M_0 contains exactly k tokens;
(2) $RG((N, M_0))$ contains exactly $k+1$ states;
(3) the set of states of $RG((N, M_0))$ is $\{(k_{ab}, k_{ba}) \in \mathbb{N}^2 \mid k_{ab} + k_{ba} = k\}$.

Proof. We first prove that (1) implies (2) and (3).

The case $k = m+n-1$ follows from the proof of Theorem 1: all the markings of the form $(k_{ab}, m+n-1-k_{ab})$ with $k_{ab} \in [0, m+n-1]$ are reachable. By the preservation of the total number of tokens through firings, these markings are distinct and their amount is thus $n+m$, proving the claim for $k = m+n-1$.

In the following, let us denote by $S_\perp$ the set of all these markings (i.e. with exactly $m+n-1$ tokens).

The case $k = \ell \cdot (m+n-1)$ for some positive integer ℓ is deduced similarly: denoting M_0 as any sum of ℓ markings $M_1 \ldots M_\ell$ such that each M_i corresponds to some distribution of k tokens over the two places, firing sequences are allowed in each (N, M_i) that lead to all the markings of $S_\perp$, i.e. $S_\perp$ is the set of states of $RG((N, M_i))$ for each i and all these RG's differ only by the choice of the initial state. All these sequences, obtained from all $i \in [1, \ell]$, are allowed independently (sequentially as well as in a shuffle) in (N, M_0). Thus, all the markings of the form $(k_{ab}, k - k_{ab})$ with $k_{ab} \in [0, k]$ are mutually reachable, describing $k+1$ distinct markings, which correspond to all the possibilities of distributing k tokens over the two places.

Now, let us consider $k > m + n - 1$, denoting the initial marking as $M_0 = (u + u', v + v')$ such that $u + v = \ell \cdot (m+n-1)$, $\ell \in \mathbb{N}_{>0}$, and u', v' are non-negative integers with $m+n-1 > u'+v' \geq 1$. From the above, all the markings of the form $M + (u', v')$, where M belongs to $RG(N, (u, v))$ are reachable from M_0, describing $u + v + 1$ distinct markings. Other markings can be reached by firing the tokens of u' and v': for each $x \in [0; u' - 1]$ and each $y \in [0; v' - 1]$, the markings $(x + n, k - x - n)$ and $(k - y - m, y + m)$ are reachable, from which we may fire b and a, respectively, leading to markings $(x, k - x)$ and $(k - y, y)$, and all these markings are distinct. Thus, we reach at least $u + v + 1 + u' + v' = k + 1$ distinct markings, which describe all the possible distributions of k.

We deduce that (1) implies (2) and (3). Now, assuming (2), and from the reasoning above, M_0 cannot have strictly less nor strictly more than k tokens, implying (1). Finally, (3) describes all the $k+1$ distributions of the k tokens over the two places, all of these markings being mutually reachable by reversibility, hence (1) and (2) are obtained. □

In Theorem 1, we provided a criterion for the cyclic solvability of a given word. In the next theorem, we abstract the word by a Parikh vector, which provides less accurate information on the behaviour of the process. This result investigates the possible WMG-solvable LTS for this vector.

Theorem 2 (WMG-solvable reversible binary LTS). *Let us consider* $\mu = (n, m) \geq \mathbb{1}$ *such that* $\gcd(n, m) = 1$, *and a positive integer* k. *Up to isomorphism and the choice of the initial state, when* $k \geq n + m$, *there exists a single finite WMG-solvable LTS* $(S, \rightarrow, \{a, b\}, \iota)$ *that satisfies* **b**, **c** *and* $|S| = k$, *and that contains a small cycle whose Parikh vector is* μ. *No such WMG-solvable LTS exists when* $k < n+m$. *In the particular case of* $S = \{0, 1, \dots, m+n-1\}$, *we have (up to isomorphism)* $\rightarrow = \{(i, a, i+m)|i, i+m \in S\} \cup \{(i, b, i-n)|i, i-n \in S\}$.

Proof. If a solution exists, it has the form of Fig. 1. If $k \geq n+m$, there are exactly $k-1$ tokens in the system by Lemma 1 and the reachability graph is unique up to isomorphism. From the previous results of this section, if $M_0 = n + m - 1$, then the RG is circular and contains exactly $n + m$ distinct states: all the values for i between 0 and $n + m - 1$ are reached in some order. Moreover, if we identify the states to i, i.e., the marking of $p_{a,b}$, the arcs are $\{(i, a, i+m)|0 \leq i, i+m < n+m \in S\} \cup \{(i, b, i-n)|0 \leq i, i-n < n+m \in S\}$. As a consequence, if $|S| < n + m$, there aren't enough states to close the circuit, and there is no solution. The rest of the claim immediately results from Lemma 1. □

3.2 WMG-solvable Infinite Cyclic Binary LTS

Let us consider an infinite LTS satisfying **b** and **c** with only two different labels. From the previous section, it cannot correspond to a net of the kind illustrated in Fig. 1 since $i + j$ remains constant, hence yields finitely many states. On the other hand, a net of the kind illustrated in Fig. 3, or the variant obtained by switching the roles of a and b, yields infinitely many occurrences of transition a, leading to infinitely many different reachable markings. Besides, from any state, there may only be finitely many consecutive b's. Moreover, this is the only way to obtain infinitely many cycles with Parikh vector (n, m).

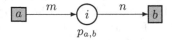

Fig. 3. A WMG solution for the infinite cyclic case.

If $n = 1$, i is the maximum number of consecutive executions of b from ι; we can then verify if the given LTS corresponds to the constructed net. Otherwise, let k and l be the Bezout coefficients corresponding to the relatively prime numbers m and n, so that $k \cdot m + l \cdot n = 1$. If $l \geq 0 \geq k$, i is the maximum number of times we may execute $a^{-k}b^l$ consecutively from ι, and we can check again if the given LTS corresponds to the constructed net (this is a direct generalisation of the case $n = 1$). Otherwise, since $-n \cdot m + m \cdot n = 0$, by adding this relation enough times to the previous one, we get $k' \cdot m + l' \cdot n = 1$ with $l' \geq 0 \geq k'$, and we apply the same idea.

4 WMG-solvable Acyclic LTS: A Geometric Approach

In what follows, we consider any acyclic LTS satisfying property **b**. First, in Subsect. 4.1, we give a geometric interpretation of WMG-solvability for acyclic LTS with only two different labels. Then, in Subsect. 4.2, we extend this result to any number of labels.

4.1 Geometric Characterisation for 2 Labels

In the following, we specialise to the WMG case the more general framework considered in [16], Theorem 2, using convex sets of $\mathbb{N}^2$. The standard definition of convex sets of $\mathbb{R}^2$ is given by the segment-inclusion property: a set $C \subseteq \mathbb{R}^2$ is convex if and only if, for any $x, y \in C$, $[x, y] \subseteq C$, where $[x, y]$ is the linear segment with extremities x and y. However, this does not work for $\mathbb{N}^2$ (nor $\mathbb{Z}^2$), as illustrated by Fig. 4: in the set $C = \{x, y, z\}$ with $x = (0, 0)$, $y = (1, 2)$ and $z = (2, 1)$, we have $[x, y] = \{x, y\}$, $[y, z] = \{y, z\}$ and $[z, x] = \{z, x\}$; hence we have the segment-inclusion property; however, clearly, this set should not be considered as convex since the node $X = (1, 1)$ is missing.

In [25], two equivalent definitions of convex sets in lattices like $\mathbb{Z}^2$ are provided, which immediately extend to $\mathbb{N}^2$:

1. either as the intersection of a convex set of $\mathbb{R}^2$ with $\mathbb{Z}^2$,
2. or as the intersection of half planes $\mathcal{L}_i$, with $\mathcal{L}_i = \{(x, y) \in \mathbb{Z}^2 | a_i \cdot x + b_i \cdot y \geq c_i$ for some $a_i, b_i, c_i \in \mathbb{Z}\}$. If the convex set is finite, we can use a finite set of such half-planes, otherwise we may need (countably) infinitely many of them (notice that infinite convex sets exist with a boundary defined by finitely many half-planes).

In order to characterise the acyclic LTS that are solvable by WMG nets with two labels, we first identify isomorphically each state s with:
Δ_s = (number of a's in any path from ι to s, number of b's in any path from ι to s) (this is coherent from Proposition 2), which amounts to consider for S a part of $\mathbb{N}^2$ containing $(0, 0)(= \iota)$. From the full reachability, S is connected (there is a directed path from $(0, 0)$ to any $(i, j) \in S$, hence an undirected path between any two states). From Keller's theorem [26] (due to determinism and persistence), full reachability and Proposition 2, we have that $(i, j) \xrightarrow{a} (i', j') \iff i' = i + 1 \wedge j' = j$ and $(i, j) \xrightarrow{b} (i', j') \iff i' = i \wedge j' = j + 1$.

Fig. 4. Non-convex set in $\mathbb{Z}^2$ with the segment-inclusion property.

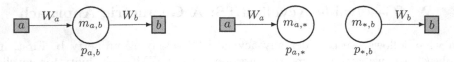

Fig. 5. General places for a WMG synthesis, with initial marking $m_{a,b} = M_0(p_{a,b})$, $m_{a,*} = M_0(p_{a,*})$ and $m_{*,b} = M_0(p_{*,b})$.

If the system is WMG-solvable, it must be defined by a finite set of places of the kind $p_{a,b}$ and $p_{b,a}$ in Fig. 5 (with $\gcd(W_a, W_b) = 1$ and $m_{a,b}, m_{b,a} \geq 0$), including the special cases $p_{*,b}$ (with $W_a = 0$, $W_b = 1$ and $m_{*,b} > 0$, the case $m_{*,b} = 0$ only serving to make b non-firable but we assumed the system weakly live) or $p_{*,a}$, and $p_{a,*}$ (with $W_b = 0$, $W_a = 1$ and $m_{*,a} = m_{a,*} = 0$) or $p_{b,*}$. For a place $p_{a,b}$, we have for each state $s = (i,j)$ that the corresponding marking is $M_s(p_{a,b}) = M_0(p_{a,b}) + i \cdot W_a - j \cdot W_b$, and since we must have $M_s(p_{a,b}) \geq 0$, this defines a 'region', both in the sense of [27] and in an intuitive geometric meaning: $R_{p_{a,b}} = \{(i,j) | M_0(p_{a,b}) + W_a \cdot i - W_b \cdot j \geq 0\}$ or, permuting the roles of a and b, $R_{p_{b,a}} = \{(i,j) | M_0(p_{b,a}) - W_a \cdot i + W_b \cdot j \geq 0\}$, i.e. in either case the intersection of $\mathbb{N}^2$ with a half plane of $\mathbb{Z}^2$. These regions will be called in the following WMG-regions. Notice that [16] considers additional regions, where $W_a < 0$ or $W_b < 0$. Each such region is convex, as well as any intersection of such regions.

We deduce the next specialisation of Theorem 2 in [16].

Theorem 3 (WMG-solvable acyclic binary LTS). *An acyclic LTS satisfying property **b** is WMG-solvable if and only if, when applied on $\mathbb{N}^2$, its set of states S is connected, convex and delimited by (i.e., it is the intersection of) a finite set of WMG-regions. A possible solution is then provided by the places corresponding to these regions.*

For any finite LTS, if it is the intersection of WMG-regions, it is the intersection of a finite set of such regions. However, the result may be extended to an infinite LTS, but then it may be necessary to specify that only a finite set of regions is allowed. This is illustrated by Fig. 6.

Note that total reachability does not arise from WMG-regions alone, as illustrated by Fig. 7: on the left, the points $\iota = (0,0)$, $(1,0)$ and $(2,1)$ form a convex set of $\mathbb{N}^2$, intersection of the WMG-regions $i - 2 \cdot j \geq 0$ and $1 - i + j \geq 0$

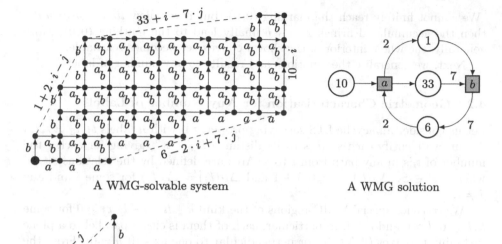

A WMG-solvable system   A WMG solution

A non-convex system: the node ∗ is missing,
hence non-WMG-solvable.

Fig. 6. Illustration of Theorem 3.

Fig. 7. Convex sets defined by WMG-regions may be non-totally reachable in $\mathbb{N}^2$.

(plus $j \geq 0$ to certify being in $\mathbb{N}^2$), but $(2,1)$ is not reachable from ι. These WMG-regions yield the WMG system on the right of the same figure.

A closer look shows that state ι corresponds to marking $(1,0)$, state $(1,0)$ to marking $(0,1)$ and $(2,1)$ to marking $(0,0)$. The latter is not reachable, but is potentially reachable in the sense of [5]. Let us recall that, from the classical state equation $M[\sigma\rangle M' \Rightarrow M' = M + C \cdot \mathbf{P}(\sigma)$ where C is the incidence matrix, and that a marking M is potentially reachable from the initial marking M_0 if $M = M_0 + C \cdot \alpha$ for some T-vector $\alpha \geq \mathbb{0}$ (non-necessarily the Parikh vector of some firing sequence). Indeed, here $C = \begin{pmatrix} -1 & 1 \\ 1 & -2 \end{pmatrix}$, and $(0,0) = (1,0) + C \cdot (2,1)$ (caution: here the vectors are to be considered as column vectors).

Another possible interpretation is to consider the net on the right of Fig. 7 as a *continuous* or *fluid* one, in the sense of [28]. In those models, a transition may be executed fractionally and reachable markings may be real vectors with no negative component. Thus, in our case, we can have the firing sequence

$$(1,0)[a\rangle(0,1)[b^{1/2}\rangle(1/2,0)[a^{1/2}\rangle(0,1/2)[b^{1/4}\rangle(1/4,0)[a^{1/4}\rangle(0,1/4)[b^{1/8}\rangle(1/8,0)\dots$$

We cannot finitely reach the marking $(0,0)$, but if we allow *limit-reachability*, then the accumulated firings $2 \cdot a + b$ finally lead to the marking $(0,0)$. More generally, the whole interior of the shown convex set becomes reachable.

Next, we generalise these notions and results to any number of labels.

4.2 Geometric Characterisation for Any Number of Labels

Let us consider an acyclic LTS satisfying property **b** with n labels $t_1, t_2, \ldots, t_n$. Again, we identify each state s to its distance $\Delta_s \in \mathbb{N}^n$, giving for each i the number of t_i's in any path from ι to s. Arcs are defined by the relations $s[t_i\rangle s'$ when $s, s' \in S$, $\Delta_{s'}(t_i) = \Delta_s(t_i) + 1$ and $\Delta_{s'}(t_j) = \Delta_s(t_j)$ for some i and any $j \neq i$.

We consider special WMG-regions of the kind $k + h \cdot x_i - l \cdot x_j \geq 0$ for some $k, h \geq 0$, $l > 0$ and $i \neq j$. In particular, each of them is either parallel to a plane including two axes (if $h > 0$), or perpendicular to one axis (if $h = 0$). From the specialisation of [16], we deduce the following.

Theorem 4 (WMG-solvable acyclic n-ary systems). *An acyclic LTS satisfying property **b** with n different labels is WMG-solvable if and only if, when applied on $\mathbb{N}^n$, its set of states S is connected, convex and delimited by (i.e., it is the intersection of) a finite set of WMG-regions. A possible solution is then provided by the places corresponding to these regions.*

However, this characterisation is less intuitively (visually) interpretable when $n > 2$. Hence it will usually be more efficient to use the general WMG synthesis procedure described in [10].

5   A Sufficient Condition of Circular WMG-solvability for Any Number of Labels

In this section, we provide a general sufficient condition for the cyclic solvability of k-ary words, for any positive integer k. This condition, embodied by the next theorem, uses binary subwords obtained by projection[2] and containing occurrences of two different labels that are contiguous somewhere in the k-ary word. The other binary subwords are not needed since they lack this contiguity and do not capture the direct causality.

Theorem 5. *Consider any word w over any finite alphabet T such that $\mathbf{P}(w)$ is prime. Suppose the following: $\forall u = w_{|t_1 t_2}$ (i.e., the projection of w on $\{t_1, t_2\}$) for some t_1, t_2 such that $t_1 \neq t_2 \in T$, and $w = (w_1 t_1 t_2 w_2)$ or $w = (t_2 w_3 t_1)$, $u = v^\ell$ for some positive integer ℓ, $\mathbf{P}(v)$ is prime, and v is cyclically solvable by a circuit. Then, w is cyclically solvable with a WMG.*

---

[2] The projection of a word $w \in A^*$ on a set $A' \subseteq A$ of labels is the maximum subword of w whose labels belong to A', noted $w_{|A'}$. For example, the projection of the word $w = \ell_1 \ell_2 \ell_3 \ell_2$ on the set $\{\ell_1, \ell_2\}$ is the word $\ell_1 \ell_2 \ell_2$.

Proof. For every such pair (t_i, t_j), $i < j$, let $C_{i,j} = ((P_{i,j}, T_{i,j}, W_{i,j}), M_{i,j})$ be a circuit solution of v for the subword $v^l = u_{i,j} = w_{|t_i t_j}$, obtained as in the construction of Theorem 1. Assuming all these nets are place-disjoint (which is always possible since the Petri net solutions are considered up to isomorphism), consider the transition-merging[3] of all these marked circuits. The result is a WMG $\mathcal{S}' = (N', M_0')$ such that $N' = (P', T, W')$ with $P' = \cup_{i,j} P_{i,j}$, $T = \cup_{i,j} T_{i,j}$, $W' = \cup_{i,j} W_{i,j}$, and $M_0' = \cup_{i,j} M_{i,j}$.

Let w be of the form aw'. We prove that a is the only transition enabled in $\mathcal{S}'$.

All the subwords of the form $w_{|a,t}$ necessarily start with a. All the input places of the transition a belong to the binary circuits defined by these subwords. Since these subwords are solvable by marked circuits which we merged together, all the input places of a are initially enabled. Now, let us suppose that another transition d is also initially enabled in $\mathcal{S}'$. Since d is not the first label of w, another label q appears in w just before the first occurrence of d. In the solution of $w_{|d,q}$, d is not initially enabled since q must occur before; hence it is not enabled in the merging either. We deduce that a is the only transition that is enabled in $\mathcal{S}'$.

Now, the same arguments apply to $w'' = w'a$ whose relevant subwords are solvable by the circuits in the same way, and we deduce that the WMG $\mathcal{S}'$ has the language $PREF(w^*)$.

Note that we did not use explicitly above the special form of u. Simply, the latter is necessary to build a circuit system $C_{i,j}$ with the language $PREF(u^*) = PREF(v^*)$. $C_{i,j}$ is a circular solution for v, but not for u unless $\ell = 1$. The fact that the merging $\mathcal{S}'$ of all the $C_{i,j}$'s yields not only a system with the adequate language $PREF(w^*)$ but a circular solution of w arises from the fact that $\mathbf{P}(w)$ is prime (by Proposition 2). We thus deduce that the WMG $\mathcal{S}'$ solves w cyclically. □

6   Synthesis of WMGs from Live Ternary LTS

In this section, we provide several conditions of WMG-solvability for a ternary LTS. We first develop a characterisation of WMG-solvability for a subclass of the cyclic ternary words in Subsect. 6.1. Then, in Subsect. 6.2, we construct two counter-examples to this condition: one for four labels with three different values in the Parikh vector, and another one for five labels with only two different values.

6.1   WMG-solvability in a Subclass of the Finite Circular Ternary LTS

First, we prove the other direction of Theorem 5, leading to a full characterisation of WMG-solvability for a special subclass of the ternary cyclic words.

The proof exploits a WMG with 3 transitions and 6 places, connecting 2 places to each pair of transitions, as illustrated in Fig. 8. In some cases, a smaller

---

[3] Also called sometimes the synchronisation on transitions.

number of places can solve the same LTS, but we do not aim here at minimising the number of nodes in a solution.

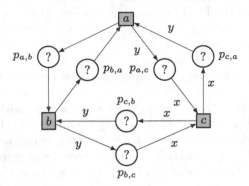

Fig. 8. A generic WMG with three labels, with minimal T-semiflow (x, x, y) and $\gcd(x, y) = 1$.

Theorem 6 (Cyclic solvability of ternary words). *Consider a ternary word w over the alphabet T with Parikh vector (x, x, y) such that $\gcd(x, y) = 1$. Then, w is cyclically solvable with a WMG if and only if $\forall u = w_{|t_1 t_2}$ such that $t_1 \neq t_2 \in T$, and $w = (w_1 t_1 t_2 w_2)$ or $w = (t_2 w_3 t_1)$, $u = v^\ell$ for some positive integer ℓ, $\mathbf{P}(v)$ is prime, and v is cyclically solvable by a circuit (i.e. a circular net).*

Proof. The right-to-left direction of the equivalence, assuming the properties on the projections, is true by Theorem 5, for the particular case that $|T| = 3$. We thus deduce the cyclic solvability.

In the rest of this proof, we consider the other direction, assuming circular solvability. If $x = y = 1$, the claim is trivially obtained since $w = t_1 t_2 t_3$, up to some permutation, and an easy marked graph solution may be found. Let us thus assume that $x \neq y$.

Let us write $T = \{a, b, c\}$. The general form of a solution has 3 transitions and 6 places (one for each ordered pair of transitions). Additional places are never necessary in the presence of a T-semiflow. Indeed, let $p_{u,v}$ be a place between transitions u and v, W_u the weight on the arc to this place and W_v the one from this place. Due to the presence of the T-semiflow $\mathbf{P}(w)$, we have $\mathbf{P}(w)(u) \cdot W_u = \mathbf{P}(w)(v) \cdot W_v$, and we may choose $W_u = \mathbf{P}(w)(v)$ as well as $W_v = \mathbf{P}(w)(u)$. We may also divide the weights around each place by their gcd. In our case, this leads to the configuration illustrated by Fig. 8. We denote by RG the reachability graph of a solution based on this net.

We show first that the projection $w_{|ab}$ of w on $\{a, b\}$ is of the form $(ab)^k$ or $(ba)^k$ for some positive integer k.

There is no pattern aab in w^2 (which allows to consider sequences on the border of two consecutive w's) because, if $M_1[a\rangle M_2[a\rangle M_3[b\rangle$, $M_1(p_{c,b}) = M_2(p_{c,b}) =$

$M_3(p_{c,b}) \geq y$ and $M_2(p_{a,b}) \geq 1$, which would also allow to perform b after the first a and RG is not circular.

If there is a pattern aac and $M_1[a\rangle M_2[a\rangle M_3[c\rangle M_4$, $M_2(p_{a,c}) \geq y$ and $M_1(p_{b,c}) = M_2(p_{b,c}) = M_3(p_{b,c}) \geq x$, hence $y < x$ otherwise M_2 already enables c and RG is not circular. Then $M_4(p_{a,b}) \geq 2$, $M_4(p_{c,b}) \geq x > y$ and $M_4(p_{c,a}) \geq x > y$, so that $M_4[b\rangle$; hence $M_4(p_{b,a}) = 0$ since otherwise we also have $M_4[a\rangle$ and RG is not circular. We thus have $M_4[ba\rangle M_5$ for some marking M_5, with $M_5(p_{b,a}) = 0$, so that M_5 does not enable a; M_5 does not enable b either since otherwise we could also perform $M_4[bb\rangle$ and again RG is not circular. Thus, we have $M_4[bac\rangle$, and then we are in a situation similar to the one after the first c. As a consequence, we must have a sequence $M_1[aa(cba)^\omega\rangle$, and RG is not circular.

Hence, in w^2 we cannot have a sequence aa, nor bb by symmetry.

Let us now assume that a pattern $ac^k a$ exists in w^2 for some $k \geq 1$. Since the first firing of a puts a token in $p_{a,b}$ and the next firing of c does not enable b, we must have $x < y$. Let us assume in the circular RG that $M_1[ac^k a\rangle M_2[\sigma\rangle M_1$. σ is not empty since it must contain x times b. It cannot end with an a, since otherwise we have a sequence aa, which we already excluded. It cannot end with a b either, since $M_1(p_{b,a}) \geq 2$ (in order to fire a twice without a b in between), so that if $M_3[b\rangle M_1[a\rangle$, $M_3(p_{b,a}) \geq 1$, we must also have $M_3[a\rangle$, and RG is not circular. Hence, σ ends with a c and for some reachable markings M_2' and M_3 we have $M_3[c\rangle M_1[ac^k\rangle M_2'[a\rangle M_2$.

We deduce that $M_2'(p_{a,b}) \geq 1$, $M_2'(p_{c,b}) \geq (k+1) \cdot x$; hence $(k+1) \cdot x < y$ otherwise M_2' also enables b and RG is not circular. Also, $M_3(p_{b,a}) \geq 2$ and $M_3(p_{c,a}) \geq 2 \cdot y - (k+1) \cdot x > y$, so that M_3 also enables a and again RG is not circular. As a consequence, we cannot have a pattern $ac^k a$, nor $bc^k b$ by symmetry, and $w_{|ab} = (ab)^k$ or $w_{|ab} = (ba)^k$ for the positive integer $k = x$. With $v = ab$ or $v = ba$, we have the adequate solvability property for $w_{|ab}$, and we can assume in the following that the sum of the tokens present in places $p_{a,b}$ and $p_{b,a}$ is 1 for all reachable markings.

Let us now suppose that we have a WMG S solving w cyclically, whose underlying net is pictured in Fig. 8. From the previous results, we can assume that $M_0(p_{a,b}) + M_0(p_{b,a}) = 1$ in S, this equality being preserved by all reachable markings. To show that $u = w_{|ac}$ has the adequate form (the case for $w_{|bc}$ is symmetrical), let us consider the circuit C_{ac}, restriction of S to $p_{a,c}, c, p_{c,a}, a$.

Let us assume in the following that u cannot be written under the form $u = v^\ell$ for some positive integer ℓ, where $\mathbf{P}(v)$ is prime and v is cyclically solvable. Since $gcd(x, y) = gcd(\mathbf{P}(w)(a), \mathbf{P}(w)(c)) = gcd(\mathbf{P}(u)(a), \mathbf{P}(u)(c)) = 1$, $\mathbf{P}(u)$ is prime and $u = v$ with $\ell = 1$, hence u is not cyclically solvable. For the net N considered, this implies the existence of some prefix σ_{ac} of u such that, for every initial marking of C_{ac} that enables the sequence u in this circuit, the marking reached by firing σ_{ac} necessarily enables both places $p_{a,c}$ and $p_{c,a}$. Indeed, Theorem 1 specifies the finite set of all possible minimal markings that allow cyclic solvability, and each such marking enables exactly one place of the circuit. Every other non-circular reachability graph is defined by some larger initial marking and contains a marking that enables both places.

Thus, for any initial marking M_0 that makes the system $\mathcal{S} = (N, M_0)$ solve w cyclically, the smallest prefix of w whose projection on $\{a, c\}$ equals σ_{ac} leads to a marking M in the WMG that enables $p_{a,c}$ and $p_{c,a}$.

Hereafter, we consider all the cases in which either a or c is enabled from M. In each case, we describe the shape of the LTS and deduce from it a reachable marking that enables two transitions, hence a contradiction.

Case $x > y$: In this case, in $\mathcal{S}$, we cannot have two consecutive c's.

- Subcase in which M enables the place $p_{a,c}$ as well as the transition a in the WMG, hence its input places $p_{b,a}$ and $p_{c,a}$. Since $M[a\rangle$, transition c is not enabled at M, implying that $p_{b,c}$ is not enabled by M. We deduce: $M(p_{a,c}) > M(p_{b,c})$. Since $p_{a,c}$ is enabled by M, the last occurrence of a transition before the next firing of c is necessarily b, implying: $M[(ab)^k c\rangle M_1$ for some integer $k \geq 1$ and some marking M_1. The inequality mentioned above is still valid at M_1, i.e. $M_1(p_{a,c}) > M_1(p_{b,c})$, and we iterate the same arguments from M_1 to deduce that the rotation w_M of w starting at M is of the form $(ab)^{k_1} c \ldots (ab)^{k_y} c$ with $\sum_{i=1..y} k_i = x$ and each k_i is positive.
- Subcase in which M enables the place $p_{c,a}$ as well as the transition c in the WMG, hence its input places $p_{a,c}$ and $p_{b,c}$. Thus, the firing of c from M cannot enable a, implying that $M(p_{c,b}) < M(p_{c,a})$ and that $M[c(ba)^k c\rangle M_1$ for some positive integer k and a marking M_1. The inequality is still valid at M_1, i.e. $M_1(p_{c,b}) < M_1(p_{c,a})$, from which we deduce that the rotation w_M of w starting at M is of the form $c(ba)^{k_1} \ldots c(ba)^{k_y}$ with $\sum_{i=1..y} k_i = x$ and each k_i is positive.

Case $x \leq y$:

- Subcase in which M enables the place $p_{a,c}$ as well as the transition a in the WMG, hence its input places $p_{c,a}$ and $p_{b,a}$. Thus, the firing of a from M cannot enable c, implying that $M(p_{b,c}) < M(p_{a,c})$ and that $M[abc^k\rangle M_1$ for some positive integer k and a marking M_1, at which the same inequality is still valid. We deduce that the rotation w_M of w starting at M is of the form $abc^{k_1} \ldots abc^{k_x}$ with $\sum_{i=1..x} k_i = y$ and each k_i is positive.
- Subcase in which M enables the place $p_{c,a}$ as well as the transition c in the WMG, hence its input places $p_{a,c}$ and $p_{b,c}$. Thus, firing one or several c's from M does not enable a, and $M(p_{c,b}) < M(p_{c,a})$, implying that $M[c^k ba\rangle M_1$ for some positive integer k and a marking M_1, at which the same inequality is still valid. We deduce that the rotation w_M of w starting at M is of the form $c^{k_1} ba \ldots c^{k_x} ba$ with $\sum_{i=1..x} k_i = y$ and each k_i is positive.

In each of the four cases developed above, we observe that each sequence of ab or ba could be seen as an atomic firing, and $w_{M|b,c}$ is obtained from $w_{M|a,c}$ by renaming each a into one b. This implies that the deletion of the initial useless tokens (also known as frozen tokens, i.e. never used by any firing) yields a system in which some reachable marking distributes the tokens in the same way in the places between c and a as in the places between c and b. This is for example the case of the marking M if it does not contain useless tokens.

We deduce that M (with or without useless tokens) enables all four places $p_{a,c}$, $p_{c,a}$, $p_{b,c}$ and $p_{c,b}$, thus enabling two transitions of the WMG at least. This contradicts the cyclic solvability of w, implying that $v = u$ is cyclically solvable by a circuit. Hence the claim.   □

6.2   Counter-Examples for 4 and 5 Labels

In Theorem 6, we provided a characterisation of cyclic WMG-solvability for ternary words w such that $\mathbf{P}(w)$ is prime with two values. However, this result does not apply to words w over 4 labels with 3 values nor 5 labels with 2 values, even if $\mathbf{P}(w)$ is prime. Indeed, Fig. 9 pictures two counter-examples: on the left, the WMG cyclically solves the word $w = aacbbdabd$ with $\mathbf{P}(w) = (3, 3, 1, 2)$, which is prime, while its projection $u = aabbab$ on $\{a, b\}$ leads to $v = u$, and $\mathbf{P}(v) = (3, 3)$ is not prime, hence is not cyclically solvable by a WMG; on the right, the WMG cyclically solves the word $w = aacbbeabd$ with $\mathbf{P}(w) = (3, 3, 1, 1, 1)$, which is prime, while its projection $u = aabbab$ on $\{a, b\}$ leads to $v = u$, and $\mathbf{P}(v) = (3, 3)$ is not cyclically solvable by a WMG.

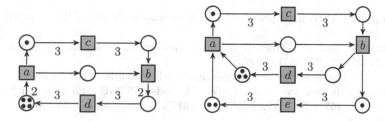

Fig. 9. The WMG on the left solves $aacbbdabd$ cyclically, and the WMG on the right solves $aacbbeabd$ cyclically.

However, presently we do not know what happens for ternary words w such that $\mathbf{P}(w)$ is prime with three values, nor when w has four labels and $\mathbf{P}(w)$ is prime with two values.

7   Conclusions and Perspectives

In this work, we specialised previous methods dedicated to the analysis and synthesis of weighted marked graphs, a well-known and useful subclass of weighted Petri nets allowing to model various real-world applications.

By restricting the size of the alphabet to 2 labels, we provided a characterisation of the WMG-solvable labelled transition systems formed of a single cycle. We also extended this investigation to finite LTS containing several cycles, and to infinite LTS.

Then, leaving out the restriction on the number of labels, we developed a geometric characterisation for acyclic LTS, using convex sets and the theory of regions; in the case circular LTS, we proposed a sufficient condition of WMG-solvability.

We exploited this sufficient condition to obtain a full characterisation of circular WMG-solvability for a subset of the possible Parikh vectors over three labels.

Finally, we proved that this condition for 3 labels does not extend to circular LTSs with 4 labels and three different Parikh values, nor with 5 labels and two Parikh values.

As perspectives of this work, we believe that relaxations of our statements may lead to other characterisations of WMG-solvable LTS, together with efficient algorithms for their analysis and synthesis.

Acknowledgements. We would like to thank the anonymous referees for their involvement and useful suggestions.

References

1. Murata, T.: Petri nets: properties, analysis and applications. Proc. IEEE **77**(4), 541–580 (1989)
2. Teruel, E., Silva, M.: Structure theory of equal conflict systems. Theoret. Comput. Sci. **153**(1&2), 271–300 (1996)
3. Hujsa, T., Devillers, R.: On liveness and deadlockability in subclasses of weighted Petri nets. In: van der Aalst, W., Best, E. (eds.) PETRI NETS 2017. LNCS, vol. 10258, pp. 267–287. Springer, Cham (2017). https://doi.org/10.1007/978-3-319-57861-3_16
4. Desel, J., Esparza, J.: Free Choice Petri Nets. Cambridge Tracts in Theoretical Computer Science, vol. 40. Cambridge University Press, New York (1995)
5. Teruel, E., Colom, J.M., Silva, M.: Choice-free Petri nets: a model for deterministic concurrent systems with bulk services and arrivals. IEEE Trans. Syst. Man Cybern. Part A **27**(1), 73–83 (1997)
6. Hujsa, T., Delosme, J.-M., Munier-Kordon, A.: On the reversibility of well-behaved weighted choice-free systems. In: Ciardo, G., Kindler, E. (eds.) PETRI NETS 2014. LNCS, vol. 8489, pp. 334–353. Springer, Cham (2014). https://doi.org/10.1007/978-3-319-07734-5_18
7. Commoner, F., Holt, A., Even, S., Pnueli, A.: Marked directed graphs. J. Comput. Syst. Sci. **5**(5), 511–523 (1971)
8. Teruel, E., Chrzastowski-Wachtel, P., Colom, J.M., Silva, M.: On weighted T-systems. In: Jensen, K. (ed.) ICATPN 1992. LNCS, vol. 616, pp. 348–367. Springer, Heidelberg (1992). https://doi.org/10.1007/3-540-55676-1_20
9. Best, E., Hujsa, T., Wimmel, H.: Sufficient conditions for the marked graph realisability of labelled transition systems. Theoret. Comput. Sci. **750**, 101–116 (2017)
10. Devillers, R., Hujsa, T.: Analysis and synthesis of weighted marked graph Petri nets. In: Khomenko, V., Roux, O.H. (eds.) PETRI NETS 2018. LNCS, vol. 10877, pp. 19–39. Springer, Cham (2018). https://doi.org/10.1007/978-3-319-91268-4_2

11. Delosme, J.M., Hujsa, T., Munier-Kordon, A.: Polynomial sufficient conditions of well-behavedness for weighted join-free and choice-free systems. In: 13th International Conference on Application of Concurrency to System Design, pp. 90–99, July 2013
12. Hujsa, T., Delosme, J.M., Munier-Kordon, A.: Polynomial sufficient conditions of well-behavedness and home markings in subclasses of weighted Petri nets. ACM Trans. Embed. Comput. Syst. **13**(4s), 141:1–141:25 (2014)
13. Barylska, K., Best, E., Erofeev, E., Mikulski, L., Piatkowski, M.: On binary words being Petri net solvable. In: Proceedings of the International Workshop on Algorithms & Theories for the Analysis of Event Data, ATAED 2015, Brussels, Belgium, pp. 1–15 (2015)
14. Barylska, K., Best, E., Erofeev, E., Mikulski, L., Piatkowski, M.: Conditions for Petri net solvable binary words. Trans. Petri Nets Other Models Concurr. **11**, 137–159 (2016)
15. Erofeev, E., Barylska, K., Mikulski, L., Piatkowski, M.: Generating all minimal Petri net unsolvable binary words. In: Proceedings of the Prague Stringology Conference 2016, Prague, Czech Republic, pp. 33–46 (2016)
16. Erofeev, E., Wimmel, H.: Reachability graphs of two-transition Petri nets. In: Proceedings of the International Workshop on Algorithms & Theories for the Analysis of Event Data 2017, Zaragoza, Spain, pp. 39–54 (2017)
17. Best, E., Devillers, R.: Synthesis and reengineering of persistent systems. Acta Inf. **52**(1), 35–60 (2015)
18. Hujsa, T., Delosme, J.M., Munier-Kordon, A.: On liveness and reversibility of equal-conflict Petri nets. Fundamenta Informaticae **146**(1), 83–119 (2016)
19. Hujsa, T.: Contribution to the study of weighted Petri nets. Ph.D. thesis, Pierre and Marie Curie University, Paris, France (2014)
20. Devillers, R., Erofeev, E., Hujsa, T.: Synthesis of weighted marked graphs from constrained labelled transition systems. In: Proceedings of the International Workshop on Algorithms & Theories for the Analysis of Event Data, Bratislava, Slovakia, pp. 75–90 (2018)
21. Crespi-Reghizzi, S., Mandrioli, D.: A decidability theorem for a class of vector-addition systems. Inf. Process. Lett. **3**(3), 78–80 (1975)
22. Devillers, R.: Products of transition systems and additions of Petri nets. In: Desel, J., Yakovlev, A. (eds.) Proceedings of 16th International Conference on Application of Concurrency to System Design (ACSD 2016), pp. 65–73 (2016)
23. Devillers, R.: Factorisation of transition systems. Acta Informatica **55**, 339–362 (2017)
24. Best, E., Erofeev, E., Schlachter, U., Wimmel, H.: Characterising Petri net solvable binary words. In: Kordon, F., Moldt, D. (eds.) PETRI NETS 2016. LNCS, vol. 9698, pp. 39–58. Springer, Cham (2016). https://doi.org/10.1007/978-3-319-39086-4_4
25. Doignon, J.P.: Convexity in cristallographical lattices. J. Geom. **3**(1), 71–85 (1973)
26. Keller, R.M.: A fundamental theorem of asynchronous parallel computation. In: Feng, T. (ed.) Parallel Processing. LNCS, vol. 24, pp. 102–112. Springer, Heidelberg (1975). https://doi.org/10.1007/3-540-07135-0_113
27. Badouel, E., Bernardinello, L., Darondeau, P.: Petri Net Synthesis. Springer, Heidelberg (2015). https://doi.org/10.1007/978-3-662-47967-4
28. David, R., Alla, H.: Discrete, Continuous, and Hybrid Petri Nets, 2nd edn. Springer, Heidelberg (2010). https://doi.org/10.1007/978-3-642-10669-9

Evaluating Conformance Measures in Process Mining Using Conformance Propositions

Anja F. Syring[1], Niek Tax[2], and Wil M. P. van der Aalst[1,2,3(✉)]

[1] Process and Data Science (Informatik 9), RWTH Aachen University,
52056 Aachen, Germany
wvdaalst@pads.rwth-aachen.de
[2] Architecture of Information Systems, Eindhoven University of Technology,
Eindhoven, The Netherlands
[3] Fraunhofer Institute for Applied Information Technology FIT,
Sankt Augustin, Germany

Abstract. Process mining sheds new light on the relationship between process models and real-life processes. Process discovery can be used to learn process models from event logs. Conformance checking is concerned with quantifying the *quality* of a business process model in relation to event data that was logged during the execution of the business process. There exist different categories of conformance measures. *Recall*, also called fitness, is concerned with quantifying how much of the behavior that was observed in the event log fits the process model. *Precision* is concerned with quantifying how much behavior a process model allows for that was never observed in the event log. *Generalization* is concerned with quantifying how well a process model generalizes to behavior that is possible in the business process but was never observed in the event log. Many recall, precision, and generalization measures have been developed throughout the years, but they are often defined in an ad-hoc manner without formally defining the desired properties up front. To address these problems, we formulate *21 conformance propositions* and we use these propositions to evaluate current and existing conformance measures. The goal is to trigger a discussion by clearly formulating the challenges and requirements (rather than proposing new measures). Additionally, this paper serves as an overview of the conformance checking measures that are available in the process mining area.

Keywords: Process mining · Conformance checking · Evaluation measures

1 Introduction

Process mining [2] is a fast growing discipline that focuses on the analysis of event data that is logged during the execution of a business process. Events in such

© Springer-Verlag GmbH Germany, part of Springer Nature 2019
M. Koutny et al. (Eds.): ToPNoC XIV, LNCS 11790, pp. 192–221, 2019.
https://doi.org/10.1007/978-3-662-60651-3_8

an event log contain information on what was done, by whom, for whom, where, when, etc. Such event data are often readily available from information systems that support the execution of the business process, such as ERP, CRM, or BPM systems. *Process discovery*, the task of automatically generating a process model that accurately describes a business process based on such event data, plays a prominent role in process mining. Throughout the years, many process discovery algorithms have been developed, producing process models in various forms, such as Petri nets, process trees, and BPMN.

Event logs are often incomplete, i.e., they only contain a sample of all possible behavior in the business process. This not only makes process discovery challenging; it is also difficult to assess the quality of the process model in relation to the log. Process discovery algorithms take an event log as input and aim to output a process model that satisfies certain properties, which are often referred to as the four quality dimensions [2] of process mining: (1) *recall*: the discovered model should allow for the behavior seen in the event log (avoiding "non-fitting" behavior), (2) *precision*: the discovered model should not allow for behavior completely unrelated to what was seen in the event log (avoiding "underfitting"), (3) *generalization*: the discovered model should generalize the example behavior seen in the event log (avoiding "overfitting"), and (4) *simplicity*: the discovered model should not be unnecessarily complex. The simplicity dimension refers to Occam's Razor: "one should not increase, beyond what is necessary, the number of entities required to explain anything". In the context of process mining, this is often operationalized by quantifying the complexity of the model (number of nodes, number of arcs, understandability, etc.). We do *not* consider the simplicity dimension in this paper, since we focus on *behavior* and abstract from the actual model representation. Recall is often referred to as *fitness* in process mining literature. Sometimes fitness refers to a combination of the four quality dimensions. To avoid later confusion, we use the term recall which is commonly used in pattern recognition, information retrieval, and (binary) classification. Many conformance measures have been proposed throughout the years, e.g., [2, 4, 6, 12–15, 24, 25, 27, 32, 33].

So far it remains an open question whether existing measures for recall, precision, and generalization measure what they are aiming to measure. This motivates the need for a formal framework for conformance measures. Users of existing conformance measures should be aware of seemingly obvious quality issues of existing approaches and researchers and developers that aim to create new measures should be clear on what conformance characteristics they aim to support. To address this open question, this paper evaluates state-of-the-art conformance measures based on 21 propositions introduced in [3]. This paper supported by a detailed publicly available report detailing the evaluations of existing techniques [29].

The remainder is organized as follows. Section 2 discusses related work. Section 3 introduces basic concepts and notations. The rest of the paper is split into two parts where the first one discusses the topics of recall and precision (Sect. 4) and the second part is dedicated to generalization (Sect. 5). In both parts, we introduce the corresponding conformance propositions and provide an overview of existing conformance measures. Furthermore, we discuss our findings of validating existing these measures on the propositions. Additionally, Sect. 4 demonstrates the importance of the propositions on several baseline conformance measures, while Sect. 5 includes a discussion about the different points of view on generalization. Section 6 concludes the paper.

2   Related Work

In early years, when process mining started to gain in popularity and the community around it grew, many process discovery algorithms were developed. But at that time there was no standard method to evaluate the results of these algorithms and to compare them to the performance of other algorithms. Based on this, Rozinat et al. [28] called on the process mining community to develop a standard framework to evaluate process discovery algorithms. This led to a variety of fitness/recall, precision, generalization and simplicity notions [2]. These notions can be quantified in different ways and there are often trade-offs between the different quality dimensions. As shown using generic algorithms assigning weights to the different quality dimensions [10], one quickly gets degenerate models when leaving out one or two dimensions. For example, it is very easy to create a simple model with perfect recall (i.e., all observed behavior fits perfectly) that has poor precision and provides no insights.

Throughout the years, several conformance measures have been developed for each quality dimension. However, it is unclear whether these measures actually measure what they are supposed to. An initial step to address the need for a framework to evaluate conformance measures was made in [30]. Five so-called *axioms* for precision measures were defined that characterize the desired properties of such measures. Additionally, [30] showed that none of the existing precision measures satisfied all of the formulated axioms. In comparison to [30] Janssenswillen et al. [19] did not rely on qualitative criteria, but quantitatively compared existing recall, precision and generalization measures under the aspect of feasibility, validity and sensitivity. The results showed that all recall and precision measures tend to behave in a similar way, while generalization measures seemed to differ greatly from each other. In [3] van der Aalst made a follow-up step to [30] by formalizing recall and generalization in addition to precision and by extending the precision requirements, resulting in a list of 21 conformance propositions. Furthermore, [3] showed the importance of probabilistic conformance measures that also take into account trace probabilities in process models. Beyond that, [30] and [3] motivated the process mining community to develop new precision measures, taking the axioms and propositions as a design criterion, resulting in the measures among others the measures that are proposed in [26] and in [8]. Using the 21 propositions of [3] we evaluate state-of-the-art recall (e.g. [4,5,16,23,26,27,34]), precision (e.g. [4,13,16,17,23,26,27,31]) and generalization (e.g. [4,13,16]) measures.

This paper uses the mainstream view that there are at least four quality dimensions: fitness/recall, precision, generalization, and simplicity [2]. We deliberately do not consider simplicity, since we focus on behavior only (i.e., not the model representation). Moreover, we treat generalization separately. In a controlled experiment one can assume the existence of a so-called "system model". This model can be simulated to create a synthetic event log used for discovery. In this setting, conformance checking can be reduced to measuring the similarity between the discovered model and the system model [9,20]. In terms of the well-known confusion matrix, one can then reason about true positives, false positives, true negatives, and false negatives. However, without a system model and just an event log, it is not possible to find false positives (traces possible in the model but not in reality). Hence, precision cannot be determined in the traditional way. Janssenswillen and Depaire [18] conclude in their evaluation of state-of-the-art conformance measures that none of the existing approaches reliably measures this similarity. However, in this paper, we follow the traditional view on the quality dimensions and exclude the concept of the system from our work.

Whereas there are many fitness/recall and precision measures there are fewer generalization measures. Generalization deals with future cases that were not yet observed. There is no consensus on how to define generalization and in [19] it was shown that there is no agreement between existing generalization metrics. Therefore, we cover generalization in a separate section (Sect. 5). However, as discussed in [2] and demonstrated through experimentation [10], one cannot leave out the generalization dimension. The model that simply enumerates all the traces in the log has perfect fitness/recall and precision. However, event logs cannot be assumed to be complete, thus proving that a generalization dimension is needed.

3   Preliminaries

A *multiset* over a set X is a function $B : X \to \mathbb{N}$ which we write as $[a_1^{w_1}, a_2^{w_2}, \ldots, a_n^{w_n}]$ where for all $i \in [1, n]$ we have $a_i \in X$ and $w_i \in \mathbb{N}^*$. $\mathbb{B}(X)$ denotes the set of all multisets over set X. For example, $[a^3, b, c^2]$ is a multiset over set $X = \{a, b, c\}$ that contains three a elements, one b element and two c elements. $|B|$ is the number of elements in multiset B and $B(x)$ denotes the number of x elements in B. $B_1 \uplus B_2$ is the sum of two multisets: $(B_1 \uplus B_2)(x) = B_1(x) + B_2(x)$. $B_1 \setminus B_2$ is the difference containing all elements from B_1 that do not occur in B_2. Thus, $(B_1 \setminus B_2)(x) = max\{B_1(x) - B_2(x), 0\}$. $B_1 \cap B_2$ is the intersection of two multisets. Hence, $(B_1 \cap B_2)(x) = min\{B_1(x), B_2(x)\}$. $[x \in B \mid b(x)]$ is the multiset of all elements in B that satisfy some condition b. $B_1 \subseteq B_2$ denotes that B_1 is contained in B_2, e.g., $[a^2, b] \subseteq [a^2, b^2, c]$, but $[a^2, b^3] \not\subseteq [a^2, b^2, c^2]$ and $[a^2, b^2, c] \not\subseteq [a^3, b^3]$.

Process mining techniques focus on the relationship between observed behavior and modeled behavior. Therefore, we first formalize event logs (i.e., observed behavior) and process models (i.e., modeled behavior). To do this, we consider a very simple setting where we only focus on the control-flow, i.e., sequences of activities.

3.1 Event Logs

The starting point for process mining is an event log. Each *event* in such a log refers to an *activity* possibly executed by a *resource* at a particular *time* and for a particular *case*. An event may have many more attributes, e.g., transactional information, costs, customer, location, and unit. Here, we focus on control-flow. Therefore, we only consider activity labels and the ordering of events within cases.

Definition 1 (Traces). $\mathcal{A}$ *is the universe of* activities. *A trace* $t \in \mathcal{A}^*$ *is a sequence of activities.* $\mathcal{T} = \mathcal{A}^*$ *is the universe of traces.*

Trace $t = \langle a, b, c, d, a \rangle$ refers to 5 events belonging to the same case (i.e., $|t| = 5$). An event log is a collection of cases each represented by a trace.

Definition 2 (Event Log). $\mathcal{L} = \mathbb{B}(\mathcal{T})$ *is the universe of event logs. An* event log $l \in \mathcal{L}$ *is a finite multiset of observed traces.* $\tau(l) = \{t \in l\} \subseteq \mathcal{T}$ *is the set of traces appearing in* $l \in \mathcal{L}$. $\overline{\tau}(l) = \mathcal{T} \setminus \tau(l)$ *is the complement of the set of non-observed traces.*

Event log $l = [\langle a, b, c \rangle^5, \langle b, a, d \rangle^3, \langle a, b, d \rangle^2]$ refers to 10 cases (i.e., $|l| = 10$). Five cases are represented by the trace $\langle a, b, c \rangle$, three cases are represented by the trace $\langle b, a, d \rangle$, and two cases are represented by the trace $\langle a, b, d \rangle$. Hence, $l(\langle a, b, d \rangle) = 2$.

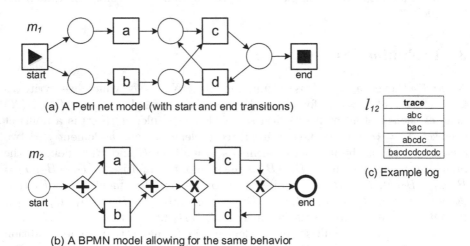

(a) A Petri net model (with start and end transitions)

l_{12}	trace
	abc
	bac
	abcdc
	bacdcdcdcdc

(c) Example log

(b) A BPMN model allowing for the same behavior

Fig. 1. Two process models m_1 and m_2 allowing for the same set of traces ($\tau(m_1) = \tau(m_2)$) with an example log l_{12} (c).

3.2   Process Models

The behavior of a process model m is simply the set of traces allowed by m. In our definition, we will abstract from the actual representation (e.g. Petri nets or BPMN).

Definition 3 (Process Model). $\mathcal{M}$ *is the set of process models. A process model $m \in \mathcal{M}$ allows for a set of traces $\tau(m) \subseteq \mathcal{T}$. $\overline{\tau}(m) = \mathcal{T} \setminus \tau(m)$ is the complement of the set of traces allowed by model $m \in \mathcal{M}$.*

A process model $m \in \mathcal{M}$ may abstract from the real process and leave out unlikely behavior. Furthermore, this abstraction can result in $\tau(m)$ allowing for traces that cannot happen (e.g., particular interleavings or loops).

We distinguish between *representation* and *behavior* of a model. Process model $m \in \mathcal{M}$ can be represented using a plethora of modeling languages, e.g., Petri nets, BPMN models, UML activity diagrams, automata, and process trees. Here, we abstract from the actual representation and focus on behavioral characteristics $\tau(m) \subseteq \mathcal{T}$.

Figure 1(a) and (b) show two process models that have the same behavior: $\tau(m_1) = \tau(m_2) = \{\langle a, b, c \rangle, \langle a, c, b \rangle, \langle a, b, c, d, c \rangle, \langle b, a, c, d, c \rangle, \dots \}$. Figure 1(c) shows a possible event log generated by one of these models $l_{12} = [\langle a, b, c \rangle^3, \langle b, a, c \rangle^5, \langle a, b, c, d, c \rangle^2, \langle b, a, c, d, c, d, c, d, c, d, c \rangle^2]$.

The behavior $\tau(m)$ of a process model $m \in \mathcal{M}$ can be of infinite size. We use Fig. 1 to illustrate this. There is a "race" between a and b. After a and b, activity c will occur. Then there is a probability that the process ends or d can occur. Let $t_{a,k} = \langle a, b \rangle \cdot (\langle c, d \rangle)^k \cdot \langle c \rangle$ be the trace that starts with a and where d is executed k times. $t_{b,k} = \langle b, a \rangle \cdot (\langle c, d \rangle)^k \cdot \langle c \rangle$ is the trace that starts with b and where d is executed k times. $\tau(m_1) = \tau(m_2) = \bigcup_{k \geq 0}\{t_{a,k}, t_{b,k}\}$. Some examples are given in Fig. 1(c).

Since any log contains only a finite number of traces, one can never observe all traces possible in m_1 or m_2.

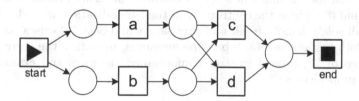

Fig. 2. A process model m_3 discovered based on log $l_3 = [\langle a, b, c \rangle^5, \langle b, a, d \rangle^3, \langle a, b, d \rangle^2]$.

3.3   Process Discovery

A discovery algorithm takes an event log as input and returns a process model. For example, the model m_3 in Fig. 2 could have been discovered based on event log $l_3 = [\langle a, b, c \rangle^5, \langle b, a, d \rangle^3, \langle a, b, d \rangle^2]$. Ideally, the process model should capture the (dominant) behavior observed but it should also generalize without becoming too imprecise. For example, the model allows for trace $t = \langle b, a, c \rangle$ although this was never observed.

Definition 4 (Discovery Algorithm). *A discovery algorithm can be described as a function disc $\in \mathcal{L} \to \mathcal{M}$ mapping event logs onto process models.*

We abstract from concrete discovery algorithms. Over 100 discovery algorithms have been proposed in literature [2]. Merely as a reference to explain basic notions, we define three simple, but extreme, algorithms: $disc_{ofit}$, $disc_{ufit}$, and $disc_{nfit}$. Let $l \in \mathcal{L}$ be a log. $disc_{ofit}(l) = m_o$ such that $\tau(m_o) = \tau(l)$ produces an overfitting model that allows only for the behavior seen in the log. $disc_{ufit}(l) = m_u$ such that $\tau(m_u) = \mathcal{T}$ produces an underfitting model that allows for any behavior. $disc_{nfit}(l) = m_n$ such that $\tau(m_n) = \overline{\tau}(l)$ produces a non-fitting model that allows for all behavior *not* seen in the log.

4   Recall and Precision

Many recall measures have been proposed in literature [2, 4, 6, 12–15, 24, 25, 27, 32, 33]. In recent years, also several precision measures have been proposed [7, 30]. Only few generalization measures have been proposed [4]. The goal of this paper is to evaluate these quality measures. To achieve this, in the following the propositions introduced in [3] are applied to existing conformance measures.

The notion of recall and precision are well established in the process mining community. Definitions are in place and there is an agreement on what these two measures are supposed to measure. However, this is not the case for generalization. There exist different points of view on what generalization is supposed to measure. Depending on these, existing generalization measures might greatly differ from each other.

To account for the different levels of maturity in recall, precision and generalization and to address the controversy in the generalization area, the following section will solely handle recall and precision while Sect. 5 focuses on generalization. Both sections establish baseline measures, introduce the corresponding propositions of [3], present existing conformance measures and evaluate them using the propositions.

4.1   Baseline Recall and Precision Measures

We assume the existence of two functions: $rec()$ and $prec()$ respectively denoting recall and precision. Both take a log and model as input and return a value between 0 and 1. The higher the value, the better.

Definition 5 (Recall). *A recall measure rec $\in \mathcal{L} \times \mathcal{M} \to [0, 1]$ aims to quantify the fraction of observed behavior that is allowed by the model.*

Definition 6 (Precision). *A precision measure $prec \in \mathcal{L} \times \mathcal{M} \to [0,1]$ aims to quantify the fraction of behavior allowed by the model that was actually observed.*

If we ignore frequencies of traces, we can simply count fractions of traces yielding the following two simple measures.

Definition 7 (Trace-Based L2M Precision and Recall). *Let $l \in \mathcal{L}$ and $m \in \mathcal{M}$ be an event log and a process model. Trace-based L2M precision and recall are defined as follows:*

$$rec_{TB}(l,m) = \frac{|\tau(l) \cap \tau(m)|}{|\tau(l)|} \qquad prec_{TB}(l,m) = \frac{|\tau(l) \cap \tau(m)|}{|\tau(m)|} \qquad (1)$$

Since $|\tau(l)|$ is bounded by the size of the log, $rec_{TB}(l,m)$ is well-defined. However, $prec_{TB}(l,m)$ is undefined when $\tau(m)$ is unbounded (e.g., in case of loops).

One can argue, that the frequency of traces should be taken into account when evaluating conformance which yields the following measure. Note that it is not possible to define frequency-based precision based on a process model that does not define the probability of its traces. Since probabilities are specifically excluded from the scope of this paper, the following approach only defines frequency-based recall.

Definition 8 (Frequency-Based L2M Recall). *Let $l \in \mathcal{L}$ and $m \in \mathcal{M}$ be an event log and a process model. Frequency-based L2M recall is defined as follows:*

$$rec_{FB}(l,m) = \frac{|[t \in l \mid t \in \tau(m)]|}{|l|} \qquad (2)$$

4.2   A Collection of Conformance Propositions

In [3], 21 *conformance propositions* covering the different conformance dimensions (except simplicity) were given. In this section, we focus on the general, recall and precision propositions introduced in [3]. We discuss the generalization propositions separately, because they reason about unseen cases not yet recorded in the event log. Most of the conformance propositions have broad support from the community, i.e., there is broad consensus that these propositions should hold. These are marked with a "+". More controversial propositions are marked with a "0" (rather than a "+").

General Propositions. The first two propositions are commonly accepted; the computation of a quality measure should be deterministic (**DetPro⁺**) and only depend on behavioral aspects (**BehPro⁺**). The latter is a design choice. We deliberately exclude simplicity notions.

Proposition 1 (DetPro⁺). *rec(), prec(), gen() are deterministic functions, i.e., the measures $rec(l,m)$, $prec(l,m)$, $gen(l,m)$ are fully determined by $l \in \mathcal{L}$ and $m \in \mathcal{M}$.*

Proposition 2 (BehPro$^+$). *For any $l \in \mathcal{L}$ and $m_1, m_2 \in \mathcal{M}$ such that $\tau(m_1) = \tau(m_2)$: $rec(l, m_1) = rec(l, m_2)$, $prec(l, m_1) = prec(l, m_2)$, and $gen(l, m_1) = gen(l, m_2)$, i.e., the measures are fully determined by the behavior observed and the behavior described by the model (representation does not matter).*

Recall Propositions. In this subsection, we consider a few *recall propositions*. $rec \in \mathcal{L} \times \mathcal{M} \rightarrow [0, 1]$ aims to quantify the fraction of observed behavior that is allowed by the model. Proposition **RecPro1$^+$** states that extending the model to allow for more behavior can never result in a lower recall. From the definition follows, that this proposition implies **BehPro$^+$**. Recall measures violating **BehPro$^+$** also violate **RecPro1$^+$** which is demonstrated as follows:

For two models m_1, m_2 with $\tau(m_1) = \tau(m_2)$ it follows from **RecPro1$^+$** that $rec(l, m_1) \leq rec(l, m_2)$ because $\tau(m_1) \subseteq \tau(m_2)$. From **RecPro1$^+$** follows that $rec(l, m_2) \leq rec(l, m_1)$ because $\tau(m_2) \subseteq \tau(m_1)$. Combined, $rec(l, m_2) \leq rec(l, m_1)$ and $rec(l, m_1) \leq rec(l, m_2)$ gives $rec(l, m_2) = rec(l, m_1)$, thus, recall measures that fulfill **RecPro1$^+$** are fully determined by the behavior observed and the behavior described by the model, i.e., representation does not matter.

Proposition 3 (RecPro1$^+$). *For any $l \in \mathcal{L}$ and $m_1, m_2 \in \mathcal{M}$ such that $\tau(m_1) \subseteq \tau(m_2)$: $rec(l, m_1) \leq rec(l, m_2)$.*

Similarly to **RecPro1$^+$**, it cannot be the case that adding fitting behavior to the event logs, lowers recall (**RecPro2$^+$**).

Proposition 4 (RecPro2$^+$). *For any $l_1, l_2, l_3 \in \mathcal{L}$ and $m \in \mathcal{M}$ such that $l_2 = l_1 \uplus l_3$ and $\tau(l_3) \subseteq \tau(m)$: $rec(l_1, m) \leq rec(l_2, m)$.*

Similarly to **RecPro2$^+$**, one can argue that adding non-fitting behavior to event logs should not be able to increase recall (**RecPro3^0**). However, one could also argue that recall should not be measured on a trace-level, but should instead distinguish between non-fitting traces by measuring the *degree* in which a non-fitting trace is still fitting. Therefore, unlike the previous propositions, this requirement is debatable as is indicated by the "0" tag.

Proposition 5 (RecPro3^0). *For any $l_1, l_2, l_3 \in \mathcal{L}$ and $m \in \mathcal{M}$ such that $l_2 = l_1 \uplus l_3$ and $\tau(l_3) \subseteq \overline{\tau}(m)$: $rec(l_1, m) \geq rec(l_2, m)$.*

For any $k \in \mathbb{N}$: $l^k(t) = k \cdot l(t)$, e.g., if $l = [\langle a, b \rangle^3, \langle c \rangle^2]$, then $l^4 = [\langle a, b \rangle^{12}, \langle c \rangle^8]$. We use this notation to enlarge event logs without changing the original distribution. One could argue that this should not influence recall (**RecPro4^0**), e.g., $rec([\langle a, b \rangle^3, \langle c \rangle^2], m) = rec([\langle a, b \rangle^{12}, \langle c \rangle^8], m)$. On the other hand, larger logs can provide more confidence that the log is indeed a representative sample of the possible behavior. Therefore, it is debatable whether the size of the event log should have influence on recall as indicated by the "0" tag.

Proposition 6 (RecPro4^0). *For any $l \in \mathcal{L}$, $m \in \mathcal{M}$, and $k \geq 1$: $rec(l^k, m) = rec(l, m)$.*

Finally, we provide a proposition stating that recall should be 1 if all traces in the log fit the model (**RecPro5$^+$**). As a result, the empty log has recall 1 for any model. Based on this proposition, $rec(l, disc_{ofit}(l)) = rec(l, disc_{ufit}(l)) = 1$ for any log l.

Proposition 7 (RecPro5$^+$). *For any $l \in \mathcal{L}$ and $m \in \mathcal{M}$ such that $\tau(l) \subseteq \tau(m)$: $rec(l, m) = 1$.*

Precision Propositions. Precision ($prec \in \mathcal{L} \times \mathcal{M} \to [0, 1]$) aims to quantify the fraction of behavior allowed by the model that was actually observed. Initial work in the area of checking requirements of conformance checking measures started with [30], where five axioms for precision measures were introduced. The precision propositions that we state below partly overlap with these axioms, but some have been added and some have been strengthened. Axiom 1 of [30] specifies **DetPro$^+$** for the case of precision, while we have generalized it to the recall and generalization dimension. Furthermore, **BehPro$^+$** generalizes axiom 4 of [30] from its initial focus on precision to also cover recall and generalization. **PrecPro1$^+$** states that removing behavior from a model that does not happen in the event log cannot lead to a lower precision. From the definition follows, that this proposition implies **BehPro$^+$**. Precision measures violating **BehPro$^+$** also violate **PrecPro1$^+$**. Adding fitting traces to the event log can also not lower precision (**PrecPro2$^+$**). However, adding non-fitting traces to the event log should not change precision (**PrecPro3^0**).

Proposition 8 (PrecPro1$^+$). *For any $l \in \mathcal{L}$ and $m_1, m_2 \in \mathcal{M}$ such that $\tau(m_1) \subseteq \tau(m_2)$ and $\tau(l) \cap (\tau(m_2) \setminus \tau(m_1)) = \emptyset$: $prec(l, m_1) \geq prec(l, m_2)$.*

This proposition captures the same idea as axiom 2 in [30], but it is more general. Axiom 2 only put this requirement on precision when $\tau(l) \subseteq \tau(m_1)$, while **PrecPro1$^+$** also concerns the situation where this does not hold.

Proposition 9 (PrecPro2$^+$). *For any $l_1, l_2, l_3 \in \mathcal{L}$ and $m \in \mathcal{M}$ such that $l_2 = l_1 \uplus l_3$ and $\tau(l_3) \subseteq \tau(m)$: $prec(l_1, m) \leq prec(l_2, m)$.*

This proposition is identical to axiom 5 in [30].

Proposition 10 (PrecPro3^0). *For any $l_1, l_2, l_3 \in \mathcal{L}$ and $m \in \mathcal{M}$ such that $l_2 = l_1 \uplus l_3$ and $\tau(l_3) \subseteq \overline{\tau}(m)$: $prec(l_1, m) = prec(l_2, m)$.*

One could also argue that duplicating the event log should not influence precision because the distribution remains the same (**PrecPro4^0**), e.g., $prec([\langle a, b \rangle^{20}, \langle c \rangle^{20}], m) = prec([\langle a, b \rangle^{40}, \langle c \rangle^{40}], m)$. Similar to (**RecPro3^0**) and (**RecPro4^0**), the equivalents on the precision side are tagged with "0".

Proposition 11 (PrecPro4^0). *For any $l \in \mathcal{L}$, $m \in \mathcal{M}$, and $k \geq 1$: $prec(l^k, m) = prec(l, m)$.*

If the model allows for the behavior observed and nothing more, precision should be maximal (**PrecPro5$^+$**). One could also argue that if all modeled behavior was observed, precision should also be 1 (**PrecPro6^0**). The latter proposition is debatable because it implies that the non-fitting behavior cannot influence perfect precision, as indicated by the "0" tag. Consider for example extreme cases where the model covers just a small fraction of all observed behavior (or even more extreme situations like $\tau(m) = \emptyset$). According to **PrecPro5$^+$** and **PrecPro6^0**, $rec(l, disc_{ofit}(l)) = 1$ for any log l.

Proposition 12 (PrecPro5$^+$). *For any $l \in \mathcal{L}$ and $m \in \mathcal{M}$ such that $\tau(m) = \tau(l)$: $prec(l, m) = 1$.*

Proposition 13 (PrecPro6^0). *For any $l \in \mathcal{L}$ and $m \in \mathcal{M}$ such that $\tau(m) \subseteq \tau(l)$: $prec(l, m) = 1$.*

4.3   Evaluation of Baseline Conformance Measures

To illustrate the presented propositions and justify their formulation, we evaluate the conformance measures defined as baselines in Sect. 4.1. Note that these 3 baseline measures were introduced to provide simple examples that can be used to discuss the propositions. We conduct this evaluation under the assumption that $l \neq [\ ]$, $\tau(m) \neq \emptyset$ and $\langle \rangle \notin \tau(m)$.

General Propositions. Based on the definition of rec_{TB} and rec_{FB} it is clear that all measures can be fully determined by the log and the model. Consequently, **DetPro$^+$** hold for these two baseline conformance measures. However, $prec_{TB}$ is undefined when $\tau(m)$ is unbound and, therefore, non-deterministic.

The behavior of the model is defined as sets of traces $\tau(m)$, which abstracts from the representation of the process model itself. Therefore, all recall and precision baseline conformance measures fulfill **BehPro$^+$**.

Recall Propositions. Considering measure rec_{TB}, it is obvious that **RecPro1$^+$** holds if $\tau(m_1) \subseteq \tau(m_2)$, because the intersection between $\tau(m_2)$ and $\tau(l)$ will always be equal or bigger to the intersection of $\tau(m_1)$ and $\tau(l)$. The **RecPro2$^+$** proposition holds for rec_{TB}, if $l_2 = l_1 \uplus l_3$ and $\tau(l_3) \subseteq \tau(m)$, because the additional fitting behavior is added to the nominator as well as the denominator of the formula: $|\tau(l_1) \cap \tau(m)| + |\tau(l_3)|)/(|\tau(l_1)| + |\tau(l_3)|$. This can never decrease recall. Furthermore, **RecPro3^0** propositions holds for rec_{TB} since adding unfitting behavior cannot increase the intersection between traces of the model and the log if $l_2 = l_1 \uplus l_3$ and $\tau(l_3) \subseteq \overline{\tau}(m)$. Consequently, only the denominator of the formula grows, which decreases recall. Similarly, we can show that these two proposition hold for rec_{FB}.

Duplication of the event log cannot affect rec_{TB}, since it is defined based on the set of traces and not the multiset. The proposition also holds for rec_{FB} since nominator and denominator of the formula will grow in proportion. Hence, **RecPro4^0** holds for both baseline measures. Considering rec_{TB},

RecPro5$^+$ holds, since $\tau(l) \cap \tau(m) = \tau(l)$ if $\tau(l) \subseteq \tau(m)$ and consequently $|\tau(l) \cap \tau(m)| / |\tau(l)| = |\tau(l)| / |\tau(l)| = 1$. The same conclusions can be drawn for rec_{FB}.

Precision Propositions. Consider proposition **PrecPro1$^+$** together with $prec_{TB}$. The proposition holds, since removing behavior from the model that does not happen in the event log will not affect the intersection between the traces of the model and the log: $\tau(l) \cap \tau(m_2) = \tau(l) \cap \tau(m_1)$ if $\tau(m_1) \subseteq \tau(m_2)$ and $\tau(l) \cap (\tau(m_2) \setminus \tau(m_1)) = \emptyset$. At the same time the denominator of the formula decreases, which can never decrease precision itself. **PrecPro2$^+$** also holds for $prec_{TB}$, since the fitting behavior increases the intersection between traces of the model and the log, while the denominator of the formula stays the same. Furthermore, **PrecPro3^0** holds for $prec_{TB}$, since unfitting behavior cannot affect the intersection between traces of the model and the log.

Duplication of the event log cannot affect $prec_{TB}$, since it is defined based on the set of traces and not the multiset, i.e. **PrecPro4^0** holds.

Considering $prec_{TB}$, **PrecPro5$^+$** holds, since $\tau(l) \cap \tau(m) = \tau(m)$ if $\tau(m) = \tau(l)$ and consequently $|\tau(l) \cap \tau(m)| / |\tau(m)| = |\tau(m)| / |\tau(m)| = 1$. Similarly, **PrecPro6^0** holds for $prec_{TB}$.

4.4 Existing Recall Measures

The previous evaluation of the simple baseline measures shows that the recall measures fulfill all propositions and the baseline precision measure only violates one proposition. However, the work presented in [30] demonstrated for precision, that most of the existing approaches violate seemingly obvious requirements. This is surprising compared to the results of our simple baseline measure. Inspired by [30], this paper takes a broad look at existing conformance measures with respect to the previously presented propositions. In the following section, existing recall and precision measures are introduced, before they will be evaluated in Sect. 4.6.

Causal Footprint Recall (rec_A). Van der Aalst et al. [5] introduce the concept of the footprint matrix, which captures the relations between the different activities in the log. The technique relies on the principle that if activity a is followed by b but b is never followed by a, then there is a causal dependency between a and b. The log can be described using four different relations types. In [2] it is stated that a footprint matrix can also be derived for a process model by generating a complete event log from it. Recall can be measured by counting the mismatches between both matrices. Note that this approach assumes an event log which is complete with respect to the directly follows relations.

Token Replay Recall (rec_B). Token replay measures recall by replaying the log on the model and counting mismatches in the form of missing and remaining tokens. This approach was proposed by Rozinat and van der Aalst [27]. During replay, four types of tokens are distinguished: p the number of *produced* tokens, c the number of *consumed* tokens, m the number of *missing* tokens that had to be added because a transition was not enabled during replay and r the number of *remaining* tokens that are left in the model after replay. In the beginning, a token is produced in the initial place. Similarly, the approach ends by consuming a token from the final place. The more missing and remaining tokens are counted during replay the lower recall: $rec_B = \frac{1}{2}(1 - \frac{m}{c}) + \frac{1}{2}(1 - \frac{r}{p})$ Note that the approach assumes a relaxed sound workflow net, but it allows for duplicate and silent transitions.

Alignment Recall (rec_C). Another approach to determine recall was proposed by van der Aalst et al. [4]. It calculates recall based on alignments, which detect process deviations by mapping the steps taken in the event log to the ones of the process model. This map can contain three types of steps (so-called moves): *synchronous* moves when event log and model agree, *log* moves if the event was recorded in the event log but should not have happened according to the model and *model* moves if the event should have happened according to the model but did not in the event log. The approach uses a function that assigns costs to log moves and model moves. This function is used to compute the optimal alignment for each trace in the log (i.e. the alignment with the least cost associated).

To compute recall, the total alignment cost of the log is normalized with respect to the cost of the worst-case scenario where there are only moves in the log and in the model but never together. Note, that the approach assumes an accepting Petri net with an initial and final state. However, it allows for duplicate and silent transitions in the process model.

Behavioral Recall (rec_D). Goedertier et al. [16] define recall according to its definition in the data mining field using true positive (TP) and false negative (FN) counters. $TP(l, m)$ denotes the number of true positives, i.e., the number of events in the log that can be parsed correctly in model m by firing a corresponding enabled transition. $FN(l, m)$ denotes the number of false negatives, i.e., the number of events in the log for which the corresponding transition that was needed to mimic the event was not enabled and needed to be force-fired. The recall measure is defined as follows: $rec_D(l, m) = \frac{TP(l,m)}{TP(l,m)+FN(l,m)}$.

Projected Recall (rec_E). Leemans et al. [23] developed a conformance checking approach that is also able to handle big event logs. This is achieved by projecting the event log as well as the model on all possible subsets of activities of size k. The behavior of a projected log and projected model is translated into the minimal deterministic finite automata (DFA)[1]. Recall is calculated by checking the fraction of the behavior that is allowed for by the minimal log-automaton that is also allowed for by the minimal model-automaton for each projection and by averaging the recall over each projection.

---

[1] Every regular language has a unique minimal DFA according to the Myhill–Nerode theorem.

Continued Parsing Measure (rec_F). This continued parsing measure was developed in the context of the heuristic miner by Weijters et al. [34]. It abstracts from the representation of the process model by translating the Petri net into a causal matrix. This matrix defines input and output expressions for each activity, which describe possible in- and output behavior. When replaying the event log on the causal matrix, one has to check whether the corresponding input and output expressions are activated and therefore enable the execution of the activity. To calculate the continued parsing measure the number of events e in the event log, as well as the number of missing activated input expressions m and remaining activated output expressions r are counted. Note, that the approach allows for silent transitions in the process model but excludes duplicate transitions.

Eigenvalue Recall (rec_G). Polyvyanyy et al. [26] introduce a framework for the definition of language quotients that guarantee several properties similar to the propositions introduced in [3]. To illustrate this framework, they apply it in the process mining context and define a recall measure. Hereby they rely on the relation between the language of a deterministic finite automaton (DFA) that describes the behavior of the model and the language of the log. In principle, recall is defined as in Definition 7. However, the measure is undefined if the language of the model or the log are infinite. Therefore, instead of using the cardinality of the languages and their intersection, the measure computes their corresponding eigenvalues and sets them in relation. To compute these eigenvalues, the languages have to be irreducible. Since this is not the case for the language of event logs, Polyvyanyy et al. [26] introduce a short-circuit measure over languages and proved that it is a deterministic measure over any arbitrary regular language.

4.5 Existing Precision Measures

Soundness ($prec_H$). The notion of soundness as defined by Greco et al. [17] states that a model is precise if all possible enactments of the process have been observed in the event log. Therefore, it divides the number of unique traces in the log compliant with the process model by the number of unique traces through the model. Note, that this approach assumes the process model in the shape of a workflow net. Furthermore, it is equivalent to the baseline precision measure $prec_T B$.

Simple Behavioral Appropriateness ($prec_I$). Rozinat and van der Aalst [27] introduce simple behavioral appropriateness to measure the precision of process models. The approach assumes that imprecise models enable a lot of transitions during replay. Therefore, the approach computes the mean number of enabled transitions x_i for each unique trace i and puts it in relation to the visible transitions T_V in the process model. Note, that the approach assumes a sound workflow net. However, it allows for duplicate and silent transitions in the process model.

Advanced Behavioral Appropriateness ($prec_J$). In the same paper, Rozinat and van der Aalst [27] define advanced behavioral appropriateness. This approach abstracts from the process model by describing the relation between activities of both the log and model with respect to whether these activities *follow* and/or *precede* each other. Hereby they differentiate between *never, sometimes* and *always* precede/follow relations. To calculate precision the set of sometimes followed relations of the log S_F^l and the model S_F^m are considered, as well as their sometimes precedes relations S_P^l and S_P^m. The fraction of sometimes follows/precedes relations of the model which are also observed by the event log defines precision. Note, that the approach assumes a sound workflow net. However, it allows for duplicate and silent transitions in the process model.

ETC-one/ETC-rep ($prec_K$) and ETC-all ($prec_L$). Munoz-Gama and Carmona [25] introduced a precision measure which constructs an automaton that reflects the states of the model which are visited by the event log. For each state, it is evaluated whether there are activities which were allowed by the process model but not observed by the event log. These activities are added to the automaton as so-called escaping edges. Since this approach is not able to handle unfitting behavior, [7] and [4] extended the approach with a preprocessing step that aligned the log to the model before the construction of the automaton. Since it is possible that traces result in multiple optimal alignments, there are three variations of the precision measure. One can randomly pick one alignment and construct the alignment automaton based on it (ETC-one), select a representative set of multiple alignments (ETC-rep) or use all optimal alignments (ETC-all). For each variation, [4] defines an approach that assigns appropriate weights to the edges of the automaton. Precision is then computed by comparing for each state of the automaton, the weighted number of non-escaping edges to the total number of edges.

Behavioral Specificity ($prec_M$) and Behavioral Precision ($prec_N$). Goedertier et al. [16] introduced a precision measure based on the concept of negative events that is defined based on the concept of a confusion matrix as used in the data mining field. In this confusion matrix, the induced negative events are considered to be the ground truth and the process model is considered to be a prediction machine that predicts whether an event can or cannot occur. A negative event expresses that at a certain position in a trace, a particular event cannot occur. To induce the negative events into an event log, the traces are split in subsequences of length k. For each event e in the trace, it is checked whether another event e_n could be a negative event. Therefore the approach searches whether the set of subsequences contains a similar sequence to the one preceding e. If no matching sequence is found that contains e_n at the current position of e, e_n is recorded as a negative event of e. To check conformance the log, that was induced with negative events, is replayed on the model.

For both measures, the log that was induced with negative events is replayed on the model. Specificity and precision are measured according to their data mining definition using true positive (TP), false positive (FP) and true negative (TN) counts.

Goedertier et al. [16] ($prec_M$) defined behavioral specificity precision as $prec_M(l, m) = \frac{TN(l,m)}{TN(l,m)+FP(l,m)}$, i.e., the ratio of the induced negative events that were also disallowed by m. More recently, De Weerdt et al. [33] gave an inverse definition, called behavioral precision ($prec_N$), as the ratio of behavior that is allowed by m that does not conflict an induced negative event, i.e. $prec_N(l, m) = \frac{TP(l,m)}{TP(l,m)+FP(l,m)}$.

Weighted Negative Event Precision ($prec_O$). Van den Broucke et al. [31] proposed an improvement to the approach of Goedertier et al. [16], which assigns weights to negative events. These weights indicate the confidence of the negative events actually being negative. To compute the weight, the approach takes the sequence preceding event e and searches for the matching subsequences in the event log. All events that have never followed such a subsequence are identified as negative events for e and their weight is computed based on the length of the matching subsequence. To calculate precision the enhanced log is replayed on the model, similar to the approach introduced in [33]. However, instead of increasing the counters by 1 they are increased by the weight of the negative event. Furthermore, van den Broucke et al. [31] also introduced a modified trace replay procedure which finds the best fitting firing sequence of transitions, taking force firing of transitions as well as paths enabled by silent transitions into account.

Projected Precision ($prec_P$). Along with projected recall (rec_E) Leemans et al. [23] introduce projected precision. To compute precision, the approach creates a DFA which describes the conjunction of the behavior of the model and the event log. The number of outgoing edges of $DFA(m|_A)$ and the conjunctive automaton $DFAc(l, m, A)$ are compared. Precision is calculated for each subset of size k and averaged over the number of subsets.

Anti-alignment Precision ($prec_Q$). Van Dongen et al. [13] propose a conformance checking approach based on anti-alignments. An anti-alignment is a run of a model which differs from all the traces in a log. The principle of the approach assumes that a very precise model only allows for the observed traces and nothing more. If one trace is removed from the log, it becomes the anti-alignment for the remaining log.

Therefore, trace-based precision computes an anti-alignment for each trace in the log. Then the distance d between the anti-alignment and the trace σ is computed. This is summed up for each trace and averaged over the number of traces in the log. The more precise a model, the lower the distance. However, the anti-alignment used for trace-based precision is limited by the length of the removed trace $|\sigma|$. Therefore, log-based precision uses an anti-alignment between

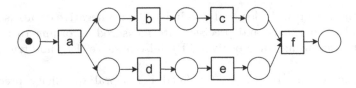

Fig. 3. A process model $m4$.

the model and the complete log which has a length which is much greater than the traces observed in the log. Anti-alignment precision is the weighted combination of trace-based and log-based anti-alignment precision. Note, that the approach allows for duplicate and silent transitions in the process model.

Eigenvalue Precision ($prec_R$). Polyvyanyy et al. [26] also define a precision measure along with the Eigenvalue recall (rec_G). For precision, they rely on the relation between the language of a deterministic finite automaton (DFA) that describes the behavior of the model and the language of the log. To overcome the problems arising with infinite languages of the model or log, they compute their corresponding eigenvalues and set them in relation. To compute these eigenvalues, the languages have to be irreducible. Since this is not the case for the language of event logs, Polyvyanyy et al. [26] introduce a short-circuit measure over languages and proof that it is a deterministic measure over any arbitrary regular language.

4.6 Evaluation of Existing Recall and Precision Measures

Several of the existing precision measures are not able to handle non-fitting behavior and remove it by aligning the log to the model. We use a baseline approach for the alignment, which results in a deterministic event log: l is the original event log, which is aligned in a deterministic manner. The resulting event log l' corresponds to unique paths through the model. We use l' to evaluate the propositions.

Evaluation of Existing Recall Measures. The previously presented recall measures are evaluated using the corresponding propositions. The results of the evaluation are displayed in Table 1. To ensure the readability of this paper, only the most interesting findings of the evaluation are addressed in the following section. For more details, we refer to [29].

The evaluation of the *causal footprint recall measure* (rec_A) showed that it is deterministic and solely relies on the behavior of the process model. However, the measure violates several propositions such as **RecPro1$^+$**, **RecPro3^0**, and **RecPro5$^+$**. These violations are caused by the fact that recall records every difference between the footprint of the log and the model. Behavior that is described by the model but not observed in the event log has an impact on recall, although Definition 5 states otherwise. To illustrate this, consider m_4 in

Table 1. Overview of the recall propositions that hold for the existing measures (under the assumption that $l \neq [\,]$, $\tau(m) \neq \emptyset$ and $\langle\rangle \notin \tau(m)$): $\checkmark$ means that the proposition holds for any log and model and $\times$ means that the proposition does not always hold.

Proposition	Name	rec_A	rec_B	rec_C	rec_D	rec_E	rec_F	rec_G
1	**DetPro$^+$**	$\checkmark$	$\times$	$\checkmark$	$\times$	$\checkmark$	$\times$	$\checkmark$
2	**BehPro$^+$**	$\checkmark$	$\times$	$\checkmark$	$\times$	$\checkmark$	$\checkmark$	$\checkmark$
3	**RecPro1$^+$**	$\times$	$\times$	$\checkmark$	$\times$	$\checkmark$	$\checkmark$	$\checkmark$
4	**RecPro2$^+$**	$\checkmark$	$\checkmark$	$\checkmark$	$\checkmark$	$\checkmark$	$\checkmark$	$\checkmark$
5	**RecPro3^0**	$\times$	$\times$	$\times$	$\times$	$\times$	$\times$	$\checkmark$
6	**RecPro4^0**	$\checkmark$	$\checkmark$	$\checkmark$	$\checkmark$	$\checkmark$	$\checkmark$	$\checkmark$
7	**RecPro5$^+$**	$\times$	$\checkmark$	$\checkmark$	$\checkmark$	$\checkmark$	$\times$	$\checkmark$

Fig. 3, event log $l_4 = [\langle a,b,c,d,e,f \rangle, \langle a,b,d,c,e,f \rangle]$ and **RecPro5$^+$**. The traces in l_4 perfectly fit process model m_4. The footprint of l_4 is shown in Table 2(b). Comparing it to the footprint of m_4 in Table 2(a) shows mismatches although l_4 is perfectly fitting. These mismatches are caused by the fact that the log does not show all possible behavior of the model and, therefore, the footprint cannot completely detect the parallelism of the model. Consequently 10 of 36 relations of the footprint represent mismatches: $rec_A(l_4, m_4) = 1 - \frac{10}{36} = 0.72 \neq 1$. Van der Aalst mentions in [2] that checking conformance using causal footprints is only meaningful if the log is complete in term of directly followed relations. Moreover, the measure also includes precision and generalization aspects, next to recall.

Table 2. The causal footprints of m_4 (a), l_4 (b). Mismatching relations are marked in red

(a)   (b)

```
     a  b  c  d  e  f         a  b  c  d  e  f
a  #  →  #  →  #  #      a  #  →  #  #  #  #
b  ←  #  →  ||  ||  #      b  ←  #  →  →  #  #
c  #  ←  #  ||  ||  →      c  #  ←  #  ||  →  #
d  ←  ||  ||  #  →  #      d  #  ←  ||  #  →  #
e  #  ||  ||  ←  #  →      e  #  #  ←  ←  #  →
f  #  #  ←  #  ←  #      f  #  #  #  #  ←  #
```

In comparison, recall based on *token replay* (rec_B) depends on the path taken through the model. Due to duplicate activities and silent transitions, multiple paths through a model can be taken when replaying a single trace. Different paths can lead to different numbers of produced, consumed, missing and remaining tokens. Therefore, the approach is neither deterministic nor independent from the structure of the process model and, consequently, violates **RecPro1$^+$**. The *continued parsing measure* (rec_F) builds on a similar replay principle as token-based replay and also violates **DetPro$^+$**. However, the approach translates the process model into a causal matrix and is therefore independent of its structure.

Table 1 also shows that most measures violate **RecPro3^0**. This is caused by the fact, that we define non-fitting behavior in this paper on a trace level: traces either fit the model or they do not. However, the evaluated approaches measure non-fitting behavior on an event level. A trace consists of fitting and non-fitting events. In cases where the log contains traces with a large number of deviating events, recall can be improved by adding non-fitting traces which contain several fitting and only a few deviating events. To illustrate this, consider *token replay* (rec_B), process model m_5 in Fig. 4, $l_5 = [\langle a, b, f, g\rangle]$ and $l_6 = l_5 \uplus [\langle a, d, e, f, g\rangle]$. The log l_5 is not perfectly fitting and replaying it on the model results in 6 produced and 6 consumed tokens, as well as 1 missing and 1 remaining token. $rec_B(l_5, m_5) = \frac{1}{2}(1 - \frac{1}{6}) + \frac{1}{2}(1 - \frac{1}{6}) = 0.833$. Event log l_6 was created by adding non-fitting behavior to l_5. Replaying l_6 on m_5 results in $p = c = 13$, $r = m = 2$ and $rec_B(l_7, m_6) = \frac{1}{2}(1 - \frac{2}{13}) + \frac{1}{2}(1 - \frac{2}{13}) = 0.846$. Hence, the additional unfitting trace results in proportionally more fitting events than deviating ones which improves recall: $rec_B(l_6, m_6) < rec_B(l_7, m_6)$.

To overcome the problems arising with the differences between trace-based and event-based fitness, one could alter the definition of **RecPro3^0** by requiring, that the initial log l_1 only contains fitting behavior ($\tau(l_1) \subseteq \tau(m)$). However, to stay within the scope of this paper, we decide to use the propositions as defined in [3] and keep this suggestion for future work.

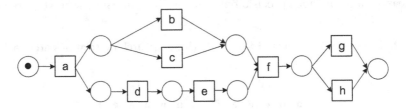

Fig. 4. Petri net m_5

Evaluation of Existing Precision Measures. The previously presented precision measures are evaluated using the corresponding propositions. The results of the evaluation are displayed in Table 3. To ensure the readability of this paper, only the most interesting findings of the evaluation are addressed in the following section. For more details, we refer to [29].

The evaluation showed that several measures violate the determinism **Det-Pro$^+$** proposition. For example, the soundness measure ($prec_H$) solely relies on the number of unique paths of the model $|\tau(m)|$ and unique traces in the log that comply with the process model $|\tau(l) \cap \tau(m)|$. Hence, precision is not defined if the model has infinite possible paths. Additionally to **DetPro$^+$**, behavioral

Table 3. Overview of the precision propositions that hold for the existing measures (under the assumption that $l \neq [\ ]$, $\tau(m) \neq \emptyset$ and $\langle\rangle \notin \tau(m)$): $\sqrt{}$ means that the proposition holds for any log and model and $\times$ means that the proposition does not always hold.

Prop	Name	$prec_H$	$prec_I$	$prec_J$	$prec_K$	$prec_L$	$prec_M$	$prec_N$	$prec_O$	$prec_P$	$prec_Q$	$prec_R$
1	**DetPro$^+$**	$\times$	$\times$	$\times$	$\times$	$\sqrt{}$	$\times$	$\times$	$\times$	$\sqrt{}$	$\sqrt{}$	$\sqrt{}$
2	**BehPro$^+$**	$\sqrt{}$	$\times$	$\times$	$\times$	$\times$	$\times$	$\times$	$\times$	$\sqrt{}$	$\sqrt{}$	$\sqrt{}$
8	**PrecPro1$^+$**	$\sqrt{}$	$\times$	$\times$	$\times$	$\times$	$\times$	$\times$	$\times$	$\times$	$\sqrt{}$	$\sqrt{}$
9	**PrecPro2$^+$**	$\sqrt{}$	$\times$	$\sqrt{}$	$\times$	$\times$	$\times$	$\times$	$\times$	$\times$	$\times$	$\sqrt{}$
10	**PrecPro3^0**	$\sqrt{}$	$\times$	$\times$	$\times$	$\times$	$\times$	$\times$	$\times$	$\times$	$\times$	$\sqrt{}$
11	**PrecPro4^0**	$\sqrt{}$	$\sqrt{}$	$\sqrt{}$	$\sqrt{}$	$\sqrt{}$	$\sqrt{}$	$\sqrt{}$	$\sqrt{}$	$\sqrt{}$	$\sqrt{}$	$\sqrt{}$
12	**PrecPro5$^+$**	$\sqrt{}$	$\times$	$\sqrt{}$	$\times$	$\times$	$\sqrt{}$	$\sqrt{}$	$\sqrt{}$	$\sqrt{}$	$\sqrt{}$	$\sqrt{}$
13	**PrecPro6^0**	$\sqrt{}$	$\times$	$\sqrt{}$	$\times$	$\times$	$\sqrt{}$	$\sqrt{}$	$\sqrt{}$	$\sqrt{}$	$\sqrt{}$	$\sqrt{}$

specificity (rec_M) and behavioral precision (rec_N) also violate **BehPro$^+$**. If during the replay of the trace duplicate or silent transitions are encountered, the approach explored which of the available transitions enables the next event in the trace. If no solution is found, one of the transitions is randomly fired, which can lead to different recall values for traces with the same behavior.

Table 3 shows that simple behavioral appropriateness ($prec_I$) violates all but one of the propositions. One of the reason is that it relies on the average number of enabled transitions during replay. Even when the model allows for all exactly observed behavior (and nothing more), precision is not maximal when the model is not strictly sequential. Advanced behavioral appropriateness ($prec_J$) overcomes these problems by relying on follow relations. However, it is not deterministic and depends on the structure of the process model.

The results presented in [30] show that ETC precision ($prec_K$ and $prec_L$), weighted negative event precision ($prec_O$) and projected precision ($prec_P$) violate **PrecPro1$^+$**. Additionally, all remaining measures aside from anti-alignment precision ($prec_Q$) and eigenvalue precision ($prec_R$) violate the proposition. The proposition states that removing behavior from a model that does not happen in the event log cannot lower precision. Consider, projected precision ($prec_P$) and a model with a length-one-loop. We remove behavior from the model by restricting the model to only execute the looping activity twice. This changes the DFA of the model since future behavior now depends on how often the looping activity was executed: the DFA contains different states for each execution. If these states show a low local precision, overall precision decreases. Furthermore, [30] showed that ETC precision ($prec_K$ and $prec_L$), projected precision ($prec_P$) and anti-alignment precision ($prec_Q$) also violate **PrecPro2$^+$**.

In general, looking at Table 3 shows that all precision measures, except for soundness ($prec_H$) and eigenvalue precision ($prec_R$) violate **PrecPro3^0**, which states that adding unfitting behavior to the event log should not change precision. However, for example, all variations of the ETC-measure ($prec_K$, $prec_L$) align the log before constructing the alignment automaton. Unfitting behavior

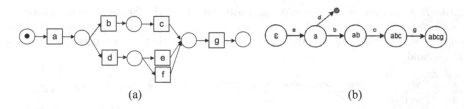

Fig. 5. Petri net m_6 (a) and the alignment automaton describing the state space of $\sigma = \langle a, b, c, g \rangle$ (b)

can be aligned to a trace that was not seen in the log before and introduce new states to the automaton. Consider process model m_6, together with trace $\sigma = \langle a, b, c, g \rangle$ and its alignment automaton displayed in Fig. 5. Adding the unfitting trace $\langle a, d, g \rangle$ could result in the aligned trace $\langle a, d, e, g \rangle$ or $\langle a, d, f, g \rangle$. Both aligned traces introduce new states into the alignment automaton, alter the weights assigned to each state and, consequently, change precision. Weighted negative precision ($prec_O$) also violates this proposition. The measure accounts for the number of negative events that actually could fire during trace replay (FP). These false positives are caused by behavior that is shown in the model but not observed in the log. As explained in the context of **RecPro3⁰**, although the trace is not fitting when considered as a whole, certain parts of the trace can fit the model. These parts can possibly represent the previously missing behavior in the event log that leads to the wrong classification of negative events. Adding these traces will, therefore, lead to a decrease in false positives and changes precision. $FP(l_1, m) > FP(l_2, m)$ and $\frac{TP(l_1,m)}{(TP(l_1,m)+FP(l_1,m))} < \frac{TP(l_2,m)}{(TP(l_2,m)+FP(l_2,m))}$.

Table 3 shows that $prec_I$, $prec_K$ and $prec_L$ violate proposition **PrecPro6⁰**, which states that if all modeled behavior was observed, precision should be maximal and unfitting behavior cannot effect precision. $prec_I$ only reports maximal precision if the model is strictly sequential and both ETC measures ($prec_K$ and $prec_L$) can run into problems with models containing silent or duplicate transitions

The ETC ($prec_K$, $prec_L$) and anti-alignment measures ($prec_Q$) form a special group of measures as they are unable to handle unfitting behavior without pre-processing unfitting traces and aligning them to the process model. Accordingly, we evaluate the conformance measure based on this aligned log. The evaluation of **PrecPro3⁰** and the ETC measure ($prec_K$, $prec_L$) is an example of the alignment of the log resulting in a violation. However, there are also cases where the proposition only holds because of this alignment. Consider, for example, anti-alignment precision ($prec_Q$) and proposition **PrecPro6⁰**. By definition, an anti-alignment will always fit the model. Consequently, when computing the distance between the unfitting trace and the anti-alignment it will never be minimal. However, after aligning the log, it exactly contains the modeled behavior, precision is maximal and the proposition holds.

5   Generalization

Generalization is a challenging concept to define, in contrast to recall and precision. As a result, there are different viewpoints within the process mining community on what generalization precisely means. The main reason for this is, that generalization needs to reason about behavior that was not observed in the event log and establish its relation to the model.

The need for a generalization dimension stems from the fact that, given a log, a model can be fitting and precise, but be overfitting. The algorithm that simply creates a model m such that $\tau(m) = \{t \in l\}$ is useless because it is simply enumerating the event log. Consider an unknown process. Assume we observe the first four traces $l_1 = [\langle a, b, c \rangle, \langle b, a, c \rangle, \langle a, b, d \rangle, \langle b, a, d \rangle]$. Based on this we may construct the model m_3 in Fig. 2 with $\tau(m_3) = \{\langle a, b, c \rangle, \langle b, a, c \rangle, \langle a, b, d \rangle, \langle b, a, d \rangle\}$. This model allows for all the traces in the event log and nothing more. However, because the real underlying process in unknown, this model may be overfitting event log l_1. Based on just four example traces we cannot be confident that the model m_3 in Fig. 2 will be able to explain future behavior of the process. The next trace may as well be $\langle a, c \rangle$ or $\langle a, b, b, c \rangle$. Now assume that we observe the same process for a longer time and consider the first 100 traces (including the initial four): $l_2 = [\langle a, b, c \rangle^{25}, \langle b, a, c \rangle^{25}, \langle a, b, d \rangle^{25}, \langle b, a, d \rangle^{25}]$. After observing 100 traces, we are more confident that model m_3 in Fig. 2 is the right model. Intuitively, the probability that the next case will have a trace not allowed by m_3 gets smaller. Now assume that we observe the same process for an even longer time and obtain the event log $l_2 = [\langle a, b, c \rangle^{53789}, \langle b, a, c \rangle^{48976}, \langle a, b, d \rangle^{64543}, \langle b, a, d \rangle^{53789}]$. Although we do not know the underlying process, intuitively, the probability that the next case will have a trace not allowed by m_3 is close to 0. This simple example shows that recall and precision are not enough for conformance checking. We need a generalization notion to address the risk of overfitting example data.

It is difficult to reason about generalization because this refers to unseen cases. Van der Aalst et al. [4] was the first to quantify generalization. In [4], each event is seen as an observation of an activity a in some state s. Suppose that state s is visited n times and that w is the number of different activities observed in this state. Suppose that n is very large and w is very small, then it is unlikely that a new event visiting this state will correspond to an activity not seen before in this state. However, if n and w are of the same order of magnitude, then it is more likely that a new event visiting state s will correspond to an activity not seen before in this state. This reasoning is used to provide a generalization metric. This estimate can be derived under the Bayesian assumption that there is an unknown number of possible activities in state s and that probability distribution over these activities follows a multinomial distribution.

It is not easy to develop an approach that accurately measures generalization. Therefore, some authors define generalization using the notion of a "system" (i.e., a model of the real underlying process). The system refers to the real behavior of the underlying process that the model tries to capture. This can also include the context of the process such as the organization or rules. For example, employees of a company might exceptionally be allowed to deviate from the defined process model in certain situations [20]. In this view, *system fitness* measures the fraction of the behavior of the system that is captured by the model and *system precision* measures how much of the behavior of the model is part of the system. Buijs et al. [11] link this view to the traditional understanding of generalization. They state that both system fitness and system precision are difficult to obtain under the assumption that the system is unknown. Therefore, state-of-the-art discovery algorithms assume that the process model discovered from an event log does not contain behavior outside of the system. In other words, they assume system precision to be 1. Given this assumption, system fitness can be seen as generalization [11]. Janssenswillen et al. [20] agree that in this comparison between the system and the model, especially the system fitness, in fact is what defines generalization. Furthermore, Janssenswillen and Depaire [18] demonstrated the differences between the traditional and the system-based view on conformance checking by showing that state-of-the-art conformance measures cannot reliably assess the similarity between a process model and the underlying system.

Although capturing the unobserved behavior by assuming a model of the system is a theoretically elegant solution, practical applicability of this solution is hindered by the fact that is often impossible to retrieve full knowledge about the system itself. Furthermore, [3] showed the importance of trace probabilities in process models. To accurately represent reality, the system would also need to include probabilities for each of its traces. However, to date, there is only one conformance measure that can actually support probabilistic process models [22]. This approach uses the Earth Movers' distance which measures the effort to transform the distributions of traces of the event log into the distribution of traces of the model.

Some people would argue that one should use cross-validation (e.g., k-fold checking). However, this is a very different setting. Cross validation aims to estimate the quality of a discovery approach and not the quality of a given model given an event log. Of course, one could produce multiple process models using fragments of the event log and compare them. However, such forms of cross-validation evaluate the quality of the discovery technique and are unrelated to generalization.

For these reasons, we define generalization in the traditional sense.

Definition 9 Generalization. *A generalization measure $gen \in \mathcal{L} \times \mathcal{M} \to [0, 1]$ aims to quantify the probability that new unseen cases will fit the model.*[2]

This definition assumes that a process generates a stream of newly executed cases. The more traces that are fitting and the more redundancy there is in the event, the more certain one can be that the next case will have a trace that fits

---

[2] Note that the term "probability" is used here in an informal manner. Since we only have example observations and no knowledge of the underlying (possibly changing) process, we cannot compute such a probability. Of course, unseen cases can have traces that have been observed before.

the model. Note that we deliberately do not formalize the notion of probability, since in real-life we cannot know the real process. Also phenomena like concept drift and contextual factors make it unrealistic to reason about probabilities in a formal sense.

Based on this definition, we present a set of propositions. Note that we do not claim our set of propositions to be complete and invite other researchers who represent a different viewpoint on generalization to contribute to the discussion.

5.1   Generalization Propositions

Generalization ($gen \in \mathcal{L} \times \mathcal{M} \to [0, 1]$) aims to quantify the probability that new unseen cases will fit the model. This conformance dimension is a bit different than the two previously discussed conformance dimensions because it reasons about future *unseen* cases (i.e., not yet in the event log). If the recall is good and the log is complete with lots of repeating behavior, then future cases will most likely fit the model. Analogous to recall, model extensions cannot lower generalization (**GenPro1$^+$**), extending the log with fitting behavior cannot lower generalization (**GenPro2$^+$**), and extending the log with non-fitting behavior cannot improve generalization (**GenPro3^0**).

Proposition 14 (GenPro1$^+$).   *For any $l \in \mathcal{L}$ and $m_1, m_2 \in \mathcal{M}$ such that $\tau(m_1) \subseteq \tau(m_2)$: $gen(l, m_1) \leq gen(l, m_2)$.*

Similar to recall, this proposition implies **BehPro$^+$**. Generalization measures violating **BehPro$^+$** also violate **GenPro1$^+$**.

Proposition 15 (GenPro2$^+$).   *For any $l_1, l_2, l_3 \in \mathcal{L}$ and $m \in \mathcal{M}$ such that $l_2 = l_1 \uplus l_3$ and $\tau(l_3) \subseteq \tau(m)$: $gen(l_1, m) \leq gen(l_2, m)$.*

Proposition 16 GenPro3^0).   *For any $l_1, l_2, l_3 \in \mathcal{L}$ and $m \in \mathcal{M}$ such that $l_2 = l_1 \uplus l_3$ and $\tau(l_3) \subseteq \overline{\tau}(m)$: $gen(l_1, m) \geq gen(l_2, m)$.*

Duplicating the event log does not necessarily influence recall and precision. According to propositions **RecPro4^0** and **PrecPro4^0** this should have no effect on recall and precision. However, making the event log more redundant, should have an effect on generalization. For fitting logs, adding redundancy without changing the distribution can only improve generalization (**GenPro4$^+$**). For non-fitting logs, adding redundancy without changing the distribution can only lower generalization (**GenPro5$^+$**). Note that **GenPro4$^+$** and **GenPro5$^+$** are special cases of **GenPro6^0** and **GenPro7^0**. **GenPro6^0** and **GenPro7^0** consider logs where some traces are fitting and others are not. For a log where more than half of the traces is fitting, duplication can only improve generalization (**GenPro6^0**). For a log where more than half of the traces is non-fitting, duplication can only lower generalization (**GenPro7^0**).

Proposition 17 (GenPro4$^+$).   *For any $l \in \mathcal{L}$, $m \in \mathcal{M}$, and $k \geq 1$ such that $\tau(l) \subseteq \tau(m)$: $gen(l^k, m) \geq gen(l, m)$.*

Proposition 18 (GenPro5$^+$). *For any $l \in \mathcal{L}$, $m \in \mathcal{M}$, and $k \geq 1$ such that $\tau(l) \subseteq \overline{\tau}(m)$: $gen(l^k, m) \leq gen(l, m)$.*

Proposition 19 (GenPro6^0). *For any $l \in \mathcal{L}$, $m \in \mathcal{M}$, and $k \geq 1$ such that most traces are fitting ($|[t \in l \mid t \in \tau(m)]| \geq |[t \in l \mid t \notin \tau(m)]|$): $gen(l^k, m) \geq gen(l, m)$.*

Proposition 20 (GenPro7^0). *For any $l \in \mathcal{L}$, $m \in \mathcal{M}$, and $k \geq 1$ such that most traces are non-fitting ($|[t \in l \mid t \in \tau(m)]| \leq |[t \in l \mid t \notin \tau(m)]|$): $gen(l^k, m) \leq gen(l, m)$.*

When the model allows for any behavior, clearly the next case will also be fitting (**GenPro8^0**). Nevertheless, it is marked as controversial because the proposition would also need to hold for an empty event log.

Proposition 21 (GenPro8^0). *For any $l \in \mathcal{L}$ and $m \in \mathcal{M}$ such that $\tau(m) = \mathcal{T}$: $gen(l, m) = 1$.*

5.2 Existing Generalization Measures

The following sections introduce several state-of-the-art generalization measures, before they will be evaluated using the corresponding propositions.

Alignment Generalization (gen_S). Van der Aalst et al. [4] also introduce a measure for generalization. This approach considers each occurrence of a given event e as observation of an activity in some state s. The approach is parameterized by a $state_M$ function that maps events onto states in which they occurred. For each event e that occurred in state s the approach counts how many different activities w were observed in that state. Furthermore, it counts the number of visits n to this state. Generalization is high if n is very large and w is small, since in that case, it is unlikely that a new trace will correspond to unseen behavior in that state.

Weighted Negative Event Generalization (gen_T). Aside from improving the approach of Goedertier et al. [16] van den Broucke et al. [31] also developed a generalization measure based on weighted negative events. It defines allowed generalizations AG which represent events, that could be replayed without errors and confirm that the model is general and disallowed generalizations DG which are generalization events, that could not be replayed correctly. If during replay a negative event e is encountered that actually was enabled the AG value is increased by $1 - weight(e)$. Similarly, if a negative event is not enabled the DG value is increased by $1 - weight(e)$. The more disallowed generalizations are encountered during log replay the lower generalization.

Anti-alignment Generalization (gen_U). Van Dongen et al. [13] also introduce an anti-alignment generalization and build on the principle that with a generalizing model, newly seen behavior will introduce new paths between the states of the model, however no new states themselves. Therefore, they define a recovery distance d_{rec} which measures the maximum distance between the states visited by the log and the states visited by the anti-alignment γ. A perfectly generalizing model according to van Dongen et al. [13] has the maximum distance to the anti-alignment with minimal recovery distance. Similar to recall they define trace-based and log-based generalization. Finally, anti-alignment generalization is the weighted combination of trace-based and log-based anti-alignment generalization.

Table 4. An overview of the generalization propositions that hold for the measures: (assuming $l \neq [\,]$, $\tau(m) \neq \emptyset$ and $\langle\rangle \notin \tau(m)$): $\checkmark$ means that the proposition holds for any log and model and $\times$ means that the proposition does not always hold.

Proposition	Name	gen_S	gen_T	gen_U
1	**DetPro$^+$**	$\checkmark$	$\times$	$\checkmark$
2	**BehPro$^+$**	$\checkmark$	$\times$	$\times$
14	**GenPro1$^+$**	$\times$	$\times$	$\times$
15	**GenPro2$^+$**	$\times$	$\times$	$\times$
16	**GenPro3^0**	$\times$	$\times$	$\times$
17	**GenPro4$^+$**	$\checkmark$	$\checkmark$	$\checkmark$
18	**GenPro5$^+$**	$\times$	$\checkmark$	$\checkmark$
19	**GenPro6^0**	$\checkmark$	$\checkmark$	$\checkmark$
20	**GenPro7^0**	$\times$	$\checkmark$	$\checkmark$
21	**GenPro8^0**	$\times$	$\checkmark$	$\times$

5.3   Evaluation of Existing Generalization Measures

The previously presented generalization measures are evaluated using the corresponding propositions. The results of the evaluation are displayed in Table 4. To improve the readability of this paper, only the most interesting findings of the evaluation are addressed in the following section. For a detailed evaluation, we refer to [29].

Table 4 displays that alignment based generalization (gen_S) violates several propositions. Generalization is not defined if there are unfitting traces since they cannot be mapped to states of the process model. Therefore, unfitting event logs should be aligned to fit to the model before calculating generalization. Aligning a non-fitting log and duplicating it will result in more visits to each state visited by the log. Therefore, adding non-fitting behavior increases generalization and violates the propositions **GenPro3^0**, **GenPro5$^+$** and **GenPro7^0**.

In comparison, weighted negative event generalization (gen_T) is robust against the duplication of the event log, even if it contains non-fitting behavior. However, this measure violates **DetPro$^+$**, **BehPro$^+$**, **GenPro1$^+$**, **GenPro2$^+$**

and **GenPro3^0**, which states that extending the log with non-fitting behavior cannot improve generalization. However, in this approach, negative events are assigned a weight which indicates how certain the log is about these events being negative ones. Even though the added behavior is non-fitting it might still provide evidence for certain negative events and therefore increase their weight. If these events are then not enabled during log replay the value for disallowed generalizations (DG) decreases $DG(l_1, m) < DG(l_2, m)$ and generalization improves: $\frac{AG(l_1,m)}{AG(l_1,m)+DG(l_1,m)} < \frac{AG(l_2,m)}{AG(l_2,m)+DG(l_2,m)}$.

Table 4 shows that anti-alignment generalization (gen_U) violates several propositions. The approach considers markings of the process models as the basis for the generalization computation which violates the behavioral proposition. Furthermore, the measure cannot handle if the model displays behavior that has not been observed in the event log. If the unobserved model behavior and therefore also the anti-alignment introduced a lot of new states which were not visited by the event log, the value of the recovery distance increases and generalization is lowered. This clashes with propositions **GenPro1$^+$** and **GenPro8$^+$**. Finally, the approach also excludes unfitting behavior from its scope. Only after aligning the event log, generalization can be computed. As a result, the measure fulfills **GenPro5$^+$**, **GenPro6^0** and **GenPro7^0**, but violates **GenPro3^0**.

6 Conclusion

With the process mining field maturing and more commercial tools becoming available [21], there is an urgent need to have a set of agreed-upon measures to determine the quality of discovered processes models. We have revisited the 21 conformance propositions introduced in [3] and illustrated their relevance by applying them to baseline measures. Furthermore, we used the propositions to evaluate currently existing conformance measures. This evaluation uncovers large differences between existing conformance measures and the properties that they possess in relation to the propositions. It is surprising that seemingly obvious requirements are not met by today's conformance measures. However, there are also measures that do meet all the propositions.

It is important to note that we do not consider the set of propositions to be complete. Instead, we consider them to be an initial step to start the discussion on what properties are to be desired from conformance measures, and we encourage others to contribute to this discussion. Moreover, we motivate researchers to use the conformance propositions as design criteria for the development of novel conformance measures.

One relevant direction of future work is in the area of conformance propositions that have a more fine-grained focus than the trace-level, i.e., that distinguish between *almost fitting* and *completely non-fitting* behavior. Another relevant area of future work is in the direction of *probabilistic conformance measures*, which take into account branching probabilities in models, and their desired properties. Even though event logs provide insights into the probability of traces, thus far all existing conformance checking techniques ignore this point of view. In [1,3], we already showed that probabilities can be used to provide more faithful definitions of recall and precision. In [22], we moved one step further and provide the first *stochastic conformance checking technique* using the so-called Earth Movers' Distance (EMD). This conformance checking approach considers the stochastic characteristics of both the event log and the process model. It measures the effort to transform the distribution of traces of the event log into the distribution of traces of the model. This way one can overcome many of the challenges identified in this paper.

Acknowledgements. We thank the Alexander von Humboldt (AvH) Stiftung for supporting our research.

References

1. van der Aalst, W.M.P.: Mediating between modeled and observed behavior: the quest for the "Right" process. In: IEEE International Conference on Research Challenges in Information Science, RCIS 2013, pp. 31–43. IEEE Computing Society (2013)
2. van der Aalst, W.M.P.: Process Mining: Data Science in Action. Springer, Berlin (2016). https://doi.org/10.1007/978-3-662-49851-4
3. van der Aalst, W.M.P.: Relating process models and event logs: 21 conformance propositions. In: van der Aalst, W.M.P., Bergenthum, R., Carmona, J. (eds.) Workshop on Algorithms & Theories for the Analysis of Event Data, ATAED 2018, pp. 56–74. CEUR Workshop Proceedings (2018)
4. van der Aalst, W.M.P., Adriansyah, A., van Dongen, B.: Replaying history on process models for conformance checking and performance analysis. WIREs Data Mining Knowl. Discov. **2**(2), 182–192 (2012)
5. van der Aalst, W.M.P., Weijters, A.J.M.M., Maruster, L.: Workflow mining: discovering process models from event logs. IEEE Trans. Knowl. Data Eng. **16**(9), 1128–1142 (2004)
6. Adriansyah, A., van Dongen, B., van der Aalst, W.M.P.: Conformance checking using cost-based fitness analysis. In: Chi, C.H., Johnson, P. (eds.) IEEE International Enterprise Computing Conference, EDOC 2011, pp. 55–64. IEEE Computer Society (2011)
7. Adriansyah, A., Munoz-Gama, J., Carmona, J., van Dongen, B.F., van der Aalst, W.M.P.: Alignment Based Precision Checking. In: La Rosa, M., Soffer, P. (eds.) BPM 2012. LNBIP, vol. 132, pp. 137–149. Springer, Heidelberg (2013). https://doi.org/10.1007/978-3-642-36285-9_15
8. Augusto, A., Armas-Cervantes, A., Conforti, R., Dumas, M., La Rosa, M., Reissner, D.: Abstract-and-Compare: A Family of Scalable Precision Measures for Automated Process Discovery. In: Weske, M., Montali, M., Weber, I., vom Brocke, J. (eds.) BPM 2018. LNCS, vol. 11080, pp. 158–175. Springer, Cham (2018). https://doi.org/10.1007/978-3-319-98648-7_10
9. Buijs, J.C.A.M.: Flexible evolutionary algorithms for mining structured process models. Ph.D. thesis, Department of Mathematics and Computer Science (2014)

10. Buijs, J.C.A.M., van Dongen, B.F., van der Aalst, W.M.P.: On the Role of Fitness, Precision, Generalization and Simplicity in Process Discovery. In: Meersman, R., Panetto, H., Dillon, T., Rinderle-Ma, S., Dadam, P., Zhou, X., Pearson, S., Ferscha, A., Bergamaschi, S., Cruz, I.F. (eds.) OTM 2012. LNCS, vol. 7565, pp. 305–322. Springer, Heidelberg (2012). https://doi.org/10.1007/978-3-642-33606-5_19
11. Buijs, J.C.A.M., van Dongen, B.F., van der Aalst, W.M.P.: Quality dimensions in process discovery: the importance of fitness, precision, generalization and simplicity. Int. J. Coop. Inf. Syst. **23**(1), 1–39 (2014)
12. Carmona, J., van Dongen, B., Solti, A., Weidlich, M.: Conformance Checking: Relating Processes and Models. Springer, Berlin (2018). https://doi.org/10.1007/978-3-319-99414-7
13. van Dongen, B.F., Carmona, J., Chatain, T.: A Unified Approach for Measuring Precision and Generalization Based on Anti-alignments. In: La Rosa, M., Loos, P., Pastor, O. (eds.) BPM 2016. LNCS, vol. 9850, pp. 39–56. Springer, Cham (2016). https://doi.org/10.1007/978-3-319-45348-4_3
14. van Dongen, B., Carmona, J., Chatain, T., Taymouri, F.: Aligning Modeled and Observed Behavior: A Compromise Between Computation Complexity and Quality. In: Dubois, E., Pohl, K. (eds.) CAiSE 2017. LNCS, vol. 10253, pp. 94–109. Springer, Cham (2017). https://doi.org/10.1007/978-3-319-59536-8_7
15. Garcia-Banuelos, L., van Beest, N., Dumas, M., La Rosa, M., Mertens, W.: Complete and interpretable conformance checking of business processes. IEEE Trans. Softw. Eng. **44**(3), 262–290 (2018)
16. Goedertier, S., Martens, D., Vanthienen, J., Baesens, B.: Robust process discovery with artificial negative events. J. Mach. Learn. Res. **10**, 1305–1340 (2009)
17. Greco, G., Guzzo, A., Pontieri, L., Saccà, D.: Discovering expressive process models by clustering log traces. IEEE Trans. Knowl. Data Eng. **18**(8), 1010–1027 (2006)
18. Janssenswillen, G., Depaire, B.: Towards confirmatory process discovery: making assertions about the underlying system. Bus. Inf. Syst, Eng (2018)
19. Janssenswillen, G., Donders, N., Jouck, T., Depaire, B.: A comparative study of existing quality measures for process discovery. Inf. Syst. 50(1), 2:1–2:45 (2017)
20. Janssenswillen, G., Jouck, T., Creemers, M., Depaire, B.: Measuring the Quality of Models with Respect to the Underlying System: An Empirical Study. In: La Rosa, M., Loos, P., Pastor, O. (eds.) BPM 2016. LNCS, vol. 9850, pp. 73–89. Springer, Cham (2016). https://doi.org/10.1007/978-3-319-45348-4_5
21. Kerremans, M.: Gartner Market Guide for Process Mining, Research Note G00353970 (2018). www.gartner.com
22. Leemans, S.J.J., Syring, A.F., van der Aalst, W.M.P.: Earth Movers' Stochastic Conformance Checking. In: Hildebrandt, T., van Dongen, B.F., Röglinger, M., Mendling, J. (eds.) BPM 2019. LNBIP, vol. 360, pp. 127–143. Springer, Cham (2019). https://doi.org/10.1007/978-3-030-26643-1_8
23. Leemans, S.J.J., Fahland, D., van der Aalst, W.M.P.: Scalable process discovery and conformance checking. Softw. Syst. Model. **17**(2), 599–631 (2018)
24. Mannhardt, F., de Leoni, M., Reijers, H.A., van der Aalst, W.M.P.: Balanced multi-perspective checking of process conformance. Computing **98**(4), 407–437 (2016)
25. Muñoz-Gama, J., Carmona, J.: A Fresh Look at Precision in Process Conformance. In: Hull, R., Mendling, J., Tai, S. (eds.) BPM 2010. LNCS, vol. 6336, pp. 211–226. Springer, Heidelberg (2010). https://doi.org/10.1007/978-3-642-15618-2_16
26. Polyvyanyy, A., Solti, A., Weidlich, M., Di Ciccio, C., Mendling, J.: Behavioural quotients for precision and recall in process mining. Technical report, University of Melbourne (2018)

27. Rozinat, A., van der Aalst, W.M.P.: Conformance checking of processes based on monitoring real behavior. Inf. Syst. **33**(1), 64–95 (2008)
28. Rozinat A., de Medeiros A.K.A., Günther, C.W., Weijters, A.J.M.M., van der Aalst, W.M.P.: The need for a process mining evaluation framework in research and practice. In: Castellanos, M., Mendling, J., Weber, B. (eds.) Informal Proceedings of the International Workshop on Business Process Intelligence, BPI 2007, pp. 73–78. QUT, Brisbane (2007)
29. Syring, A.F., Tax, N., van der Aalst, W.M.P.: Evaluating Conformance Measures in Process Mining using Conformance Propositions (Extended Version). CoRR, arXiv:1909.02393 (2019)
30. Tax, N., Lu, X., Sidorova, N., Fahland, D., van der Aalst, W.M.P.: The imprecisions of precision measures in process mining. Inf. Process. Lett. **135**, 1–8 (2018)
31. vanden Broucke, S.K.L.M., De Weerdt, J., Vanthienen, J., Baesens, B.: Determining process model precision and generalization with weighted artificial negative events. IEEE Trans. Knowl. Data Eng. 26(8), 1877–1889 (2014)
32. De Weerdt, J., De Backer, M., Vanthienen, J., Baesens, B.: A multi-dimensional quality assessment of state-of-the-art process discovery algorithms using real-life event logs. Inf. Syst. **37**(7), 654–676 (2012)
33. De Weerdt, J., De Backer, M., Vanthienen, J., Baesens, B.: A robust f-measure for evaluating discovered process models. In: Chawla, N., King, I., Sperduti, A. (eds.) IEEE Symposium on Computational Intelligence and Data Mining, CIDM 2011, pp. 148–155. IEEE, Paris (2011)
34. Weijters, A.J.M.M., van der Aalst, W.M.P., de Medeiros, A.K.A.: Process Mining with the Heuristics Miner-algorithm. BETA Working Paper Series, WP 166, Eindhoven University of Technology, Eindhoven (2006)

Relabelling LTS for Petri Net Synthesis via Solving Separation Problems

Uli Schlachter[1,2] and Harro Wimmel[1(✉)]

[1] Department of Computing Science, Carl von Ossietzky Universität Oldenburg, Oldenburg, Germany
{uli.schlachter,harro.wimmel}@informatik.uni-oldenburg.de
[2] Institute of Networked Energy Systems, German Aerospace Center, Oldenburg, Germany

Abstract. Petri net synthesis deals with finding an unlabelled Petri net with a reachability graph isomorphic to a given usually finite labelled transition system (LTS). If there is no solution for a synthesis problem, we use label splitting. This means that we relabel edges until the LTS becomes synthesisable. We obtain an unlabelled Petri net and a relabelling function, which together form a labelled Petri net with the original, intended behaviour. By careful selection of the edges to relabel we hope to keep the alphabet of the LTS and the constructed Petri net as small as possible. Even approximation algorithms, not yielding an optimal relabelling, are hard to come by. Using region theory, we develop a polynomial heuristic based on two kinds of separation problems. These either demand distinct Petri net markings for distinct LTS states or a correspondence between the existence of an edge in the LTS and the activation of a transition under the state's marking. If any separation problem is not solvable, relabelling of edges in the LTS becomes necessary. We show efficient ways to choose those edges.

Keywords: Labelled transition systems · Petri nets · System synthesis · Regions · Separation problems · Label splitting

1 Introduction

There are two general approaches to investigate the behaviour of Petri nets [14,16]. *Analysis* is used to construct a variety of descriptions from sets of firing sequences [11] to event structures [15]. One of the most common forms for describing the sequential behaviour is the reachability graph, containing the reachable markings as states together with edges denoting transitions that fire

U. Schlachter—Supported by DFG (German Research Foundation) through grant Be 1267/15-1 ARS (Algorithms for Reengineering and Synthesis).

H. Wimmel—Supported by DFG (German Research Foundation) through grant Be 1267/16-1 ASYST (Algorithms for Synthesis and Pre-Synthesis Based on Petri Net Structure Theory).

M. Koutny et al. (Eds.): ToPNoC XIV, LNCS 11790, pp. 222–254, 2019.
https://doi.org/10.1007/978-3-662-60651-3_9

to reach one marking from another. In the reverse direction, i.e. *synthesis*, we try to build a Petri net that behaves like a given specification, e.g. a labelled transition system (LTS). If we allow *labelled* Petri nets, this is very simple: We generate one transition for each edge in the LTS, one place for each state, and connect a place and a transition with an arc (of the flow relation of the Petri net) if the corresponding state and edge are adjacent. We obtain a Petri net that is structurally isomorphic to the given LTS. This is, of course, not the output we desire since its information density is as low as that of the specifying LTS, not yielding a compact description of the behaviour.

If we want to synthesise an *unlabelled* Petri net, all edges in the LTS with the same label must be generated by a single transition. This is much more difficult but leads to a small Petri net model in terms of the number of transitions. Region theory, first introduced by Ehrenfeucht and Rozenberg [9] and well covered by Badouel, Bernadinello, and Darondeau [1], provides a good basis for tackling this problem. Unfortunately, many behaviours cannot be modelled this way. Even for a linearly ordered LTS representing e.g. the single word *abbaa* there exists no unlabelled Petri net [3].

To be able to synthesise any LTS, we must therefore use labelled Petri nets where only as few transitions as possible have a common label. This optimisation problem has been investigated by Carmona et al. [6,7] and an algorithm has been suggested for Petri nets with an a priori bound k.[1] Because 1-bounded Petri net synthesis is known to be NP-complete [20], the algorithm requires at least exponential time even for $k = 1$ unless $\mathsf{P} = \mathsf{NP}$. For larger k, similar run times are expected. The algorithm investigates sets of region candidates such that, in the best case, one of the candidates is indeed a region (equivalent to a place of the sought Petri net). If no region is found (up to bound k), a heuristic decides on how to relabel some edges for a remedy. In the worst case, this leads to the 'isomorphic' solution (which works for any k), even if there is a nice, but not k-bounded solution. Stepwise incrementing k to extend the search would require some termination criterion differentiating between good and bad solutions. The algorithm of Carmona et al. computes solutions up to language-equivalence while we will consider the stronger isomorphism equivalence in the present paper.

In the context of process mining, several algorithms for event splitting were proposed, e.g. [5,13,17]. These algorithms (typically taking event logs as input) are based on heuristics that show nice results in their setting. The heuristics check if labels appear in different contexts and relabel them if that is the case. In contrast, our proposed algorithm is based on the semantics of Petri nets. Events are only split if exact linear algebra indicates that this is necessary for Petri net solvability. Our algorithm takes an LTS as input which provides different information than an event log. When transforming the label sequences of an event log into an LTS, we can either identify common prefixes of the sequences or explicitly distinguish them by introducing non-determinism. In both cases, we

---

[1] For every place p and reachable marking M of the Petri net to be constructed, $M(p) \leq k$ holds, i.e. no place has ever more than k tokens.

would make a premature decision that labels either must or cannot stem from the same source. A comparison between the approaches is therefore difficult.

We try to obtain an approximative algorithm for relabelling with a polynomial run time by applying concepts from region theory. A *separation problem* takes one of two forms: either it is a pair of states, and a solution to it is a region/place distinguishing them by assigning different markings of the Petri net to them, or it is a pair of a state and a label *not* emanating from this state. Then a solution is a region/place assigning a lower value (number of tokens) to the marking for this state than for the consumption of any transition with that label. Separation problems can be formulated as linear inequality systems (in the integers), the solutions of which are regions. If all of them are solvable, the computed regions allow the generation of the sought (unlabelled) Petri net [1]. This Petri net is bounded, but the bound does occur explicitly anywhere. By using duality theorems like Farkas' lemma [10], we also find alternative inequality systems, that have a solution if and only if a separation problem is not solvable [4,18]. These latter inequality systems are defined in the rational numbers. Solving them takes polynomial time [12] (unlike systems over integers) and their solutions give insight into which edges should be relabelled.

In the next section, we introduce the basic concepts around labelled transition systems and Petri nets as well as a short description of synthesis and how separation problems are defined. Section 3 deals with event/state separation problems, where the occurrence of a transition with a certain label is prohibited at a certain state. The other kind of separation problem, dealing with the distinction of states, will be covered in two parts of Sect. 4, where we show that these problems need to be handled differently depending on the involvement of cycles or the lack of such cycles. In Sect. 5 we combine the results of the previous sections. Section 6 covers experimental results with regard to run time as well as the number of new labels needed to make an LTS synthesisable. Finally, we give a summary and an outlook in Sect. 7.

2 Basic Concepts

Definition 1. LTS
A *labelled transition system* (LTS) with initial state is a tuple $TS = (S, \Sigma, \rightarrow, s_0)$ with nodes S (a countable set of states), edge labels Σ (a finite set of letters), edges $\rightarrow \subseteq (S \times \Sigma \times S)$, and an initial state $s_0 \in S$. An edge $(s, t, s') \in \rightarrow$ may be written as $s[t\rangle s'$. A *walk* $\sigma \in \Sigma^*$ from s to s', written as $s[\sigma\rangle s'$, is given inductively by $s = s'$ for the empty word $\sigma = \varepsilon$ and by $\exists s'' \in S: s[w\rangle s''[t\rangle s'$ for $\sigma = wt$ with $w \in \Sigma^*$ and $t \in \Sigma$. A walk $s[\sigma\rangle s'$ is a *cycle* if and only if $s = s'$. The set $[s\rangle$ for $s \in S$ is the set of all states reachable from s, $[s\rangle = \{s' \mid \exists \sigma \in \Sigma^* : s[\sigma\rangle s'\}$. The *Parikh vector* $\mathcal{P}(\sigma)\colon \Sigma \rightarrow \mathbb{Z}$ of a word $\sigma \in \Sigma^*$ maps each letter $t \in \Sigma$ to its number of occurrences in σ, it will often be written as an element of the group spanned by Σ. The neutral element is written as $\mathbf{0}$, comparisons are done componentwise with $\lneq$ meaning "less or equal in all components, but not entirely equal". We map to $\mathbb{Z}$ here instead of $\mathbb{N}$ to be able to extend the notion of a Parikh

vector later and to handle differences of Parikh vectors more easily. The *support* of a word $\sigma \in \Sigma^*$ is the set $supp(\sigma) = supp(\mathcal{P}(\sigma)) = \{t \in \Sigma \mid \mathcal{P}(\sigma)(t) \neq 0\}$.

A *spanning tree* E of TS is a set of edges $E \subseteq \rightarrow$ such that for every $s \in S$ there is a unique walk from s_0 to s using edges in E only. This implies that E is cycle-free. A *walk in E* is a walk that uses edges in E only (and not any of $\rightarrow\backslash E$). Edges in $\rightarrow\backslash E$ are called *chords*. The *Parikh vector of a state s* in a spanning tree E is $\mathcal{P}_E(s) = \mathcal{P}(\sigma)$ where $s_0[\sigma\rangle s$ is the unique walk in E. The *Parikh vector of an edge $s[t\rangle s'$* in TS is $\mathcal{P}_E(s[t\rangle s') = \mathcal{P}_E(s) + 1t - \mathcal{P}_E(s')$, see Fig. 1 for an example. Note that Parikh vectors of edges in E always evaluate to zero; for chords the Parikh vector may even contain negative values. For a chord $s[t\rangle s'$, s and s' have a latest common predecessor r in E, $t', t'' \in T$ with $t' \neq t''$, $\sigma, \sigma' \in T^*$ with two walks $r[t'\sigma\rangle s[t\rangle s'$ and $r[t''\sigma'\rangle s'$ in E. These two walks form a *cycle in the LTS' underlying undirected graph* with the Parikh vector the same as the chord's, $\mathcal{P}_E(s) + 1t - \mathcal{P}_E(s')$. The *Parikh vector of a walk* $s_1[t_1\rangle s_2 \ldots s_n[t_n\rangle s_{n+1}$ is defined as $\mathcal{P}_E(s_1[t_1 \ldots t_n\rangle s_{n+1}) = \sum_{i=1}^n \mathcal{P}_E(s_i[t_i\rangle s_{i+1})$. Obviously, $\mathcal{P}_E(s_1[t_1 \ldots t_n\rangle s_{n+1}) = \mathcal{P}_E(s_1) + \mathcal{P}(t_1 \ldots t_n) - \mathcal{P}_E(s_{n+1})$. If the walk is a cycle (with $s_1 = s_{n+1}$), we thus find $\mathcal{P}(t_1 \ldots t_n) = \sum_{i=1}^n \mathcal{P}_E(s_i[t_i\rangle s_{i+1})$ where all non-zero Parikh vectors in the sum stem from chords. The set $\{\mathcal{P}_E(s[t\rangle s') \mid (s, t, s') \in \rightarrow\backslash E\}$ is then a generator for all Parikh vectors of cycles (the latter being linear combinations of its elements). By simple linear algebra, we can compute a basis from this generator. This *cycle base* Γ contains at most $|\Gamma| \leq |\Sigma|$ different Parikh vectors. There will be no need to distinguish between cycles in the LTS and in its underlying undirected graph at all. The *size* of a Parikh vector, $|\mathcal{P}(\sigma)|$ or $|\mathcal{P}_E(s)|$, is the length of the corresponding word, i.e. $|\sigma|$ (of the walk $s_0[\sigma\rangle s$ in E in case of $\mathcal{P}_E(s)$).

An LTS $TS = (S, \Sigma, \rightarrow, s_0)$ is *finite* if S is finite and it is *deterministic* if $s[t\rangle s' \wedge s[t\rangle s''$ implies $s' = s''$ for all $s \in S$ and $t \in \Sigma$. We call TS *reachable* if for every state $s \in S$ exists some $\sigma \in \Sigma^*$ with $s_0[\sigma\rangle s$. Reachability implies the existence of a spanning tree. Two labelled transition systems $TS_1 = (S_1, \Sigma_1, \rightarrow_1, s_{01})$ and $TS_2 = (S_2, \Sigma_2, \rightarrow_2, s_{02})$ are *isomorphic* if $\Sigma_1 = \Sigma_2$ and there is a bijection $\zeta: S_1 \rightarrow S_2$ with $\zeta(s_{01}) = s_{02}$ and $(s, t, s') \in \rightarrow_1 \Leftrightarrow (\zeta(s), t, \zeta(s')) \in \rightarrow_2$, for all $s, s' \in S_1$. □ 1

Definition 2. PETRI NETS
An *(initially marked) Petri net* is denoted as $N = (P, T, W, M_0)$ where P is a finite set of *places*, T is a finite set of *transitions*, W is the weight function $W: ((P \times T) \cup (T \times P)) \rightarrow \mathbb{N}$ specifying the *arc weights*, and M_0 is the *initial marking* (where a *marking* is a mapping $M: P \rightarrow \mathbb{N}$, indicating the number of *tokens* in each place). A transition $t \in T$ is *enabled* at a marking M, denoted by $M[t\rangle$, if $\forall p \in P: M(p) \geq W(p, t)$. *Firing* t leads from M to M', denoted by $M[t\rangle M'$, if $M[t\rangle$ and $M'(p) = M(p) - W(p, t) + W(t, p)$. This can be extended, by induction as usual, to $M[\sigma\rangle M'$ for words $\sigma \in T^*$, and $[M\rangle = \{M' \mid \exists \sigma \in T^*: M[\sigma\rangle M'\}$ denotes the set of markings reachable from M. The *reachability graph* $RG(N)$ of a Petri net N is the labelled transition system with the set of nodes $[M_0\rangle$, initial state M_0, label set T, and set of edges $\{(M, t, M') \mid M, M' \in [M_0\rangle \wedge M[t\rangle M'\}$.

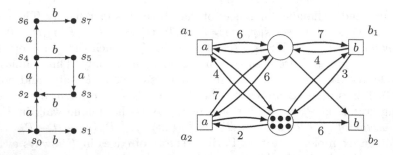

Fig. 1. An LTS (left) with a unique spanning tree (all edges except $s_3[b\rangle s_2$). States s_2 and s_3 have Parikh vectors $\mathcal{P}_E(s_2) = 1a$ and $\mathcal{P}_E(s_3) = 3a + 1b$, yielding $\mathcal{P}_E(s_3[b\rangle s_2) = \mathcal{P}_E(s_3) + 1b - \mathcal{P}_E(s_2) = 2a + 2b$ for the chord $s_3[b\rangle s_2$. The LTS is not synthesisable to an unlabelled net due to several unsolvable separation problems, e.g. the ESSP $\neg s_2[b\rangle$ with $\mathcal{P}_E(s_2) = 2a = \frac{1}{2}(\mathcal{P}_E(s_0) + \mathcal{P}_E(s_4))$ and $s_0[b\rangle$ as well as $s_4[b\rangle$. Also SSPs (s_3, s_7) and (s_2, s_5) are unsolvable due to $\mathcal{P}_E(s_3) = \mathcal{P}_E(s_7)$ and $\mathcal{P}_E(s_5) = \mathcal{P}_E(s_2) + \frac{1}{2}\mathcal{P}_E(s_3[b\rangle s_2)$. The labelled net on the right is a solution where transitions a_1 and a_2 both generate the same label a, and b_1 and b_2 produce b

A *labelled* Petri net $N = (P, T, W, M_0, \Sigma, \ell)$ is a Petri net with an additional alphabet Σ and a labelling function $\ell\colon T \to \Sigma$. We define $\ell(RG(N))$ as the LTS $([M_0\rangle, \Sigma, \{(M, \ell(t), M') \mid M, M' \in [M_0\rangle \land M[t\rangle M'\}, M_0)$. If a labelled transition system TS is isomorphic to the reachability graph $RG(N)$ of a Petri net N, or to $\ell(RG(N))$ in case of a labelled Petri net N, we say that N *PN-solves* (or simply *solves*) TS, and that TS is *synthesisable* to N. The Petri net on the right hand side of Fig. 1 is the smallest labelled Petri net solving the LTS shown on the left (with regard to the number of transitions as well as places). A (labelled) Petri net N is *k-bounded* for $k \in \mathbb{N}$ if $\forall p \in P\ \forall M \in [M_0\rangle\colon M(p) \leq k$ and it is *bounded* if it is k-bounded for some $k \in \mathbb{N}$. The reachability graph of N is *finite* if and only if N is bounded. □ 2

Definition 3. SYNTHESIS [1]
A *region* $r = (R, B, F)$ of an LTS $(S, \Sigma, \to, s_0)$ consists of three functions $R\colon S \to \mathbb{N}$, $B\colon \Sigma \to \mathbb{N}$, and $F\colon \Sigma \to \mathbb{N}$ such that for all edges $s[t\rangle s'$ in the LTS we have $R(s) \geq B(t)$ and $R(s') = R(s) - B(t) + F(t)$. This mimics the firing rule of Petri nets and makes regions essentially equivalent to places, i.e. a place p can be defined from r via $M_0(p) = R(s_0)$, $W(p, t) = B(t)$, and $W(t, p) = F(t)$ for all $t \in \Sigma$. When a Petri net is constructed from a set of regions of a reachable LTS, this implies a *uniquely defined marking* $\mathcal{M}(s)$ for each state s with $\mathcal{M}(s)(p) = M_0(p) + \sum_{t \in \Sigma} \mathcal{P}(\sigma)(t) \cdot (W(t, p) - W(p, t))$ for an arbitrary walk $s_0[\sigma\rangle s$.

The construction of a Petri net $N = (P, T, W, M_0)$ with one place in P for each region of the LTS guarantees $s[t\rangle \Rightarrow \mathcal{M}(s)[t\rangle$, but has three issues: (1) P might become infinite, (2) $\mathcal{M}$ may not be injective, and (3) $\mathcal{M}(s)[t\rangle \Rightarrow s[t\rangle$ need not hold. Failing (2) means that there are states $s, s' \in S$ with $s \neq s'$ that are identified in $RG(N)$ (leading to non-isomorphism). A *state separation problem* (SSP) is a pair $(s, s') \in S \times S$ with $s \neq s'$. A region r solves an SSP

(s, s') if $R(s) \neq R(s')$ (and thus $\mathcal{M}(s) \neq \mathcal{M}(s')$). An example of an SSP in Fig. 1 is (s_3, s_7), for which no solving region exists due to the fact that $R(s_3) = R(s_4) - B(b) + F(b) - B(a) + F(a) = R(s_4) - B(b) + F(b) - B(a) + F(a) = R(s_7)$. Failing (3) results in an edge $\mathcal{M}(s)[t\rangle$ in $RG(N)$ but not in the LTS, $\neg s[t\rangle$. An *event/state separation problem* (ESSP) is a pair $(s, t) \in S \times \Sigma$ with $\neg s[t\rangle$. A region r solves an ESSP (s, t) if $R(s) < B(t)$ (and thus $\neg \mathcal{M}(s)[t\rangle$). An example is the ESSP (s_2, b) in Fig. 1, which is not solvable due to the fact that $R(s_0) \geq B(b)$ (as $s_0[b\rangle$), $R(s_2) = R(s_0) - B(a) + F(a) < B(b)$ (as $\neg s_2[b\rangle$), and $R(s_4) = R(s_2) - B(a) + F(a) \geq B(b)$ (as $s_4[b\rangle$). Then, $R(s_0) \leq R(s_2) \leq R(s_4)$ or $R(s_0) \geq R(s_2) \geq R(s_4)$, contradicting one of the comparisons to $B(b)$. The set of all *separation problems*, $\{(s, s') \in S \times S \mid s \neq s'\} \cup \{(s, t) \in S \times \Sigma \mid \neg s[t\rangle\}$, is finite for finite LTS, and finding a solution for every separation problem solves all three issues, making the synthesis successful with $RG(N)$ isomorphic to the finite LTS.   □ 3

An SSP (s, s') can be written as a linear inequality system [4] with $\forall \gamma \in \Gamma$: $(F - B)^\mathsf{T} \cdot \gamma = 0$ (all cycles must have effect zero in a region), $(F - B)^\mathsf{T} \cdot (\mathcal{P}_E(s) - \mathcal{P}_E(s')) \neq 0$ ($\mathcal{M}(s) \neq \mathcal{M}(s')$, compared via $R(s_0)$), and $\forall s''[t\rangle s'''$: $R(s_0) + (F - B)^\mathsf{T} \cdot \mathcal{P}_E(s'') \geq B(t)$ (definition of a region, but without checking cycles). Here $F \geq \mathbf{0}$, $B \geq \mathbf{0}$ and $R(s_0) \geq 0$ are the variables, a solution forming a region solving the SSP. Note that if the first two formulas are solved, $R(s_0)$ can always be selected high enough such that the third one is also fulfilled.

An ESSP (s, t) is equivalent to a linear inequality system [18] with $\forall \gamma \in \Gamma$: $(F - B)^\mathsf{T} \cdot \gamma = 0$, $\forall s'[t'\rangle s''$: $R(s_0) + (F - B)^\mathsf{T} \cdot \mathcal{P}_E(s') \geq B(t')$, and $R(s_0) + (F - B)^\mathsf{T} \cdot \mathcal{P}_E(s) < B(t)$ ($\neg \mathcal{M}(s)[t\rangle$). With variables $F \geq \mathbf{0}$, $B \geq \mathbf{0}$, and $R(s_0) \geq 0$, a solution forms a region solving the ESSP.

If the LTS is finite, the linear inequality systems are also finite, and we may employ standard means, e.g. an ILP- or SMT-solver [8], to solve them. If an LTS is not reachable, it is not structurally isomorphic to a reachability graph, i.e. no Petri net solving it exists. For these reasons, we assume all LTS to be finite and reachable in the remainder of this paper.

3   Eliminating Unsolvable ESSPs

Let $TS = (S, \Sigma, \rightarrow, s_0)$ be an LTS. If some separation problem has no solution, synthesis fails. To remedy this situation, we adapt the LTS. In case of an unsolvable event/state separation problem (s, t), a seemingly simple modification would be to just allow the edge $s[t\rangle$ in the LTS [19], but it is unclear where this edge should lead to. And however we solve this, there is a deviation from the intended behaviour that is not visible directly in the synthesised Petri net later. When relabelling edges instead, we store away the changes to the labels, synthesise the modified LTS to an unlabelled Petri net, and then build a labelling function for the Petri net from the stored information. In this way, we obtain a labelled Petri net with the original, intended behaviour, with an underlying unlabelled Petri net modelling the modified LTS.

So, with an ESSP (s,t), how do we relabel an edge if the edge $s[t\rangle$ does not even exist? An unsolvable ESSP means that such an edge is enforced in the reachability graph by the existence of other edges with the same label. With the help of Farkas' lemma [10], a duality theorem from linear algebra, we have obtained a characterisation for solvability of ESSPs:

Theorem 1. UNSOLVABLE ESSP [18]
Let $TS = (S, \Sigma, \rightarrow, s_0)$ be an LTS with a state $s \in S$ and a transition $t \in T$. An ESSP (s,t) is not solvable if and only if $\mathcal{P}_E(s) \in \operatorname{span} \Gamma + \operatorname{convex}\{\mathcal{P}_E(s') \mid s' \in S \wedge s'[t\rangle\}$. Here, $\operatorname{span} \Gamma$ denotes the vector space spanned by the cycle base Γ and $\operatorname{convex} X$ is the convex hull of a set X of points, i.e. all weighted sums of points in X where the total weight is one.

Relabelling one of the edges contributing to a convex hull could make the ESSP solvable. A simple example is the ESSP (s_2, b) in Fig. 1. Here, $\mathcal{P}_E(s_2) = 1a = \mathbf{0} + \frac{1}{2}\mathcal{P}_E(s_0) + \frac{1}{2}\mathcal{P}_E(s_4)$, where $\mathbf{0} \in \operatorname{span} \Gamma$ and the total weight of $\frac{1}{2} + \frac{1}{2} = 1$ shows that $\mathcal{P}_E(s_2)$ lies in the convex hull of $\mathcal{P}_E(s_0)$ and $\mathcal{P}_E(s_4)$. By Theorem 1, the ESSP is not solvable. To make it solvable, one of the two edges $s_0[b\rangle$ or $s_4[b\rangle$ must be relabelled.

For the general case, let us assume we get

$$\mathcal{P}_E(s) = \sum_{\gamma \in \Gamma} k_\gamma \cdot \gamma + \sum_{i=1}^{n} k_i \cdot \mathcal{P}_E(s_i)$$

where $\{s_i \mid 1 \leq i \leq n\} \subseteq \{s' \in S \mid s'[t\rangle\}$, all $k_\gamma, k_i \in \mathbb{Q}$, $k_i > 0$ (only those s_i appear), and $\sum_{i=1}^{n} k_i = 1$ (total weight of one). We distinguish two cases then:

Case 1: $n = 1$. The missing edge $\neg s[t\rangle$ arises from a single edge $s_1[t\rangle$ plus some linear combination of cycles, i.e. $\mathcal{P}_E(s) - \mathcal{P}_E(s_1) \in \operatorname{span} \Gamma$. This coincides exactly with the characterisation of solvability of SSP from [4], i.e. *the ESSP is induced by an SSP*, which will be handled and solved in the next section. If t does not occur anywhere in span Γ, relabelling the edge $s_1[t\rangle$ to some t' will just lead to a new unsolvable ESSP (s, t') anyway, so solving the SSP instead (where we find other candidate edges for relabelling) is advised. Consider the states s_3 and s_7 in Fig. 1 as an example, where $\mathcal{P}(s_7) - \mathcal{P}(s_3) = \mathbf{0} \in \operatorname{span} \Gamma$ not only proves the SSP (s_7, s_3) unsolvable, but also the ESSP (s_7, b) due to $s_3[b\rangle$ and $\neg s_7[b\rangle$. Relabelling the b-edge $s_3[b\rangle$ to b' introduces a new ESSP (s_7, b') (again unsolvable) and is no help in solving the SSP (s_7, s_3) either.

Case 2: $n > 1$. When we select one of the s_i and relabel the edge $s_i[t\rangle$ to t', our expression for $\mathcal{P}_E(s)$ becomes invalid. Still, it is possible that $\mathcal{P}_E(s) \in \operatorname{span} \Gamma + \operatorname{convex}\{\mathcal{P}_E(s') \mid s' \in S \wedge s'[t\rangle\}$ remains true. We may find another expression representing $\mathcal{P}_E(s)$ and possibly have to relabel another edge. For an example, take a look at the unsolvable ESSP (s_2, b) with $\neg s_2[b\rangle$ in Fig. 2. We find $\mathcal{P}(s_2) = \frac{1}{2}\mathcal{P}(s_0) + \frac{1}{2}\mathcal{P}(s_4)$ but also $\mathcal{P}(s_2) = \frac{3}{4}\mathcal{P}(s_1) + \frac{1}{4}\mathcal{P}(s_5)$. $\mathcal{P}(s_2)$ remains in the convex hull after relabelling $s_0[b\rangle$ (or $s_4[b\rangle$) to b' due to the second equation.

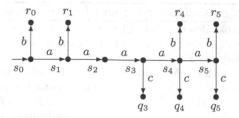

Fig. 2. An LTS with two unsolvable ESSPs, (s_2, b) and (s_3, b)

Relabelling an edge to a label that already exists in the LTS might make solvable separation problems of both kinds unsolvable again, instigating a reprocessing of already solved separation problems. In Fig. 2, relabelling, say, $s_0[b\rangle$ to c in an attempt to solve the ESSP (s_3, b) will lead to a new, unsolvable ESSP (s_2, c) with $\mathcal{P}(s_2) = \frac{1}{3}\mathcal{P}(s_0) + \frac{2}{3}\mathcal{P}(s_3)$ due to $s_0[c\rangle$, $s_3[c\rangle$ but $\neg s_2[c\rangle$. We avoid this by introducing new labels instead. To keep the alphabet size and thus the number of transitions of the Petri net low, we relabel several instances of t-edges to the same new label t'. This will not introduce new SSPs, and the ESSPs for the new t' have to be checked later anyway. In our example, relabelling both $s_0[b\rangle$ and $s_1[b\rangle$ to d makes the LTS synthesisable.

Algorithm 1 implements a heuristic strategy to deal with all ESSPs of a deterministic LTS (line 14 requires determinism). Remember that reachability graphs of unlabelled Petri nets are always deterministic, i.e. we have to remove non-determinism anyway. A trivial approach checks for each label t how often it occurs at most at any state and supplies that many different replacement labels for it. At each state, we then select for each occurrence of t one of the replacement labels, not using any of them twice. Obviously, relabelling appropriately is a difficult task that highly depends on the structure of an LTS.

The idea of the algorithm is to check for each label t_j (selected in line 4) which ESSPs are not solved yet (line 5) and provide (at most) one single replacement label for all these ESSPs (line 15). When an ESSP is selected (line 7, 8), we check it until it becomes solvable (line 9). As long as it is unsolvable, the dual characterisation of line 9 provides a solution (line 10, 11). If this solution is based on an unsolvable SSP (s, s_1), i.e. if $n = 1$ and $k_1 = 1$, we leave it unsolved (line 12). Otherwise we select one of the edges $s_\ell[t_j\rangle$ (line 13) that contribute to the solution and relabel it (line 16). If one of the ESSPs for t_j affords the replacement label (i.e. *new_label* becomes true in line 18), we record this relabelling (line 19). After changing an edge, we recompute the cycle base Γ (line 20) by simple linear algebra. After checking all ESSPs for t_j, we append the replacement label (line 23) to our list of labels to check via the outermost loop (line 4). That means the upper bound of the loop is not constant but may be incremented. After all ESSPs (except those equivalent to an SSP) have been made solvable, we return the modified LTS together with a mapping of the new labels to their originals.

There are several imaginable ways to select the edge to relabel (line 13). The solution of lines 10–11 provides coefficients k_i that show how strong the

---

Algorithm 1. An algorithm modifying an LTS to eliminate unsolvable ESSPs

---

Input: a finite, reachable, deterministic LTS $(S, \Sigma, \to, s_0)$ to be modified
 1: **index** the alphabet, i.e. $\Sigma = \{t_1, t_2, \ldots, t_m\}$ with $m \in \mathbb{N}$
 2: **compute** a spanning tree E with cycle base Γ
 3: **let** $relabel: \Sigma \to \Sigma$ with $relabel = \mathbf{id}$ initially   ▷ back-projection for new labels
 4: **for** j **from** 1 **upto** m ▷ upper bound m can grow dynamically
 5:   **let** $Q = \{s \in S \mid \neg s[t_j\rangle\}$ ▷ ESSPs to solve for t_j
 6:   **let** $new_label = $ false   ▷ notification of a needed new label
 7:   **for each** $s \in Q$
 8:     **let** $Q = Q \setminus \{s\}$
 9:     **while** $\mathcal{P}_E(s) \in \text{span}\,\Gamma + \text{convex}\{\mathcal{P}_E(s') \mid s' \in S \land s'[t_j\rangle\}$ ▷ via SMT-solver
10:       **compute** $\mathcal{P}_E(s) = \sum_{\gamma \in \Gamma} k_\gamma \cdot \gamma + \sum_{i=1}^{n} k_i \cdot \mathcal{P}_E(s_i)$ **with**
11:         $k_\gamma, k_i \in \mathbb{Q}$, $k_i > 0$, $\sum_{i=1}^{n} k_i = 1$, and $\{s_i \mid 1 \leq i \leq n\} \subseteq \{s' \in S \mid s'[t_j\rangle\}$
12:       **if** $n = 1$ **then break**   ▷ case 1 from above, leave while-loop
13:       **select** $\ell \in \{1, \ldots, n\}$ such that $|\mathcal{P}_E(s_\ell)|$ is minimal   ▷ heuristic
14:       **find** the unique $s'' \in S$ with $(s_\ell, t_j, s'') \in \to$
15:       **let** $\Sigma = \Sigma \cup \{t_{m+1}\}$ ▷ the same new label for all ESSPs with t_j
16:       **let** $\to = (\to \setminus \{(s_\ell, t_j, s'')\}) \cup \{(s_\ell, t_{m+1}, s'')\}$
17:       **let** $Q = Q \cup \{s_\ell\}$ ▷ (s_ℓ, t_j) is now an ESSP
18:       **let** $new_label = $ true
19:       **let** $relabel(t_{m+1}) = t_j$ ▷ save original label for t_{m+1}
20:       **compute** cycle base Γ ▷ $\to$ has changed
21:     **end while**
22:   **end for each**
23:   **if** new_label **then** $m = m + 1$ ▷ new transitions must also be checked
24: **end for**
25: **return** the LTS and $relabel$

---

involvement of an edge in the convex hull is, a high coefficient being a high weight in the solution. We could hope that selecting the edge with the highest weight means that we arrive faster at a solvable ESSP. Unfortunately, regular structures in the LTS often lead to random distributions of weights in the solution and too many relabelled edges in the end. Figure 3 provides an example where this kind of selection is a bad idea.

Checking the unsolvable ESSP (s_2, b), a solver provides one of several possible representations for $\mathcal{P}_E(s_2)$ (with E containing *all* edges for this LTS). Each representation must contain at least one state from $\{s_0, s_1\}$ and at least one state

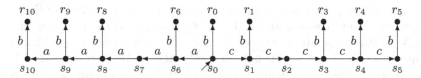

Fig. 3. A two-way comb structure starting at s_0 with unsolvable ESSPs (s_2, b) and (s_7, b)

from $\{s_3, s_4, s_5\}$; some possibilities are $\frac{1}{2}\mathcal{P}_E(s_0) + \frac{1}{2}\mathcal{P}_E(s_4)$ or $\frac{3}{4}\mathcal{P}_E(s_1) + \frac{1}{4}\mathcal{P}_E(s_5)$ or $\frac{3}{5}\mathcal{P}_E(s_1) + \frac{1}{5}\mathcal{P}_E(s_3) + \frac{1}{5}\mathcal{P}_E(s_4)$. Relabelling a single b-edge will not make the ESSP solvable, instead we have to relabel all b-edges at $\{s_0, s_1\}$ or at $\{s_3, s_4, s_5\}$. If we apply the solver repeatedly, we are forced to relabel between two and four of the b-edges to, say, b', until there are no more b-edges at the states in one of the two sets. It is likely that some states on both sides of s_2 get a b'-edge, introducing a new unsolvable ESSP (s_2, b') that will require another new label b''.

This can largely be avoided by determining how to relabel edges from the structure of the LTS and not from the specific weights computed by the solver. The easiest way to do this is to choose a state with minimal distance from the initial state (maximal distance would be ok, too). For $\frac{1}{2}\mathcal{P}_E(s_1) + \frac{1}{2}\mathcal{P}_E(s_3)$ we would select s_1. The solver is applied only twice, and the second time the b-edge at s_0 is relabelled to b'. Now, the ESSP (s_2, b) is solvable. There are new ESSPs (s_i, b') for $i \geq 2$, but all of these are immediately solvable, too, without any need for further relabelling. The ESSP (s_7, b) works analogously, so in an optimal solution the b-edges at s_0, s_1, and s_6 are relabelled to b'. Choosing the maximal distance from the initial state relabels all b-edges emanating from $\{s_3, s_4, s_5, s_8, s_9, s_{10}\}$ to b' instead. We might expect the number of states with distance d from the initial state to increase with d, thus the minimal distance strategy could lead to less relabelling overall.

Let us take a look at the complexity of Algorithm 1 now. Computing the spanning tree takes $O(|S| \cdot |\Sigma|)$ for looking at all the edges. We can also create the sets Q for each t_j (line 5) in this phase. Computing the cycle base Γ in line 20 can be done in $O(|S| \cdot |\Sigma|^3)$ (with $O(|S| \cdot |\Sigma|)$ Parikh vectors of the chords, each with $|\Sigma|$ entries). This complexity is clearly dominated by that of the main loop (lines 4–23) and can be neglected. The outer for-loop is repeated $O(|S| \cdot |\Sigma|)$ times in case every edge obtains its own label. We have $O(|S|)$ repeats in the for-each-loop; Q may grow during a run, but a state can only be added to Q if it loses its t_j-edge, i.e. only once. Solving the equation system has a worst-case time complexity[2] of $O((|S| + |\Sigma|)^{5.5} \cdot |\Sigma|^2 \cdot \log^2 |S| \cdot (\log(|S| + |\Sigma|) + \log|\Sigma| + \log\log|S|) \cdot \log(\log(|S| + |\Sigma|) + \log|\Sigma| + \log\log|S|))$. Using rough upper bounds for the logarithmic expressions, we can simplify this to $O((|S| + |\Sigma|)^{5.5} \cdot |\Sigma|^2 \cdot \log^3(|S| + |\Sigma|))$. Each time the while-loop is repeated, one Parikh vector is removed from the set defining the convex hull. This means at most $O(|S|)$ repeats. Selecting a Parikh vector of minimal size in line 13 takes $O(|S|)$ if we compute the size of the Parikh vectors during the creation of the spanning tree. Everything else aside the (re-)computation of Γ takes constant time, so the dominating factor inside the while-loop is solving the equation system. Overall, we get to $O((|S| + |\Sigma|)^{5.5} \cdot |S|^3 \cdot |\Sigma|^3 \cdot \log^3(|S| + |\Sigma|))$, i.e. a polynomial algorithm. In a typical case, we expect a lower run time though, as we usually have very

---

[2] The ellipsoid method has $O(n^6 \cdot L^2 \cdot \log L \cdot \log\log L)$ complexity while Karmarkar's algorithm has $O(n^{3.5} \cdot L^2 \cdot \log L \cdot \log\log L)$ where $n = |S| + |\Sigma|$ is the number of variables and $L = n \cdot |\Sigma| \cdot \log|S|$ is the size of the input (in bits), cf. [12].

few unsolvable ESSPs and also solving equation systems will likely be faster in the average case. Overall, we obtain

Theorem 2. MAKING ESSPs SOLVABLE VIA ALGORITHM 1 IS POLYNOMIAL
Algorithm 1 is a polynomial-time algorithm relabelling any finite, reachable, and deterministic LTS, such that all ESSPs either become solvable or are reduced to an SSP.

We have also seen that Algorithm 1 leads to optimal results for simple comb structures, which is certainly better than just giving each edge its own label.

4   Dealing with Unsolvable SSPs

Unsolvable SSPs are simpler to characterise than unsolvable ESSPs, as shown in the literature:

Theorem 3. UNSOLVABLE SSP [4]

An SSP (s, s') is not solvable if and only if $\mathcal{P}_E(s) - \mathcal{P}_E(s') \in \operatorname{span} \Gamma$, i.e. both states are in the same equivalence class regarding the cycle base.

Unfortunately, this does not mean they are easier to make solvable. Unsolvable SSPs (s, s') come in two flavors: Either $\mathcal{P}_E(s) = \mathcal{P}_E(s')$ or there are cycles in the cycle base with $\mathcal{P}_E(s) = \mathcal{P}_E(s') + \sum_{\gamma \in \Gamma} k_\gamma \cdot \gamma$.

In the former case, we need to relabel some edge on the spanning tree path from either s_0 to s or from s_0 to s' to yield $\mathcal{P}_E(s) \neq \mathcal{P}_E(s')$ and make the SSP solvable. For the latter case, Fig. 4 shows an example where relabelling on the spanning tree paths relevant to the SSP (s_0, s_1) does not suffice to make the SSP solvable. For any region $r = (R, B, F)$ we get $R(s_0) = R(s_1) - B(b) + F(b) = R(s_0) - 2B(b) + 2F(b)$ with $F(b) = B(b)$ and thus $R(s_0) = R(s_1)$. We must relabel a b-edge, but the spanning tree does not contain one.

As a consequence, it is necessary to relabel chords, if cycles (in the underlying undirected graph) are involved, but if not, we must relabel the relevant spanning tree paths. This leads to two different approaches. We start with SSPs involving cycles.

Fig. 4. An unsolvable SSP (s_0, s_1) which no relabelling in the spanning tree can solve. Spanning tree edges are drawn solid, chords are dashed edges

4.1   Unsolvable SSPs Involving Cycles

Let (s, s') be an SSP and let us assume we have a solution of the inequality system as presented in Theorem 3. We relabel an edge based on the equation $\mathcal{P}_E(s) = \mathcal{P}_E(s') + \sum_{\gamma \in \Gamma} k_\gamma \cdot \gamma$. If a cycle is involved, at least one of the k_γ is not equal to zero. Note that $\gamma \in \Gamma$ are not concrete cycles in the LTS, but Parikh vectors (with possibly negative entries). Concrete cycles in the LTS are represented by chords, where a chord $c = (q, t, q')$ also has a Parikh vector representation via the cycle base, $\mathcal{P}_E(c) = \mathcal{P}_E(q) + 1t - \mathcal{P}_E(q') = \sum_{\gamma \in \Gamma} r_\gamma \cdot \gamma$ with $r_\gamma \in \mathbb{Q}$. The walk from s_0 to q, then via c to q' and back to s_0 forms a cycle in the underlying undirected graph. Our solution can be expressed with the help of the cycle of chord c if it shares a common base vector $\gamma \in \Gamma$ with the chord's representation, i.e. if $\bigvee_{\gamma \in \Gamma} (r_\gamma \cdot k_\gamma \neq 0)$ is true. The chord can then replace the base vector in the solution via a base transformation.

If we select some chord $c = (q, t, q')$ and some edge e on its cycle to be relabelled, the effect might be cancelled out in our solution. The edge e might be used also by other cycles (and even in the opposite direction), e.g. a chord $c' = (q', t', q'')$ might also contain the edge e if e happens to be on the spanning tree path from s_0 to q'. If c and c' appear in our solution with the same weight, relabelling e will not prevent this solution. Checking such a condition for all edges e regarding all possible linear combinations of chords can be quite time consuming. This is similar to the idea of solving inequality systems directly on chords, where the cycle base with at most $|\Sigma|$ elements is replaced with the set of all chords having up to $|S| \cdot (|\Sigma| - 1) + 1$ elements. This leads to a dramatic increase in the run time. There is one way around this, as there is one edge on every chord cycle not occurring on another chord cycle: the chord edge *itself*. For SSPs involving cycles, we will therefore relabel chords themselves only.

Relabelling just one chord c might not prohibit our solution. Other chords with the same Parikh vector and even chord cycles whose Parikh vector is a multiple of $\mathcal{P}_E(c)$ can take the place of our chord c in the solution. All these chords should be relabelled in a single step to prevent the given solution. Relabelling more chords than these, on the other hand, may lead to cancellation effects again, especially if their Parikh vectors are linearly dependent but not multiples of each other.

Note that chords with the same Parikh vector (or a multiple thereof) do not necessarily have the same label. For a low count of new labels it is therefore wise to choose a set of chords to relabel that has as few as possible different chord labels.

Consider the example of Fig. 5. We ignore unsolvable ESSPs for now. The only unsolvable SSPs are (r_i, s_{i+1}) $(i \in \mathbb{N})$ with $\mathcal{P}_E(r_{2i}) - \mathcal{P}_E(s_{2i+1}) = 1c - 1a$ and $\mathcal{P}_E(r_{2i+1}) - \mathcal{P}_E(s_{2i+2}) = 1b - 1a$, both being expressible via chords. We have essentially three different Parikh vectors for chords and thus three sets of chords: $\mathcal{P}_E(r_{4i}[c\rangle r_{4i+1}) = -1a - 1b + 2c$, $\mathcal{P}_E(r_{4i+2}[c\rangle r_{4i+3}) = -2a + 1b + 1c$, and $\mathcal{P}_E(r_{4i+3}[b\rangle r_{4i+4}) = -1a + 2b - 1c = -\mathcal{P}_E(s_{4i+2}[c\rangle r_{4i+2})$. Let us now look for solutions of the dual SSP equation systems. We find $\mathcal{P}_E(r_{2i}) - \mathcal{P}_E(s_{2i+1}) = 1c - 1a = \frac{1}{3}(-1a - 1b + 2c) + \frac{1}{3}(-2a + 1b + 1c)$ but also $1c - 1a = \frac{2}{3}(-2a + 1b + 1c) -$

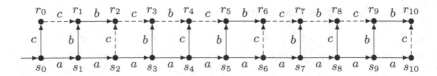

Fig. 5. An LTS where all SSPs (r_i, s_{i+1}) are unsolvable but $\mathcal{P}_E(r_i) \neq \mathcal{P}_E(s_{i+1})$. A spanning tree with solid edges is shown, chords are dashed edges

$\frac{1}{3}(-1a+2b-1c)$ and $1c-1a = \frac{2}{3}(-1a-1b+2c)+\frac{1}{3}(-1a+2b-1c)$. To make these SSPs solvable, we must invalidate all three equations and thus relabel at least two of the sets of chords. A wise choice is not to relabel the third set with $r_{4i+3}[b\rangle r_{4i+4}$ and $s_{4i+2}[c\rangle r_{4i+2}$ as this requires two new labels, say b_3 and c_3, while the other two sets introduce only one new label each: all chords $r_{4i}[c\rangle r_{4i+1}$ get relabelled to c_1, while the chords $r_{4i+2}[c\rangle r_{4i+3}$ obtain the new label c_2. After this is done, all other unsolvable SSPs also have become solvable. We cannot choose $c_1 = c_2$ as then $\mathcal{P}_E(r_{2i+1}) - \mathcal{P}_E(s_{2i+2}) = 1b-1a = \frac{2}{3}(-2a+1b+1c_1) - \frac{1}{3}(-1a-1b+2c_1)$, so the SSPs (r_{2i+1}, s_{2i+2}) remain unsolvable. This also means our solution is optimal regarding the alphabet size. Note that a different spanning tree might lead to different results. If we select all edges $(s_i, b/c, r_i)$ for $i \geq 1$ as chords, all chords will have different Parikh vectors and we eventually provide each chord with an individual label. Therefore, it is a good idea to select a spanning tree that implies an as-small-as-possible set of chord labels and/or as few as possible different Parikh vectors for the chords. So far, we have not found any fast, fail-safe method to select a good spanning tree. The ideas behind our example lead us to Algorithm 2.

Here, line 4 supplies us with sets of chords having the same Parikh vector $\mathcal{P}$ (or a rational multiple thereof) and the same label t. Only Parikh vectors of chords can yield non-empty sets, and these Parikh vectors can be generated during building the spanning tree. Chords with Parikh vector $\mathbf{0}$ do not contribute to solutions of one of our equation systems, i.e. $\mathcal{C}(\mathbf{0})(t)$ can be ignored. The loop in line 5 is executed at most $|S|^2$ times, once for each (unsolvable) SSP. If necessary, we can run the loop for each SSP, checking for solvability via line 7. The solver computes the representation of line 8. In line 9 we test whether an unsolvable SSP involves cycles, other SSPs are handled in the next section. After finding a solution $(k_\gamma)_{\gamma \in \Gamma}$, we detect the Parikh vectors of chords that contribute to the solution via their representation $(r_\gamma)_{\gamma \in \Gamma}$ in line 10–11. Amongst all such Parikh vectors of chords, we select one that requires the least number of new labels (second part of line 11). We relabel edges at lines 12–19, label by label that occurs at one of the selected chords. Changing the edge labels invalidates the cycle base, so Γ and the representation of the chords are recomputed after that (line 20). Looking at the run times, it is easy to see that line 9 is in $O(|\Sigma|)$, lines 10–11 in $O(|S|\cdot|\Sigma|^2)$, lines 12–19 in $O(|S|\cdot|\Sigma|)$, and line 20 is in $O(|S|\cdot|\Sigma|^3)$, dominating all previous parts. Adding the run time for solving the equations systems, we arrive at $O(|S| \cdot |\Sigma|^3 + |\Sigma|^{7.5} \cdot \log^4 |\Sigma|)$ or, much simplified, at

Algorithm 2. Elimination of SSPs involving cycles

Input: a finite, reachable, deterministic LTS $(S, \Sigma, \rightarrow, s_0)$ to be modified

1: **let** $Q = \{(s, s') \in S^2 \mid (s, s')$ is an unsolvable SSP$\}$
2: **let** $relabel \colon \Sigma \rightarrow \Sigma$ with $relabel = \mathbf{id}$ initially   ▷ projection for new labels
3: **compute** spanning tree E and cycle base Γ ▷ incl. representation of all chords
4: **let** $\mathcal{C}(\mathcal{P})(t) = \{(r, t, r') \in \rightarrow\backslash E \mid \exists c \neq 0 \colon c \cdot \mathcal{P}_E(r[t\rangle r') = \mathcal{P}\}$ for $\mathcal{P} \colon \Sigma \rightarrow \mathbb{Z}$ and $t \in \Sigma$
5: **for each** $(s, s') \in Q$
6:    **let** $Q = Q \setminus \{(s, s')\}$
7:    **while** $\mathcal{P}_E(s') - \mathcal{P}_E(s) \in \text{span}\,\Gamma$ ▷ via solver
8:      **compute** $\sum_{\gamma \in \Gamma} k_\gamma \cdot \gamma = \mathcal{P}_E(s') - \mathcal{P}_E(s)$ with $k_\gamma \in \mathbb{Q}$
9:      **if** $\forall \gamma \in \Gamma \colon k_\gamma = 0$ **then break end if**   ▷ no cycles involved
10:     **select** representation $(r_\gamma)_{\gamma \in \Gamma}$ of some $(r, t, r') \in \rightarrow \setminus E$ with
11:       $\bigvee_{\gamma \in \Gamma}(r_\gamma \cdot k_\gamma \neq 0) = true$ and with minimal $|\{u \in \Sigma \mid |\mathcal{C}(\mathcal{P}_E(r[t\rangle r'))(u)| > 0\}|$
12:     **for each** $u \in \Sigma$ with $\mathcal{C}(\mathcal{P}_E(r[t\rangle r'))(u) \neq \emptyset$ ▷ relabel chords
13:       **create** a new label u'
14:       **let** $\Sigma = \Sigma \cup \{u'\}$
15:       **let** $relabel(u') = u$
16:       **for each** $(q, u, q') \in \mathcal{C}(\mathcal{P}_E(r[t\rangle r'))(u)$
17:         **let** $\rightarrow = (\rightarrow \setminus \{(q, u, q')\}) \cup \{(q, u', q')\}$
18:       **end for each**
19:     **end for each**
20:     **compute** cycle base Γ and representation of chords   ▷ $\rightarrow$ has changed
21:    **end while**
22: **end for each**
23: **return** the LTS and $relabel$

$O(|S| \cdot |\Sigma|^{7.5} \cdot \log^4 |\Sigma|)$. Multiplying the factors for the two loops, $O(|S|^2)$ and $O(|S| \cdot |\Sigma|)$ (we relabel at least one chord each time we run through the while-loop and thus exclude it from further solutions), we finally arrive at $O(|S|^4 \cdot |\Sigma|^{8.5} \cdot \log^4 |\Sigma|)$. We conclude:

Theorem 4. MAKING SSPs INVOLVING CYCLES SOLVABLE IS POLYNOMIAL
Algorithm 2 is a polynomial-time algorithm relabelling any finite, reachable, and deterministic LTS, such that all SSPs (s, s') either become solvable or fulfill $\mathcal{P}_E(s) = \mathcal{P}_E(s')$.

The example of Fig. 5 shows that our algorithm can provide far better solutions than trivially relabelling each edge with its own label (also being polynomial). We make no claim about always finding an optimal solution (regarding the alphabet size).

4.2 Direct Unsolvability of SSPs

Let (s, s') be an SSP such that $\mathcal{P}_E(s) = \mathcal{P}_E(s')$ holds, i.e. it is not solvable and there is no cycle involved. Such unsolvable SSPs can be detected during the construction of the spanning tree. Whether we construct it depth-first or breadth-first, we start at the initial state and recurse through the LTS, never selecting edges leading to states that we have already seen. Due to the recursion, we know the Parikh vector of each state we encounter. For each Parikh vector, we collect the corresponding states in one set, building equivalence classes of states. For the example LTS in Fig. 6, we easily obtain classes $\{s_2, r_2\}$, $\{q_{2i+1}, r_{2i+1}\}$, and $\{s_{2i+2}, r_{2i+2}, q_{2i+2}\}$ for $i > 0$ with $\mathcal{P}_E(s_2) = \mathcal{P}_E(r_2) = 1 \cdot a + 1 \cdot b$, $\mathcal{P}_E(q_{2i+1}) = \mathcal{P}_E(r_{2i+1}) = i \cdot a + (i + 1) \cdot b$, and $\mathcal{P}_E(s_{2i+2}) = \mathcal{P}_E(r_{2i+2}) = \mathcal{P}_E(q_{2i+2}) = (i + 1) \cdot a + (i + 1) \cdot b$, while all other states quite obviously form a class of their own. Note that states that do not have the same distance from the initial state (in the spanning tree) never have the same Parikh vector.

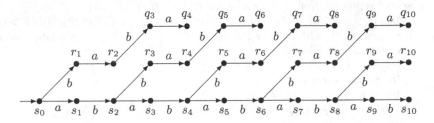

Fig. 6. An LTS where for $i > 0$ all SSPs (s_{2i}, r_{2i}), (r_{2i+1}, q_{2i+1}), (s_{2i+2}, q_{2i+2}), and (r_{2i+2}, q_{2i+2}) are unsolvable with $\mathcal{P}_E(s_{2i}) = \mathcal{P}_E(r_{2i})$, $\mathcal{P}_E(q_{2i+1}) = \mathcal{P}_E(r_{2i+1})$, and $\mathcal{P}_E(q_{2i+2}) = \mathcal{P}_E(s_{2i+2})$

Directly unsolvable SSPs exist exactly between members of the same class. It is not necessary to search the whole space $S \times S$ for them. Consider now the class $\{s_8, r_8, q_8\}$ contributing the three unsolvable SSPs (s_8, r_8), (s_8, q_8) and (r_8, q_8). If we follow the spanning tree backwards simultaneously from all three states, we see that the paths for s_8 and r_8 split up at the state s_6, while for q_8 we leave the common path already at s_4. To make the SSP (s_8, r_8) solvable we need to relabel some edge in $s_6[ab\rangle s_8$ or in $s_6[ba\rangle r_8$, any earlier edge contributes to both Parikh vectors, $\mathcal{P}_E(s_8)$ and $\mathcal{P}_E(r_8)$. To make the other two SSPs solvable, we need to change either an edge in $s_4[baba\rangle q_8$ or in $s_4[ab\rangle s_6$. After s_6, changing an edge label will only solve one of the two SSPs. A good solution relabels e.g. both $s_4[a\rangle s_5$ and $s_6[a\rangle s_7$ to a'. Obviously, the states s_4 and s_6 are important for determining which edges to relabel.

Definition 4. SPLITTER STATES
In an LTS $(S, \Sigma, \rightarrow, s_0)$ with spanning tree E, a state $s \in S$ is called a *splitter* if there is an SSP (s', s'') with $\mathcal{P}_E(s') = \mathcal{P}_E(s'')$ such that there are spanning tree paths $s[t\sigma\rangle s'$ and $s[t'\sigma'\rangle s''$ with $t \neq t'$.   □ 4

Algorithm 3 computes the splitters (set SP), the partition of equivalence classes invoked by the splitters, as well as the edge labels by which states in an equivalence class are reached from the splitters. The latter is recorded by the mapping A in line 3. For our example class $\{q_8, r_8, s_8\}$, we find $A(s_2, \{q_8, r_8, s_8\}) = \{a\}$ and $A(s_4, \{q_8, r_8, s_8\}) = \{a, b\}$. For a splitter and an

---

Algorithm 3. Splitter detection

Input: a finite, reachable, deterministic LTS $(S, \Sigma, \rightarrow, s_0)$ with spanning tree E
1: **let** EQ be the set of equivalence classes over S with respect to Parikh vectors
2: **let** $SP = \emptyset$ ▷ for collecting splitters
3: **let** $A: (S \times EQ) \rightarrow 2^{\Sigma}$, initially $\forall s, Q: A(s, Q) = \emptyset$ ▷ edges that Q uses at s
4: **let** $path: (S \times \Sigma) \rightarrow \Sigma^*$, initially $\forall s, t: path(s, t) = \varepsilon$ ▷ paths for relabelling
5: **let** $ap: S \rightarrow \Sigma^*$, initially $\forall s: ap(s) = \varepsilon$ ▷ active paths
6: **for each** $Q \in EQ$ with $|Q| > 1$ ▷ handle each class of SSPs separately
7:    **let** $R = Q$ ▷ mutable copy of Q
8:    **while** $R \neq \{s_0\}$ ▷ move backwards to initial state
9:       **let** $R' = \emptyset$ ▷ for collecting predecessors of R
10:       **for each** $s \in R$
11:          **find** $s' \in S$ and $t \in \Sigma$ with $s'[t\rangle s$ in E ▷ backwards on spanning tree
12:          **if** $|A(s', Q)| = 1$ **then**   ▷ know now that s' is splitter for Q
13:             **let** $SP = SP \cup \{s'\}$ ▷ make it a splitter
14:             **let** $t' = \text{first}(ap(s'))$ ▷ first letter of $ap(s')$
15:             **if** $path(s', t') = \varepsilon$ **then let** $path(s', t') = ap(s')$ ▷ previously found path
16:             **else let** $path(s', t') = \text{common prefix}(path(s', t'), ap(s'))$ **end if**
17:          **end if**
18:          **let** $R' = R' \cup \{s'\}$
19:          **let** $A(s', Q) = A(s', Q) \cup \{t\}$ ▷ Q can leave s' via t
20:          **let** $ap(s') = t \circ ap(s)$ ▷ prepend t to active path
21:          **if** $|A(s', Q)| > 1$ **then**   ▷ save/adapt paths at splitters
22:             **if** $path(s', t) = \varepsilon$ **then let** $path(s', t) = ap(s')$ ▷ set path for relabelling
23:             **else let** $path(s', t) = \text{common prefix}(path(s', t), ap(s'))$ **end if**
24:             **let** $ap(s') = \varepsilon$ ▷ cut active path at splitter
25:          **end if**
26:       **end for each**
27:       **let** $R = R'$ ▷ one step towards the initial state
28:    **end while**
29: **end for each**
30: **return** $SP, A, path$ ▷ note: only values at splitters are of interest

---

outgoing edge we also store (in *path*, line 4) which edges could be relabelled. In our example, we expect $path(s_4, a) = ab$ and $path(s_4, b) = baba$. In every equivalence class (loop in line 6), all states have the same distance from the initial state. By going stepwise backwards through the spanning tree (lines 8, 9, 11, 18, 27), we ensure that common predecessors of two states in the same equivalence class are found at the "same" time, inside the same call of the loop in line 10. By simply checking if we have already encountered a state for a class Q once (line 12) we can detect new splitters (perhaps some old ones, too). In line 19 we put down that a member of this equivalence class (Q) is an (indirect) successor via the edge we came through (backwards) in line 11. Line 20 extends the path, where we could relabel edges for this equivalence class, to the active state s' by prepending the edge's label t. If this is the first time we come through t to a splitter s', we just save the path, otherwise we take the common prefix with what has been stored previously (for another equivalence class, lines 22–23). The common prefix never becomes empty (at least t will remain) and it only contains edges we are allowed to relabel for every equivalence class that follows a path through s' and its t-edge. Line 24 makes sure that such paths never extend beyond the next splitter, avoiding to relabel an edge that only helps a subset of the states in Q reachable from here. When we encounter a state s' for the first time (for an equivalence class Q), we do not know yet if it "splits" this class Q. This means, we have to decide later if we need to store a path in $path(s', t)$. If we detect later that s' is a splitter for Q, the path information is still available in $ap(s')$ and is retrieved via lines 14–16 (essentially a copy of lines 22–23).

Overall, the while-loop (line 8) and the for-each-loop (line 10) combined can visit each edge (in line 11) at most once, i.e. we have $O(|S| \cdot |\Sigma|)$ runs of the inner part (lines 11–25) overall. The variable $path$ at most stores paths from one splitter to the next, i.e. each edge at most once. The variable ap can also be managed as such by not copying a path in line 20, instead extending it and pointing to the entry spot. Thus, the while-loop (line 8–28) has a run time of $O(|S| \cdot |\Sigma|)$. Multiplying this with the number of equivalence classes for the loop in line 6, $O(|S|)$, we obtain an overall complexity of $O(|S|^2 \cdot |\Sigma|)$ for the algorithm. Note that with an increasing number of equivalence classes we expect to visit a decreasing number of states per run in the while-loop, so we might see a lower complexity in the average case. We obtain:

Lemma 1. SPLITTER DETECTION
Algorithm 3 computes splitters, paths on which an edge might be relabelled, and the information by which edges states of the same equivalence class are reachable from a splitter in $O(|S|^2 \cdot |\Sigma|)$ time.

Now we use this information to decide which edges to relabel and how. When a splitter s divides an equivalence class Q, i.e. $|A(s, Q)| > 1$, we must relabel edges on some path starting at s with an edge label in $A(s, Q)$. We select a label t that occurs most often in $A(s, \cdot)$ (a maximal number of equivalence classes is split at s with the help of the t-edge). We relabel some edge from $path(s, t)$, and remove t from $A(s, \cdot)$ altogether. We repeat this until $\forall Q: |A(s, Q)| \leq 1$, so that no longer paths starting at s with different first labels have the same Parikh vectors. The main question is here if we can use the same replacement labels for different splitters.

Figure 7 helps to detect the problems that may occur. Assume that we relabel $s_3[b]r_8$ to b' to make the SSP (s_7, s_8) solvable. Then, we want to reuse the new label b' for other splitters (than s_3). We relabel $s_5[b]r_{12}$ to b' for the SSP (s_{11}, s_{12}) and also $s_0[b]r_2$ to b' to distinguish s_9 and s_{10} from s_{11} and s_{12}. We must not relabel $s_1[b]r_4$ to b' now to solve the SSP (s_3, s_4), though. This would make e.g. the SSP (s_{10}, s_{11}) unsolvable again, both states obtaining a Parikh vector of $2a + 1b + 1b' + 1c + 1d$. A similar problem occurs if we instead try to relabel a's: We change $s_0[a]r_1$, $s_1[a]r_3$, and $s_3[a]r_7$ to a', but then not $s_5[a]r_{11}$. So, the same replacement label can sometimes be used at splitters that lie "parallel" like s_3 and s_5, and sometimes for splitters with a common path like s_0, s_1, and s_3. Reuse is impossible though, if we would reach the replacement label on two

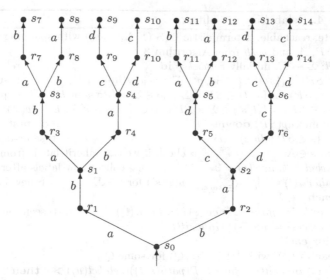

Fig. 7. An LTS where s_0 and all non-leaves s_i with odd i allow ab and ba while other non-leaves s_i allow cd and dc. Non-trivial equivalence classes of states forming SSPs are $\{s_1, s_2\}$, $\{s_3, s_4\}$, $\{s_5, s_6\}$, $\{s_7, s_8\}$, $\{s_9, s_{10}, s_{11}, s_{12}\}$, and $\{s_{13}, s_{14}\}$

paths leaving the same splitter via different edges, as for $s_0[b\rangle r_2$ and $s_0[abb\rangle r_4$ or for $s_0[a\rangle r_1$ and $s_0[bacda\rangle r_{11}$. We may relabel an a *and* a b somewhere, but this already introduces one more new label than we hoped for. Furthermore, at some splitters neither an a nor a b exists, like at s_4. We propose the following heuristic algorithm, Algorithm 4, to reduce the number of new labels.

The algorithm runs through classes X of splitters of the same depth (line 2, 5, 6). The depths are easily precomputed by a single run through the spanning tree. A depth i for a splitter s means that there are $i - 1$ more splitters on the spanning tree path from s_0 to s. (In the following, we write just *path* when we actually mean a path in the spanning tree.) Splitters of the same depth (line 6) are handled in the same run of the loop in line 5, starting with the highest depth, closest to the leaves. For splitters of the same depth, we check in the loop of lines 7–20 if they can reuse previously introduced new labels (for a higher depth). We collect all new labels occurring at states reachable from a splitter s via first label t in a set $inherit(t)$ (line 9). Then, we check if labels occur in two different sets $inherit(t)$ and $inherit(t')$ (lines 10–13), these labels enter a set *forbid*. If we would use such a label on a path starting with $s[t\rangle$, we might involuntarily identify Parikh vectors of paths starting with $s[t\rangle$ and $s[t'\rangle$. This does not happen if the new label occurs in the set $inherit(t)$ only. We relabel some edge in $s[path(s,t)\rangle$ then. All paths starting with $s[t\rangle$ (including all those already using the new label) will get one extra instance of the new label, while all other paths starting at s with a label different from t remain unchanged. In line 11 we collect the doubly used new labels in *forbid* to exclude them from the relabelling process at s. Line 12 collects the set of all new labels occurring

---

Algorithm 4. Relabelling at splitters

Input: a finite, reachable, deterministic LTS $(S, \Sigma, \rightarrow, s_0)$ with spanning tree E
and SP, A, and $path$ from Algorithm 3

1: **let** $relabel \colon \Sigma \to \Sigma$, initially $relabel = \mathbf{id}$ ▷ relabelling done
2: **let** $dp(s \in SP) = |\{s' \in SP \mid \exists \sigma, \varrho \in \Sigma^* \colon s_0[\sigma\rangle s'[\varrho\rangle s \text{ in } E\}|$ ▷ splitter depth
3: **let** $pred(s \in SP) = \{(s', t) \in SP \times \Sigma \mid \exists \sigma \in \Sigma^* \colon s'[t\sigma\rangle s \text{ in } E\}$ ▷ "predecessors"
4: **let** $dsucc(s \in SP) = \{(t, s') \in \Sigma \times SP \mid \exists \sigma \in \Sigma^* \colon s[t\sigma\rangle s' \text{ in } E \wedge dp(s') = dp(s) + 1\}$
5: **for** i **from** $\max dp(SP)$ **downto** 1   ▷ deepest splitters first
6:    **let** $X = \{s \in SP \mid dp(s) = i\}$ ▷ same-depth splitters
7:    **for each** $s \in X$ ▷ check if we can inherit labels from depth $i+1$
8:      **let** $forbid = \emptyset$, $new(s) = \emptyset$ ▷ collect new labels after s in $new(s)$
9:      **let** $inherit(t) = \bigcup_{(t,s') \in dsucc(s)} new(s')$ for $t \in \Sigma$ ▷ new labels via $s[t\rangle$
10:      **for each** $t \in \Sigma$
11:        **let** $forbid = forbid \cup (new(s) \cap inherit(t))$ ▷ seen more than once
12:        **let** $new(s) = new(s) \cup inherit(t)$
13:      **end for each**
14:      **for each** $t \in \Sigma$ with $\{t\} \subsetneq A(s, Q)$ for some Q
15:        **if** $\exists u \in inherit(t) \setminus forbid \colon \mathcal{P}(path(s, t))(relabel(u)) > 0$ **then**
16:          **relabel** one $relabel(u)$ in $s[path(s, t)\rangle$ to u
17:          **for each** Q **let** $A(s, Q) = A(s, Q) \setminus \{t\}$ **end for each**
18:        **end if**
19:      **end for each**
20:    **end for each**
21:    **while** $\exists s \in X \; \exists Q \colon |A(s, Q)| > 1$ ▷ more relabelling needed
22:      **let** $rp(s \in SP) = \{t \in \Sigma \mid \exists Q \colon \{t\} \subsetneq A(s, Q)\}$ ▷ on some $path(s, t)$
23:      **let** $occ(X \subseteq SP, t \in \Sigma) = \{s \in X \mid \exists u \in rp(s) \colon \mathcal{P}(path(s, u))(t) > 0\}$
24:      **select** $t \in \Sigma$ with maximal $|occ(X, t)|$ ▷ t can be relabelled most often for X
25:      **if** $\exists s \in SP \colon occ(X, t) = \{s\}$ **then**   ▷ renew selection if singleton
26:        **select** $t \in \Sigma$ with $s \in occ(\{s\}, t)$ and   ▷ by looking at predecessors
27:            maximal $|\{(s', u) \in pred(s) \mid u \in rp(s') \wedge \mathcal{P}(path(s', u))(t) > 0\}|$
28:      **end if**
29:      **create** new label t' and **let** $relabel(t') = t$
30:      **for each** $s \in occ(X, t)$ ▷ use new label wherever possible
31:        **select** $u \in rp(s)$ with $\mathcal{P}(path(s, u))(t) > 0$
32:        **relabel** one t in $s[path(s, u)\rangle$ to t'
33:        **for each** Q **let** $A(s, Q) = A(s, Q) \setminus \{u\}$ **end for each**
34:        **let** $new(s) = new(s) \cup \{t'\}$ ▷ allow to inherit t'
35:      **end for each**
36:    **end while**
37: **end for**
38: **return** the modified LTS and $relabel$

---

after s. When we later look at a predecessor splitter s' of s with depth decreased by one, we use this information for collecting its set of available new labels in line 9. Labels forbidden at s may be allowed at the direct predecessor splitter s' of s, since s is reachable via *one* edge leaving s' only. Thus, the whole set $new(s)$ is provided to s' and not just $new(s) \setminus forbid$.

In lines 14–19 we relabel edges with the reusable labels. If a set $A(s, Q)$ contains n labels, we must relabel (at least) $n-1$ edges on the n paths starting at s with the n different labels, eventually. One of the paths may remain unchanged. Line 14 travels through the edges t at s that require relabelling, i.e. some $A(s, Q)$ contains t and at least one other label. If we find a reusable label u, i.e. one that s inherited via t that is not forbidden (line 15), we check if its original version $relabel(u)$ occurs in $path(s, t)$. If so, we relabel (one instance of) it to u (line 16) and remove t from all $A(s, Q)$ (line 17), as we do not need to relabel more than one edge on this path.

The lines 21–36 relabel edges when we cannot reuse some replacement label and instead need a new one. We terminate when no splitter requires us to relabel more edges, i.e. all $A(s, Q)$ for all the splitters have become singletons (or are empty). Until then, we look for a label that can be relabelled at as many splitters in our active splitter set X as possible (line 23): If a label t occurs on some $path(s, u)$ that must still be handled (as $u \in rp(s)$, i.e. u still appears in some $A(s, Q)$ with more than one entry, line 22), s will occur in $occ(X, t)$. We select the label t with maximal $|occ(X, t)|$ (line 24). If this number is one, we revoke our selection and check for a label t that can be relabelled at s and at as many predecessor splitters of s as possible (lines 25–28). Note, that it is difficult to check at how many predecessors we might relabel t if $|occ(X, t)| > 1$: Other splitters in X could prevent relabelling for the predecessors of s via the inheritance mechanism (lines 7–20). Since the breadth of a tree can grow exponentially with its depth, we expect a better reuse of new labels if we go for a greedy strategy that prioritises breadth. We still may be able to reuse labels at lower depths later via the inheritance mechanism, but the label selection might not be optimal for it.

After choosing t, we create a replacement label t' for it (line 29). The following loop (lines 30–35) tries to relabel an edge for as many states in X as possible (but for s only if we revoked the first selection of t). In each instance, we look for some $path(s, u)$ that still requires relabelling, and relabel one t in it to t'. Again, the reference to this path is removed from any $A(s, Q)$, and additionally, we put t' in $new(s)$, so it may be passed on to splitters of lower depth later.

Regarding time complexity, let us take a look at some individual checks first. $A(s, Q)$ is accessed for membership and size (lines 14, 17, 21, 22, 33). This can easily be realised in time $O(1)$ with an array of size $|\Sigma|$ for the potential members and a counter for the size. Let us define a mapping $\alpha: S \times \Sigma \rightarrow 2^{EQ}$ with $\alpha(s, t) = \{Q \in EQ \mid t \in A(s, Q) \wedge |A(s, Q)| > 1\}$. This mapping α can be computed initially in $O(|S|^2 \cdot |\Sigma|)$ and then adapted in $O(1)$ every time some $A(s, Q)$ is modified. This also allows us to build rp (line 22) in $O(|S| \cdot |\Sigma|)$ in the initial phase of the algorithm and adapt it in $O(1)$ for each change of some $A(s, Q)$. Checks $\mathcal{P}(path(s, u))(t) > 0$ (lines 15, 23, 27, 31) can be precomputed yielding mappings $\beta_{s,u}: \Sigma \rightarrow S \cup \{\bot\}$ which supply the first state at which an edge labelled t occurs in $s[path(s, u)\rangle$ (or $\bot$ if t does not occur). This takes $O(|S|^2 \cdot |\Sigma|)$ time initially but reduces each single test to $O(1)$. This also allows us to relabel edges (lines 16, 32) in $O(1)$. For computing a single instance $occ(X, t)$

(line 23) we need to look through $rp(s)$ for each $s \in X$ taking $O(|S| \cdot |\Sigma|)$ time. Each set operation in lines 9, 11, 12 takes time according to the cardinality of the sets, which is in $O(|S| \cdot |\Sigma|)$ (the maximal number of new labels). The big union of line 9 takes amortised the same time as every splitter can only occur in one such union and the loops of line 5 and 7 in composition are run exactly $|SP|$ times. The initial computations of the depth dp, predecessor splitters $pred$, and direct successor splitters $dsucc$ (lines 2–4) can be done during a single depth-first search through the spanning tree in $O(|S| \cdot |\Sigma|)$.

Now let us take a look at the loops. Lines 10–13 have a run time of $O(|S| \cdot |\Sigma|^2)$, $O(|\Sigma|)$ for the loop itself and $O(|S| \cdot |\Sigma|)$ for the body. The loop in line 14 itself takes $O(|\Sigma|)$ (the test being in $O(1)$), the if-construct $\exists u$ in the body uses $O(|S| \cdot |\Sigma|)$, and the inner for-each-loop takes $O(|S|)$ for the number of equivalence classes. Adding the latter two, we get $O(|S| \cdot |\Sigma|)$ for the body and $O(|S| \cdot |\Sigma|^2)$ for the whole loop. Since the loops of line 5 and 7 in composition are run $O(|S|)$ times, we get $O(|S|^2 \cdot |\Sigma|^2)$ for the front part of the algorithm upto line 20. The while loop of line 21 can be entered at most $O(|S| \cdot |\Sigma|)$ times during the run of the algorithm, since each time one label $t \in \Sigma$ is eliminated from $A(s, \cdot)$. For the lines 22–28 we obtain $O(|S| \cdot |\Sigma|^2)$ for the computation of $occ(X, \cdot)$ and $O(|S| \cdot |\Sigma|)$ for the second select statement in lines 26–27, everything else takes constant time. The loop of lines 30–35 is entered $O(|S|)$ times, the selection (line 31) uses $O(|\Sigma|)$ and line 33 $O(|S|)$ time, the rest is in $O(1)$. Together, the loop takes $O(|S| \cdot (|S| + |\Sigma|))$. The second part of the algorithm is then in $O(|S|^3 \cdot |\Sigma|^3)$, which is also the complexity of the whole algorithm.

From the above we conclude:

Theorem 5. MAKING DIRECT SSPS SOLVABLE
Algorithm 4 is a polynomial-time algorithm relabelling any finite, reachable, and deterministic LTS, such that all SSPs (s, s') with $\mathcal{P}_E(s) = \mathcal{P}_E(s')$ become solvable.

For the example of Fig. 6 our algorithm yields the optimal result. It will first introduce one new label, say a', replacing a at the edge $s_8[a\rangle s_9$ or $r_9[a\rangle r_{10}$. Since s_8 is the deepest splitter (with no other at the same depth), the algorithm uses the second selection (line 26), but both a and b can be relabelled at equally many predecessor splitters. The selection of a is therefore random. At each previous splitter, starting with s_6, we inherit the label a' via the lower branch. Thus, the first part of the algorithm now replaces each a in $s_6[a\rangle s_7$, $s_4[a\rangle s_5$, $s_2[a\rangle s_4$, and $s_0[a\rangle s_1$ successively by a'.

The example of Fig. 7 is not solved optimally. For splitter depth 3 ($\{s_3, s_4, s_5, s_6\}$) we might relabel $s_3[b\rangle r_8$, $s_5[b\rangle r_{12}$ to b' and $s_4[c\rangle r_9$, $s_6[c\rangle r_{13}$ to c'. For the splitters s_1 and s_2 we inherit b' on the left and c' on the right path. Since c' does not occur in $path(s_1, b) = ba$, it is unusable here. With $path(s_1, a) = ab$ we relabel $r_3[b\rangle s_3$ to b'. On the other side, $r_6[c\rangle s_6$ gets relabelled to c' for similar reasons. At the splitter s_0 we inherit b' and c' both from both sides, so these labels are forbidden. Instead, we create a new label and relabel say $s_0[a\rangle r_1$ to a'.

The minimal possible number of new labels for this LTS is two. We relabel $s_0[b\rangle r_2$ to b' (instead of using a'). This introduces one b' in $s_0[b'acdab\rangle s_{11}$. While it is theoretically possible that this path is now equivalent to some path on the left side of the tree, this does not actually happen here: The only path on the left containing exactly one b' is $s_0[abab'ab\rangle s_7$, and it does not contain any c or d.

If we increase the depth of the LTS (by adding ab-ba-structures to s_7, s_9, s_{11}, and s_{13} as well as cd-dc-structures at s_8, s_{10}, s_{12}, and s_{14}), our algorithm yields the optimal result. Now, cd-dc-structures also occur after s_3, and we get equivalent paths if we relabel $s_0[b\rangle r_2$ to b'. Indeed, if we build a family of such LTS canonically, the number of required new labels is exactly the maximal splitter depth, unless the latter is three. Each two consecutive sets of splitters of same depth can share their new labels, and each set needs two of them. For splitter depth one (s_0) we only need one new label, which can be shared with depth two in case of an even maximal splitter depth.

We also obtain optimal results for LTS where every c is replaced by a and every d by b. In this case, the number of required new labels is always exactly the maximal splitter depth.

5   Combining SSP and ESSP

In this section, we combine all algorithms presented in Sects. 3 and 4 to synthesise a labelled solution of an LTS. The following theorem states, if we combine all algorithms, there will be no negative side effects.

Theorem 6. SEPARATION PROBLEMS CANNOT BE MADE UNSOLVABLE
None of the Algorithms 1, 2, and 4 can make a solvable separation problem unsolvable again.

Proof: Assume an unsolvable SSP (s, s') or an unsolvable ESSP (s, t) with $\neg s[t\rangle$ after the application of one of the algorithms. The dual inequality system according to Theorems 3 or 1 then has a solution. Note that every row in this inequality system corresponds to one label. Let u be any label and $relabel^{-1}(u)$ be the set of labels used as replacement for u (including u itself) by the algorithm. The dual inequality system before applying the algorithm contains a single row for u with entries u_i (i going over the columns). For $v \in relabel^{-1}(u)$ let v_i be the corresponding entries after applying the algorithm, then obviously $\sum_{v \in relabel^{-1}(u)} v_i = u_i$. Since our solution solves all rows $v \in relabel^{-1}(u)$, it also solves the row for u in the prior system. As this holds for all labels u, the dual inequality system prior to applying the algorithm has a solution, i.e. the SSP (s, s') or the ESSP (s, t) was also unsolvable before the algorithm was used.

$\square$ 6

This means we are essentially free to apply the algorithms in any order. Note however, that relabelling can create completely new separation problems. If we relabel an edge $s[t\rangle$ to t', we obtain ESSPs (s, t) with now $\neg s[t\rangle$ and (s', t') for all $s' \in S$ with $\neg s'[t'\rangle$ that may not have existed before. If Algorithm 1 has not yet been applied to these ESSPs, we have to invoke this algorithm once again.

Assume that some SSP (s, s') is directly unsolvable and unsolvable via cycles at the same time. Using Algorithm 2 will not change any edges on the spanning tree, i.e. (s, s') will still be directly unsolvable afterwards. Using Algorithm 4 first, on the other hand, can affect the unsolvability via cycles, too, since here the solution of the dual SSP equation system depends on both kinds of edges, spanning tree edges and chords. So attacking the (easier) directly unsolvable SSPs first might avoid some of the SSPs involving cycles at all. The same is true with regard to directly unsolvable SSPs and ESSPs. When we look at the big differences regarding time complexity, it becomes clear that we should apply the Algorithms 3 and 4 first.

We also should take on the ESSPs last, the main reasons being the heavy time complexity for these problems and the possibility of creating new ESSPs by relabelling (but no new SSPs). Still, ESSPs and SSPs involving cycles can influence each other, and it is more likely that making such an SSP solvable relieves us of an ESSP, than the other way around. The SSP algorithm relabels chords and cycles, which have a strong presence in the dual ESSP equation system. Relabelling due to an ESSP only seems to randomly affect unsolvable SSPs involving cycles. Figure 8 shows an example LTS where dealing with any single of the SSPs (s_i, q_{i+1}) or (q_i, r_{i+1}) for $0 \le i \le 9$ leads to immediately relabelling all (dashed) chord labels a to some new a'. When this is done, we have obtained a solvable LTS with three labels. Attacking ESSPs first, on the other hand, will also lead to relabelling edges on the spine. The result according to our algorithm used seven different labels.

We should also note, though, that the knowledge that all ESSPs are solvable, could greatly reduce the search for unsolvable SSPs. If for two states s, s' we have $s[a\rangle$ but $\neg s'[a\rangle$, (s, s') is a solvable SSP. An unsolvable SSP (s, s') would imply an unsolvable ESSP (s', a). Therefore, it is only necessary to solve an equation system for an SSP (s, s') if s and s' allow exactly the same set of labels at their edges. Eliminating some equation systems for SSPs without solving them is possible at the cost of doing all ESSPs first.

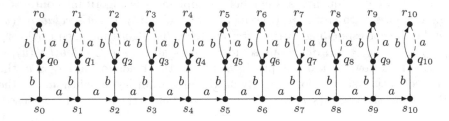

Fig. 8. In this LTS each set $\{s_i, q_{i+1}, r_{i+2}\}$ for $0 \le i \le 8$ forms three unsolvable SSPs (s_i, q_{i+1}), (s_i, r_{i+2}), and (q_{i+1}, r_{i+2}) as well as two unsolvable ESSPs (q_i, a) and (r_{i+1}, b). Additionally, (q_0, r_1), (s_9, q_{10}), (q_9, a), (q_{10}, a), and (r_{10}, b) are also unsolvable separation problems.

6   Experimental Results

We implemented the proposed algorithms in a tool written in Java.[3] Inequality systems were solved via the SMTInterpol library [8]. In this section, the performance of this implementation is evaluated.

For our experiments, we measure two parameters: the number of labels in the relabelled LTS and the run time of the algorithm. We start by looking at the number of labels in Sect. 6.1 and afterwards we investigate the run time of the algorithm in Sect. 6.2.

Finding informative, realistic benchmarks proved difficult. To measure the quality of our algorithm, we want to compare its result with the optimum, which requires structural analysis of the LTS (brute force methods are prohibitive) and therefore regular structures, not random ones. We also want scalable LTS to prevent random flukes and get an impression about the run times. Known Petri net models are either unlabelled (their reachability graphs do not require relabelling) or their reachability graphs are so large that they cannot be handled. Other models we have found were not scalable. We therefore rely on synthetic benchmarks shown in this paper.

6.1   Number of Labels

For each experiment, 100 breadth-first spanning trees were constructed, and the best of them (with a minimal number of different Parikh vectors of chords) was then passed on as input to our algorithm. To conduct the experiments, families of LTS of different sizes were constructed (the LTS shown in this paper allow for canonical extensions). The sizes are measured by the depth of the state farthest from the initial state, so the number of states might grow exponentially with the depth (and does indeed so for the trees). In the LTS of the two-way comb (Fig. 3) every fourth tooth in the comb was left out. The quadratic grid is shown in Fig. 9. Note that transitions on an axis of the coordinate system have a direction opposite to other transitions with the same label. The ab-tree is structurally identical to the tree in Fig. 7, but all instances of c and d are replaced by a and b, respectively. Finally, we use an LTS (Fig. 10) structurally identical to Fig. 5, but with differences in the labelling. These differences allow us to optimally relabel using only one new label, but we have to find a specific spanning tree for this to happen.

Optimal values for the alphabet size where found by analysing the LTS (cf. Table 1). For the abcd-trees we do not know the optimal sizes for depths greater than eight. There are many possibilities for making direct SSP solvable, and it is unclear how to measure the number of ESSP instances that remain unsolved thereafter. A brute force algorithm prohibits itself, as the number of combinations of edges to relabel will grow exponentially in the number of states, which is already 125 for depth ten.

---

[3] The tool is available at https://github.com/CvO-Theory/apt-relabelling.

Table 1. Measurement of the alphabet size after the various stages of the algorithm: initially ($|\Sigma|$), after solving direct SSP (dSSP), cyclic SSP (cSSP), and ESSP, i.e. at the end. "Minimum" shows the optimal value where known. The first three columns give the name of the LTS family, the size of the LTS (by depth, deepest node in a breadth-first spanning tree), and the number of states of the LTS

Name	Depth	$|S|$	$|\Sigma|$	After dSSP	After cSSP	After ESSP	Minimum
twig	10	26	2	3	3	3	3
(Fig. 6)	100	296	2	3	3	3	3
	1000	2996	2	3	3	3	3
abcd-tree	4	13	4	6	6	6	6
(Fig. 7)	6	29	4	7	7	7	6
	8	61	4	8	8	9	8
	10	125	4	9	9	11	?
	20	4093	4	14	14	18	?
ab-tree	4	13	2	4	4	4	4
(Fig. 7 with	6	29	2	5	5	5	5
$c \mapsto a, d \mapsto b$)	8	61	2	6	6	6	6
	10	125	2	7	7	7	7
	20	4093	2	12	12	12	12
cycle-path 1	10	20	3	3	6	13	11
(Fig. 5)	100	200	3	3	23	113	101
	200	400	3	3	38	220	201
cycle-path 2	10	20	3	3	5	8	4
(Fig. 10)	100	200	3	3	5	113	4
	200	400	3	3	6	220	4
quadratic grid	6	16	2	3	7	7	4
(Fig. 9)	8	25	2	3	11	13	4
	10	36	2	3	13	15	4
	20	121	2	4	17	27	4
	30	256	2	3	21	46	4
	40	441	2	4	29	69	4
knobby comb	10	27	2	2	3	3	3
(Fig. 8)	100	297	2	2	3	3	3
	1000	2997	2	2	3	3	3
two-way comb	6	20	3	3	3	4	4
(Fig. 3)	10	34	3	3	3	5	5
	14	48	3	3	3	6	6
	410	1434	3	3	3	105	105

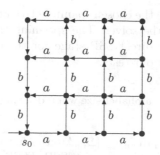

Fig. 9. Quadratic grid structure (of depth 6)

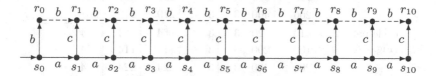

Fig. 10. An LTS with the optimal spanning tree for relabelling. Since $s_1[ac\rangle r_2$ and $s_1[cb\rangle r_2$, a and b must have the same effect in a Petri net solution. Then, (s_1, r_0) is an unsolvable SSP. Relabelling just $s_0[b\rangle r_0$ to b' will make all separation problems solvable

In many cases, optimal or near optimal values for the alphabet size were obtained in the experiments. Optimal values occur for twig, ab-tree, knobby comb, and two-way comb for all LTS sizes. Only one of the three kinds of problems, direct SSP, SSP involving cycles, or ESSP needs to be solved here, and our algorithm handles that nicely. All newly created ESSP instances are immediately solvable in these examples. Since the three algorithms for the three kinds of problems act independently, we do not expect optimal results when more than one kind of problem occurs. The abcd-trees (Fig. 7) contain inherent ESSP instances, e.g. starting at the LTS of depth eight we get $\frac{1}{2}\mathcal{P}_E(s_2) + \frac{1}{2}\mathcal{P}_E(s_{14}) = \mathcal{P}_E(s_5)$ with c- and d-edges at s_2 and s_{14} but not at s_5. Only if we manage to relabel say a c-edge at s_2 and a d-edge at s_{14}, we can prevent an unsolvable ESSP for s_5. For depth eight, we determined a way to relabel edges for the direct SSP instances that also makes all ESSPs solvable, but our algorithm does not find it. With higher depths, we do not even know whether such a way exists, as the number of unsolvable ESSP instances increases sharply.

The quadratic grid and both cycle-path LTS classes involve solving both ESSPs and SSPs based on cycles. For these examples, it might make a difference if we solve ESSPs first (as far as that is possible). Table 2 shows the results for the alternative order of the algorithms. It might be worth investing in the run time for this reversed order of solving, if the alphabet sizes became smaller this way. Unfortunately, as we can see, this is not the case. If the results have changed at all, they probably have become worse. Note though, that there is some randomness here. Different runs with the same LTS may provide different alphabet sizes depending on the chosen spanning tree and the computed solutions for inequality systems by the solver. The spanning tree choice even influences the alphabet size after the direct SSP stage for the quadratic grid, with sizes from three to five letters in our tests.

Table 2. Measurement of the alphabet size when delaying SSPs involving cycles until after ESSPs

Name	Depth	$|S|$	$|\Sigma|$	After dSSP	After ESSP	After cSSP	Minimum
cycle-path 1	10	20	3	3	13	13	11
(Fig. 5)	100	200	3	3	120	120	101
	200	400	3	3	219	219	201
cycle-path 2	10	20	3	3	4	4	4
(Fig. 10)	100	200	3	3	114	114	4
	200	400	3	3	224	224	4
quadratic grid	6	16	2	3	7	7	4
(Fig. 9)	8	25	2	3	12	12	4
	10	36	2	4	23	23	4
	20	121	2	5	32	32	4
	30	256	2	5	62	62	4
	40	441	2	5	116	116	4

The cycle-paths (Figs. 5 and 10) are of special interest. They have – despite their identical graph structure – very different optimal values for the alphabet size. Indeed, in the second case it suffices to relabel a single edge, $s_0[b\rangle r_0$. In the first case, the ESSP instances enforce at least one relabelled edge for every cycle. So, finding an optimal result is not just a graph theoretical problem, we definitely need to look at the concrete labelling in the LTS, too. This also explains why it would be hard to show that relabelling is NP-complete, as the usual NP-complete candidates from graph theory cannot easily be used for a polynomial reduction.

Note, that the direct SSP algorithm does not detect the difference of the two example LTS classes. There are no directly unsolvable SSP instances at all. The algorithm for SSPs involving cycles does not relabel the single edge $s_0[b\rangle r_0$ as it is not a chord. We have tried to combine both algorithms. While SSPs involving cycles in general cannot be made solvable by relabelling the spanning tree, there may be instances where this works. Such an instance (s, s') can be fed into the direct SSP algorithm. Note that s and s' may have different depth and the direct SSP algorithm demands the same depth for all "identifiable" states. Say s is deeper in the spanning tree than s', then we can find the ancestor s'' of s with the same depth as s' and give the set $\{s'', s'\}$ (if $s'' \neq s'$) as an additional "equivalence class" to the direct SSP algorithm. The latter will then try to distinguish the two states via relabelling the spanning tree. If this does not make the SSP (s, s') solvable, we can still use the algorithm for cyclic SSPs. Unfortunately, the results became even worse for all tested examples. For cycle-path 2, additional relabelled edges always occurred along the upper spine, where relabelling is not necessary.

Table 2 shows an optimal result for the cycle-path 2 of depth ten. Here, indeed the edge $s_0[b\rangle r_0$ was selected for relabelling (with medium probability, depending on the spanning tree) due to some solution of an ESSP, but this probability decreases with growing size of the LTS. Closer inspection reveals the reason being the inequality system solver, which randomly provides different solutions. Sometimes, the (dual) solution found for the unsolvable ESSP (s_2, b) is $\mathcal{P}_E(s_2) = \frac{1}{2}\mathcal{P}_E(s_0) + \frac{1}{2}\mathcal{P}_E(r_3) + \frac{1}{2}(bb - ac) + \frac{3}{2}(a - b)$. The spanning tree contains the path $s_0[aac\rangle r_2[b\rangle r_3$ and $(bb - ac)$ and $(a - b)$ represent two Parikh vectors of the cycles. Here, $s_0[b\rangle r_0$ gets relabelled. Other solutions occur, e.g. $\mathcal{P}_E(s_2) = \frac{1}{2}\mathcal{P}_E(r_0) + \frac{1}{2}\mathcal{P}_E(r_2) + \frac{1}{2}(bb - ac) + \frac{3}{2}(a - b)$, where a different edge must be selected. Solving an ESSP (s_i, b) with higher index i first leads to more solutions relabelling an edge different from $s_0[b\rangle r_0$.

6.2 Run Time

We measured the run time of the algorithm for some families of LTS. Just as in the previous section, for each experiment, 100 breadth-first spanning trees (BFS) were constructed. The tree with a minimal number of different Parikh vectors of chords was then passed as input to our algorithm. The time to select a spanning tree is included in the run time of the algorithm. To increase precision, each measurement was repeated three times and the best result was used.

The resulting run times (in seconds) for increasing depths of LTS can be seen in Figs. 11 and 12 on a logarithmic scale. In all cases, the run time of the algorithm seems to be sub-exponential in the size of the input LTS (the size of abcd-tree grows exponentially with increasing depth). For a better estimation, we tried to apply the formula $x = \lim_{n\to\infty} \log_2(\frac{time(2n)}{time(n)})$ to each family of LTS, where n is the number of states. If the formula converges, the average time complexity should lie in $O(n^x)$. Unfortunately, we only have finitely many data points, and the limit x cannot even be extrapolated from our data. While the above logarithm typically yields values between 2.3 and 3, the increment rate still seems to be nearly linear at the high end of our data points. So, we can only guess that the average case time complexity of our algorithm is at least cubic, and likely quite a bit higher.

The curves for cycle-path 1 and cycle-path 2 contain a lot of randomness due to spanning tree selection and choices of the solver. Still, the "true" curves can easily be guessed and cycle-path 2 is clearly the easier problem of the two. This is no surprise, as cycle-path 1 has one unsolvable ESSP per cycle at the beginning. In cycle-path 2 these problems do not exist at first, they can only be created by solving SSPs.

Figure 12 shows a few examples for comparisons between different algorithms. In the upper left picture, we show our proposed heuristic compared to a brute

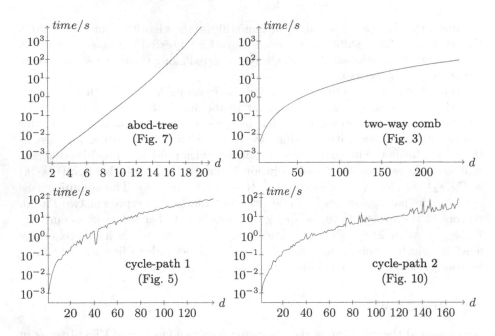

Fig. 11. Run time versus depth d for four of our LTS families on a logarithmic time scale. The abcd-tree grows exponentially with its depth, thus we get a linear curve. The two cycle-path families allow for lots of random effects, still the sub-exponential curve is clearly visible.

force algorithm for finding the optimal solution. The brute force algorithm performs very poorly and has an (at least) exponential increase in run time for larger inputs. In contrast, the heuristic algorithm performs a lot better, even for large inputs. The family twig was chosen for a comparison with the brute force approach, because here the brute force algorithm showed the best results. For other families, even less data points could be measured in the available time.

The effect of alternative orders for solving separation problems is examined in the lower left part for the knobby comb family of LTS. The figure shows that solving SSPs involving cycles first leads to smaller run times than solving ESSPs first. This effect was also observed for most other families. One exception is the cycle-path 2 family, where solving ESSPs first had a small positive influence on the resulting run time.

In the lower right, we show the times for the quadratic grid family. This figure also presents the performance of the algorithm when spanning trees are not constructed in a breadth-first manner, but randomly. The performance of these two variants of the algorithm is initially similar, but then the random spanning tree variant quickly becomes more than 15 times slower. In all considered families, true random spanning trees either had no noticeable effect or negatively

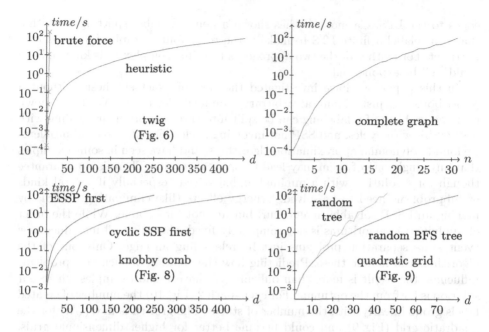

Fig. 12. Comparisons for different approaches. The brute force curve is typical for all our examples. The difference between ESSP first and our standard of solving cyclic SSP first is usually negligible, except for the knobby comb. Using true random spanning trees instead of breadth-first ones is often detrimental, and never better. The complete graph (with n states and all edges having the same label) shows the effect of randomly removing non-determinism. An optimal solution would take linear time.

influenced the run time of the algorithm, compared to breadth-first spanning (BFS) trees.

Finally, the upper right picture shows our result for the complete graph with n states where all edges have the same label. Since this LTS is non-deterministic, we cannot use it as input for our algorithm directly. We therefore use an alphabet with n labels and relabel, at each state, the emanating edges to the n different labels randomly. We obtain the expected polynomial run time, but there is a better solution that works in linear time: We order all states in a circle and relabel an edge to either z or $\bar{z}$ if it skips exactly z states counted clockwise. It is relabelled to $\bar{z}$ if and only if it skips or reaches the initial state. We have yet to find a good algorithm for removing non-determinism in general.

7   Concluding Remarks

Synthesising a small model like a Petri net from a large LTS or a set of observable processes can be done in more than one way. For Process Mining [21], Badouel and Schlachter [2] have constructed an incremental over-approximation algorithm. This allows behaviour that has not been observed essentially by adding edges to the LTS. Carmona [6] has shown a heuristic label splitting algorithm that can relabel a finite LTS to make it the reachability graph of a k-bounded Petri net. For neither of the two approaches the time complexity is known, both could well be exponential.

In this paper, we have investigated the case of exact synthesis without a priori bound k, just aiming at arbitrary, bounded Petri nets. We have shown that the problem of relabelling can be split into three subproblems dealing with ESSP, SSP with cycles, and SSP not involving cycles. For all three subproblems we found polynomial approximation algorithms, and have seen in some examples that any single one of them may lead to optimal results. There is no guarantee though for a solution with a minimal alphabet size, especially if several kinds of subproblems need to be solved. Synergy effects (like eliminating an SSP by making an ESSP solvable) may occur, but are not foreseeable. While the order of applying the algorithms is not completely fixed, they may all introduce new event/state separation problems just by relabelling an edge. Only one of the algorithms can handle these. Predicting how the creation of such new problems influences the result is more than difficult. We have found examples where our solution is far from the optimum. For cycle-path 2 (Fig. 10) the number of created labels grows linearly with the number of states. The growth is in $O(\sqrt{n})$ for the quadratic grid (Fig. 9) and could become better for higher dimensional grids. For both families though, the optimum is constant in the number of states.

As randomising factors, the choice of a spanning tree (with usually exponential possibilities in the number of states) and the solver for inequality systems come in. For the latter, we have seen the results to depend e.g. on the input order of the variables and the inequalities. Removing the choice of the spanning tree from our algorithm was unsuccessful. Randomly picking a number of spanning trees and choosing the best of them also did not work well, likely because there often are only a few good ones among an exponentially growing number (in the size of the LTS). Finding a good spanning tree might well be an NP-hard problem.

We believe that the presented algorithm can be adapted to language-equivalence instead of isomorphism. The idea is to leave SSPs (s, s') unsolvable if s and s' allow edges with exactly the same labels. The successors of s and s' will then also form pairwise unsolvable SSPs with the same property and synthesis will identify these pairs of states. We obtain a labelled Petri net whose reachability graph is bisimilar to the original LTS with a possibly smaller alphabet than for the isomorphic solution.

References

1. Badouel, E., Bernardinello, L., Darondeau, P.: Petri Net Synthesis. TTCS, 339 p. Springer, Heidelberg (2015). https://doi.org/10.1007/978-3-662-47967-4
2. Badouel, É., Schlachter, U.: Incremental process discovery using Petri net synthesis. Fundamenta Informaticae **154**(1–4), 1–13 (2017). https://doi.org/10.3233/FI-2017-1548

3. Barylska, K., Best, E., Erofeev, E., Mikulski, Ł., Piątkowski, M.: On binary words being Petri net solvable. In: Carmona, J., Bergenthum, R., van der Aalst, W. (eds) ATAED 2015, pp. 1–15 (2015). http://ceur-ws.org/Vol-1371
4. Best, E., Devillers, R., Schlachter, U.: A graph-theoretical characterisation of state separation. In: Steffen, B., Baier, C., van den Brand, M., Eder, J., Hinchey, M., Margaria, T. (eds.) SOFSEM 2017. LNCS, vol. 10139, pp. 163–175. Springer, Cham (2017). https://doi.org/10.1007/978-3-319-51963-0_13
5. vanden Broucke, S.K.L.M., De Weerdt, J.: Fodina: a robust and flexible heuristic process discovery technique. Decis. Support Syst. **100**, 109–118 (2017). https://doi.org/10.1016/j.dss.2017.04.005
6. Carmona, J.: The label splitting problem. In: Jensen, K., van der Aalst, W.M., Ajmone Marsan, M., Franceschinis, G., Kleijn, J., Kristensen, L.M. (eds.) Transactions on Petri Nets and Other Models of Concurrency VI. LNCS, vol. 7400, pp. 1–23. Springer, Heidelberg (2012). https://doi.org/10.1007/978-3-642-35179-2_1
7. Carmona, J., Cortadella, J., Kishinevsky, M., Kondratyev, A., Lavagno, L., Yakovlev, A.: A symbolic algorithm for the synthesis of bounded Petri nets. In: van Hee, K.M., Valk, R. (eds.) PETRI NETS 2008. LNCS, vol. 5062, pp. 92–111. Springer, Heidelberg (2008). https://doi.org/10.1007/978-3-540-68746-7_10
8. Christ, J., Hoenicke, J., Nutz, A.: SMTInterpol: an interpolating SMT solver. In: Donaldson, A., Parker, D. (eds.) SPIN 2012. LNCS, vol. 7385, pp. 248–254. Springer, Heidelberg (2012). https://doi.org/10.1007/978-3-642-31759-0_19. https://ultimate.informatik.uni-freiburg.de/smtinterpol/
9. Ehrenfeucht, A., Rozenberg, G.: Partial 2-structures, Part I: basic notions and the representation problem, and Part II: state spaces of concurrent systems. Acta Informatica **27**(4), 315–368 (1990)
10. Farkas, G.: Über die Theorie der einfachen Ungleichungen. Journal für die Reine und Angewandte Mathematik **124**, 1–27 (1902)
11. Hack, M.H.T.: Petri Net Languages. Computation Structures Memo 124, Project MAC, MIT (1975)
12. Karmarkar, N.: A new polynomial-time algorithm for linear programming. Combinatorica **4**, 373–395 (1984). https://doi.org/10.1007/BF02579150
13. Lu, X., Fahland, D., van den Biggelaar, F.J.H.M., van der Aalst, W.M.P.: Handling duplicated tasks in process discovery by refining event labels. In: La Rosa, M., Loos, P., Pastor, O. (eds.) BPM 2016. LNCS, vol. 9850, pp. 90–107. Springer, Cham (2016). https://doi.org/10.1007/978-3-319-45348-4_6
14. Murata, T.: Petri nets: properties, analysis and applications. Proc. IEEE **77**, 541–580 (1989)
15. Nielsen, M., Plotkin, G., Winskel, G.: Petri nets, event structures and domains, part I. Theoret. Comput. Sci. **13**(1), 85–100 (1981). https://doi.org/10.1016/0304-3975(81)90112-2
16. Reisig, W.: Petri Nets. EATCS Monographs on Theoretical Computer Science, vol. 4. Springer, Heidelberg (1985). https://doi.org/10.1007/978-3-642-69968-9
17. de San Pedro, J., Cortadella, J.: Discovering duplicate tasks in transition systems for the simplification of process models. In: La Rosa, M., Loos, P., Pastor, O. (eds.) BPM 2016. LNCS, vol. 9850, pp. 108–124. Springer, Cham (2016). https://doi.org/10.1007/978-3-319-45348-4_7
18. Schlachter, U., Wimmel, H.: A geometric characterisation of event/state separation. In: Khomenko, V., Roux, O.H. (eds.) PETRI NETS 2018. LNCS, vol. 10877, pp. 99–116. Springer, Cham (2018). https://doi.org/10.1007/978-3-319-91268-4_6

19. Schlachter, U.: Over-approximative Petri net synthesis for restricted subclasses of nets. In: Klein, S.T., Martín-Vide, C., Shapira, D. (eds.) LATA 2018. LNCS, vol. 10792, pp. 296–307. Springer, Cham (2018). https://doi.org/10.1007/978-3-319-77313-1_23
20. Tredup, R., Rosenke, C.: Towards completely characterizing the complexity of boolean nets synthesis. ArXiv https://arxiv.org/abs/1806.03703v3 (2018)
21. van der Aalst, W.M.P.: Process Mining: Discovery, Conformance and Enhancement of Business Processes, p. 352. Springer, Heidelberg (2011). https://doi.org/10.1007/978-3-642-19345-3. ISBN 978-3642193446

Author Index

Printed in the United States
By Bookmasters